Monographs of the Physiological Society No. 32

Salt glands in birds and reptiles

(Overleaf)

Wandering albatross (*Diomedea exulans*) photographed by Mr Ederic Slater, A.R.P.S., A.F.I.A.P., on its nesting territory on Macquarie Island in the Subantarctic. These birds spend most of their lives at sea, feed mainly on squid and other marine invertebrates and have well-developed salt glands. Note that this bird has a drop of secretion at the tip of the beak. (From CSIRO Division of Wildlife Research, Canberra, Australia, Report 1972-4, Plate 8.)

Frontispiece

SALT GLANDS IN BIRDS AND REPTILES

M. PEAKER and J. L. LINZELL

Department of Physiology, Agricultural Research Council
Institute of Animal Physiology, Babraham, Cambridge

CAMBRIDGE UNIVERSITY PRESS

Cambridge

London · New York · Melbourne

Published by the Syndics of the Cambridge University Press
The Pitt Building, Trumpington Street, Cambridge CB2 1RP
Bentley House, 200 Euston Road, London NW1 2DB
32 East 57th Street, New York, N.Y. 10022, USA
296 Beaconsfield Parade, Middle Park, Melbourne 3206, Australia

Library of Congress Catalogue Card Number: 74–12966

ISBN: 0 521 20629 4

First published 1975

Printed in Great Britain
at the University Printing House, Cambridge
(Euan Phillips, University Printer)

CONTENTS

The Plates are between pp. 52 and 53.

PREFACE

Our aim in writing this monograph has been to provide a comprehensive but personal account of salt glands. Since the discovery of their function, these glands have been of great interest to biologists from a wide range of disciplines. However, papers on salt glands are scattered through a great many different journals. For example, in the last decade the 266 papers of which we are aware were published in 84 different journals quite apart from articles in books and proceedings of symposia. It is hardly surprising therefore that some authors appear to be unaware of the work of others; as in all fields a few are reluctant to give due credit to the work of others but in this subject it is difficult for the most diligent and generous newcomer to know what has been done previously.

A danger of the comprehensive approach is that the result can be little more than an annotated bibliography – both dull and uninspiring. We naturally hope that we have avoided this pitfall, and only attempted a wide coverage because, at the start, it seemed that the total number of papers we would have to consider did not appear too forbidding for an account of the present state of knowledge to be presented in a readable form. We sincerely hope that the section headings and index will enable readers to find what they require without their having to plough through the whole work!

We should point out that we have restricted our coverage to the cranial salt glands of birds and reptiles, although the same descriptive term is being increasingly applied to organs which serve a similar function in other animals and plants, for example, the rectal gland of elasmobranchs and glands in certain insects.

We must thank Dr D. W. Fawcett and Dr J. R. McLean for allowing us to reproduce unpublished micrographs, as well as the numerous publishers and authors whose published figures are reproduced. Like all authors we have been helped by a number of people who have given freely of their time; the following gave us scientific advice on topics of which we had considerably less knowledge than they have: Professor E. C. Amoroso, Mr R. W. Ash, Dr R. M. C. Dawson, Dr M. H. Evans, Dr J. B. Furness, Dr Ann Hanwell, Dr Catherine Hebb, Dr (now Professor) R. D. Keynes, Dr D. B. Lindsay, Dr J. R. McLean, Mrs Stephanie Peaker, Dr J. S. Perry, Dr B. P. Setchell, Dr Ann Silver, Dr M. W. Smith, Dr Marthe Vogt and Dr F. B. P. Wooding.

We are particularly grateful to Professor R. D. Keynes, lately Director of the Institute, for encouraging us in our work on salt glands; to Dr B. P. Setchell for assisting us with the chapter on the early history of nasal glands, and to Dr Ann Hanwell for reading the entire manuscript. We have also benefited from the expertise of L. G. Jarvis with histology, A. L. Gallup with photography, and also wish to thank Mrs Dorothy George for typing and Miss Maureen Hamon for assistance in preparing the manuscript. It would have been impossible to write the book without the Institute's excellent library and its devoted librarian, Mr D. W. Butcher, and assistant librarian, Miss Wendy M. Reynolds. We wish to thank Miss Maureen Hamon sincerely for assistance with the proofs and index, and our colleagues at Cambridge University Press for their valuable assistance in the preparation of the book.

A number of papers have appeared since the book was completed; those that have reached us are included in a short addendum.

M. P.
J. L. L.

Babraham,
October 1974

1 HISTORICAL INTRODUCTION

Knowledge of the existence of salt glands in birds and reptiles is recent. The function of avian nasal glands was discovered by Knut Schmidt-Nielsen, C. Barker Jörgensen and Humio Osaki in the Double-crested Cormorant (*Phalacrocorax auritus*), and the first account of their work was published in 1957, as an abstract in *Federation Proceedings*; in the following year a full account was published in the *American Journal of Physiology.*

The discovery was made as a result of asking a simple question. How do oceanic birds survive at sea for long periods without fresh water and do they drink sea water? The background to the problem cannot be better put than in the introduction to their paper:

It is well known that sea water is toxic to man and most other mammals. Marine mammals, such as seals and whales, face a special problem since they have no access to fresh water and must either abstain or drink sea water. The evidence that has been collected indicates rather strongly that marine mammals obtain so much water with their food that the normal losses by evaporation, feces and urine can be adequately covered, and that salts as well as urea can be excreted without difficulties. Hence, these animals should have no need for drinking sea water, and there is no reason to believe that they do.

Marine birds have a problem similar to that of marine mammals. Many birds remain at sea for extended periods, far removed from any possible sources of fresh water. Ornithologists have discussed the problem of water supply for oceanic birds, and it has frequently been stated that they do indeed drink sea water. It has even been stated that 'Most birds are known to drink salt water in preference to fresh; indeed captive gulls may die without it'. However, ornithological literature also expresses the opposite view-point, that marine birds do not need to drink.

It would be extremely difficult to prove whether oceanic birds regularly use sea water as a normal water supply. If the food consists of fish the water content of the food is some 80 %, and the salt concentration is appreciably lower than that of sea water. Birds that feed on invertebrates (for example, petrels and penguins) have a more serious problem because these organisms are in osmotic equilibrium with sea water and hence have a high salt content.

While mammals need a considerable quantity of water for urea excretion, birds have no such problem of nitrogen excretion. They excrete insoluble uric acid or urates which require only small amounts of water. In birds the focus of attention must therefore be on the salt metabolism. Furthermore, their water requirement may be higher than that of marine mammals because of problems of heat regulation. Aquatic mammals probably do not evaporate water for heat dissipation, while in birds the heat loss through the well insulated body surface is small, and the respiratory tract, including the air sacs, may play a major role in the heat regulation through evaporation from their moist surfaces.

In the course of investigations of the salt and water balance of marine birds (cormorants) we obtained no evidence in support of drinking sea water, but it was found that their salt excretion is very unusual. Glands in the nasal region can secrete a highly concentrated sodium chloride solution, and this extrarenal route for elimination of salt plays a major role in excretion.

Briefly, for we shall be dealing with these aspects in other chapters, Schmidt-Nielsen and his colleagues found that following the administration of sea water to cormorants by stomach tube, a substantial proportion of the salt was excreted extrarenally.

It was observed that birds subjected to a salt load secrete a clear, water-like fluid from the nasal cavity. The secretion accumulates at the tip of the beak, from which the birds shake the drops with a sudden jerk of the head. The nasal secretion was observed only under osmotic load. It did not occur in fasting birds or in birds fed fish.

The concentrations of sodium and chloride in the nasal secretion were always high and since its rate of flow was appreciable the total amount of salt eliminated this way must be of major importance in the salt balance. The mechanism is unique among excretory processes of higher vertebrates...[The composition of nasal fluid is shown in Table 1.1]. The concentrations usually were rather constant and in most samples sodium and chloride concentrations were almost identical and between 500 and 600 mEq/l. The potassium concentration was always low. Magnesium and sulfate, which are present in appreciable amounts in sea water, were virtually absent from nasal secretion. Protein is apparently absent.

Hence, the nasal secretion is a practically pure solution of sodium chloride, with a concentration twice as high as the maximum renal concentration. The rate of elimination is so high that with continuous secretion the entire sodium content of the body could be eliminated in roughly 10 hours.

This, together with later work (Fänge, Schmidt-Nielsen & Robinson, 1958), clearly showed that the gland responsible is the nasal gland, which although opening into the nasal cavity is usually situated, in marine birds, subcutaneously between the eyes. The group then went on to make important observations on the control of secretion by the nasal gland and they, together with McFarland (1959; 1960*b*, *c*), extended their work to other marine birds (Schmidt-Nielsen & Fänge, 1958*c*; Schmidt-Nielsen & Sladen, 1958). These early studies are

Table 1.1. *Typical composition of nasal fluid in the gull,* Larus argentatus (modified from Schmidt-Nielsen, 1960)

	(mM)
Sodium	718
Potassium	24
Calcium+magnesium	1
Chloride	720
Bicarbonate	13
Sulphate	0.35

described by Schmidt-Nielsen in reviews published in 1960, 1963 and 1965, as well as in a popular article in *Scientific American* in 1959; these are required reading for anyone interested in any aspect of salt-gland function.

This important discovery had another major effect. In one stroke it ended centuries of speculation, based on detailed and voluminous anatomical studies, on the function of the nasal glands in birds. The glands were first described in the duck at a meeting of anatomists in Amsterdam on 14 November 1665, *Supra oculi orbitam glandula est magna, lata, longa, rubra, carnosa ad utrumque oculi canthum extensa, hinc cum annexis membranosis tendinibus, totum implebat circulum. Glandula pressa saccum flavescentum crassum insipidum dabat.* Their discovery has often been attributed to Caspar Commelin (sometimes called Comelin or Commelinum), but in fact he was what might now be described as minute secretary of the college, and because no names appear with individual contributions it is impossible to decide which member of the Amsterdam Private College actually demonstrated his findings at the meetings. Therefore the discovery may have been made by Blasius, Veen, Ruy(s)ch, Quina, Boddens, Cordes, Swammerdam, Godtke or even the English anatomist Matthew Slade. The reference to the rare account of these meetings is shown in the reference list under Anon (1667) in the form employed in the *British Museum Catalogue.*

Schmidt-Nielsen and his colleagues must have been surprised and delighted at their result and at once appreciated its importance. It had been known for some time that the nasal glands are much larger in marine birds and even that their size changes when the salinity of the drinking water is altered (Chapter 9) but the generally accepted view was that the secretion from the glands flushes sea water out of the nostrils in order to protect the mucosa of the nasal cavity (Stresemann, 1928; Marples, 1932; Technau, 1936)!

We should point out that the term 'salt gland' was used by Schmidt-Nielsen and his colleagues in the Department of Zoology at Duke University in North Carolina because they not only discovered salt glands in birds but in marine reptiles as well (Chapter 12). However most reptilian salt glands are not homologous with the nasal gland of marine birds so a general descriptive term was adopted. Schmidt-Nielsen (1960) continued, '...this designation then denotes any gland in the head region of marine birds and reptiles which, irrespective of anatomic origin, has an osmoregulatory function and secretes highly hypertonic sodium chloride solutions'. At this time of course salt-secreting cranial glands had not been discovered in terrestrial reptiles and birds and few would disagree with including these glands in the same category.

The discovery of salt glands came at a time when research into the comparative aspects of salt and water metabolism was booming. It seems possible that this was largely due to the availability of commercial flame photometers which permitted the hitherto difficult and tedious determination of sodium and potassium in biological fluids as a matter of routine. Such studies were also of course continuations of research and interest aroused earlier by such doyens in this field as August Krogh and Homer Smith, but which was interrupted by the second world war. With such a background, work carried out in the 1950s, often by Schmidt-Nielsen and his colleagues, led to major advances being made in our knowledge of salt and water regulation in vertebrates. Finally, when J. Wendell Burger and W. N. Hess discovered that the rectal gland of cartilaginous fishes is also a salt-secreting gland in 1960, major extrarenal avenues of salt and water exchange were evident in most classes of vertebrates.

It is sad that the size and complexity of the scientific world usually precludes the passage of knowledge from one discipline to another and that the formal manner of scientific literature tends to inhibit the publication of casual observations which may give vital clues to other workers. For example, following the discovery of salt glands two eminent zoologists and field naturalists realized that they had observed nasal secretion in the wild and their letters were published in *Nature*:

Salt Excretion in Marine Birds

In connexion with recent research on salt excretion in marine birds the following observation which I made some twenty-five years [1933] ago may be of interest. While cruising in a yacht far from land in the northern North Sea, I often used to watch lesser black-backed gulls as they soared above the ship or were perched on the mast head. I was very puzzled to account for a regular rain of drops of water which each bird shook at intervals from the tip of its beak. This was evidently salt excretion in action, and I know of no similar published account.

F. S. Russell

The Laboratory, Citadel Hill,
Plymouth

May I add a further observation to F. S. Russell's interesting communication under this title? More than thirty years ago [before 1928] I noticed a similar thing in the giant petrel or stinker (*Macronectes giganteus*) of the southern oceans, and published* an account of it, though not in a scientific paper, in 1951: 'When [a flying stinker] is right alongside and you can see him plainly, nearly always you will notice that he has a drop on his nose; hanging from the tip of the hook at the end of his beak is a drop of something – whether water or some other fluid I know not.' This, too, was evidently salt secretion in action.

L. Harrison Matthews

Zoological Society of London,
Regent's Park, London, N. W. 1.

It is interesting to speculate whether, if the anatomists and naturalists had been able to share their information at this time, the true function of nasal glands would have been deduced 25 years earlier. However it is clear that their interest in and wide knowledge of salt metabolism meant that Schmidt-Nielsen and his colleagues at once appreciated the importance of their discovery. Another spur to the study of salt glands (the rate of growth of the literature on this topic is shown in Fig. 1.1) was the realization (Scothorne, 1958*b*, 1959*a*, *b*, *c*, *d*) that it is not essential to have truly marine birds as experimental animals. Even the domestic duck and goose are suitable experimental animals because the glands in these birds, although smaller than in truly marine forms, will immediately secrete a hypertonic sodium chloride solution in response to the administra-

* Matthews, L. Harrison (1951) *Wandering Albatross*. Macgibbon and Kee: London.

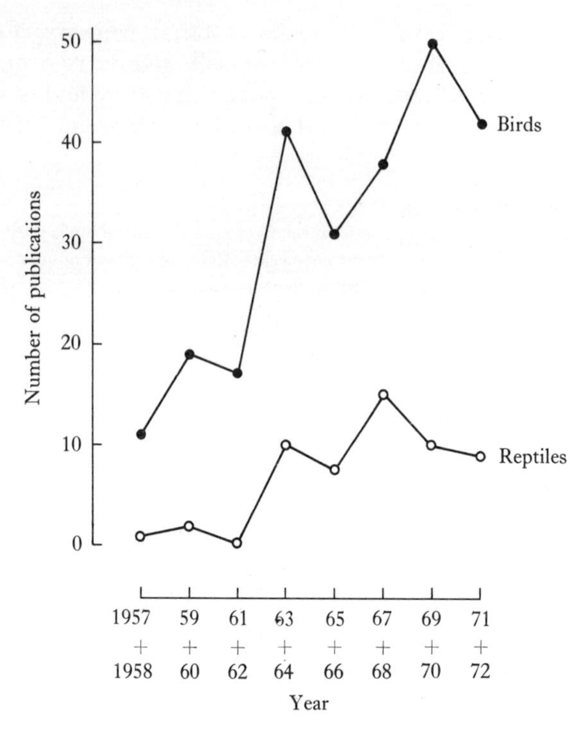

Fig. 1.1. Graph showing the growth of the literature on salt glands in birds and reptiles (preliminary communications, reviews, chapters but not theses are included).

tion of salt water. The advantages of this discovery to the laboratory worker are obvious. It should be pointed out that the domestic duck and goose referred to throughout are the domesticated varieties of *Anas platyrhynchos* and *Anser anser* respectively.

PART ONE · AVIAN SALT GLANDS

2 STRUCTURE OF SALT GLANDS

About 150 years ago was the start of the heyday of comparative anatomy, when many species were being discovered and dissected. Jacobson (1813) and Nitzsch (1820) first described in detail the anatomy of nasal glands in birds. Later a British surgeon travelling in the tropics gave the first account in English.

In 1828, when dissecting the head of the Albatross, I observed, imbedded in a bony cavity, situated immediately over the orbit, a gland, which was covered above by a dense fascia. The cavity to which the gland corresponded was of a semilunar form, and situated over each orbit: at the anterior part of this cavity or depression a small portion was left membranous, excepting a minute orifice, permitting the passage of what seemed to me to be an excretory duct, but the course of which I lost soon after it had penetrated this membrane. The floor of the cavity was perforated by numerous minute foramina, probably for the passage of nutrient vessels to the substance of the gland. This gland is found in most, if not in all, the aquatic birds, but varies in them, both with respect to exact situation or extent...

In July, 1832, during a passage to New South Wales, a capture of an Albatross afforded me an opportunity of again dissecting this gland, with the view of ascertaining, if possible, whether an excretory duct actually existed. I found the gland of a hard granulated substance and pale colour, consisting of numerous, distinct, minute oval bodies, and on being cut it is found to be abundantly nourished by blood-vessels; the nerves supplying it came from the minute foramina seen on the floor of the cavity, and are distributed in and about the substance of the gland...Not finding any duct when I came to the anterior portion of the gland, which was visible, I commenced a further research, by laying open the bony plate which covered the olfactory part of the upper mandible, continuing the destruction of this portion of the bill, on one side, to the nostril. I then found that the gland was continued for a short distance further, under the bone, towards the nostril, situated rather above and anterior to the nasal portion of the orbit. On pursuing my dissection further, I found a nerve (a branch of the fifth pair?) passing down the thin bony plate, at the interior part of the upper mandible; this I traced until it entered the skull, and thus found it had no connexion with the gland, as its first appearance led me to suspect; but close to and under it was another appearance, which could be distinctly traced, emerging from the gland; it was about a line, or rather more, in thickness, and, tracing its course, I found it proceeded in a straight direction, and then had an almost imperceptible inclination upwards, until I lost it among the cellular substance of the upper mandible, (to which it was attached partially, if not entirely) rather more than an inch from the base of the bill; the length of the duct as far as traced, was one inch and seven-eighths. I made a drawing of the preparation.

An interesting subject next for inquiry is, what this gland secretes, and what is its use in that situation ? which at present cannot be answered.

From, *Wanderings in New South Wales, Batavia, Pedir Coast, Singapore and China*, vol. II, by George Bennett, Esq. FLS, Fellow of the Royal College of Surgeons, &c. Richard Bentley: London, 1834. [The drawing referred to was demonstrated at a meeting of The Zoological Society of London in 1834.]

There is no shortage of papers describing the gross anatomy, histology and histochemistry of the nasal glands in birds. Those seeking detailed information may consult: Jacobson (1813); Nitzsch (1820); Bennett (1834*a*, *b*); Jobert (1869); Gaupp (1888); Ganin (1890); Wetsheloff (1900); Cords (1904); Marples (1932); Mihálik (1932); Technau (1936); Webb (1957); Fänge, Schmidt-Nielsen & Osaki (1958); Fänge, Schmidt-Nielsen & Robinson (1958); Schmidt-Nielsen & Fänge (1958*a*); Scothorne (1958*a*); Bang & Bang (1959); Schmidt-Nielsen (1959; 1960); Lange & Staaland (1965); Staaland (1967*b*); McLelland, Moorhouse & Pickering (1968); Ash, Pearce & Silver (1969); Cottle & Pearce (1970).

Histology and histochemistry

(Further more specialized references are given in other chapters.)

Marples (1932); Fänge, Schmidt-Nielsen & Osaki (1958); Scothorne (1958*b*; 1959*a*, *b*, *d*; 1960); Schmidt-Nielsen (1959; 1960); McFarland (1960*a*); Goodge (1961); Natochin & Krestinskaya (1961); Zaks & Sokolova (1961); Bernard & Wynn (1963*b*); Ellis, Goertemiller, DeLellis & Kablotsky (1963); Benson & Phillips (1964); Ellis (1965); Buxton, Hally & Scothorne (1966); Spannhof & Jürss (1967); Dulzetto (1967); Staaland (1967*b*); McLelland *et al.* (1968); Burock, Kühnel & Petry (1969); Kühnel, Burock & Petry (1969); Kühnel, Petry & Burock (1969).

Ultrastructure

Doyle (1960); Scothorne & Hally (1960); Fawcett (1962); Komnick (1963*a*, *b*, *c*; 1964; 1965); Komnick & Komnick (1963); Dulzetto (1965); Ernst & Ellis (1969); Komnick & Kniprath (1970); Levine, Higgins & Barrnett (1972).

Gross anatomy

The nasal glands are usually situated in or around the orbit. The different positions have been classified by Technau (1936) and are shown in Fig. 2.1. Marine birds fall into all these categories and it seems that the position is related to the size of the head, position of the eyes etc., which in turn are determined by feeding habits and other factors. However in most marine birds the gland is above the eye and there are depressions in the skull where the glands are situated. The presence of these led Marples (1932) to suppose that large nasal glands were present in the extinct, presumably aquatic birds, *Hesperornis* and *Ichthyornis.* The glands in extant birds can be so extensive that they actually cover the surface of the skull above the eyes and meet in the mid-line (Fig. 2.2); in such forms there is usually a supra-orbital foramen through which blood vessels reach part of the gland.

There are one or two ducts on each side, which enter the nasal cavity. Thus the secretion flows out of the nostrils or is thrown out by a characteristic shake of the bird's head. In many birds (e.g. pelicans, domestic geese and ducks) there are ridges along the edge of the beak which prevent nasal fluid from flowing into the mouth; petrels, which spend many months at sea but only rarely settle on the water, have tubular extensions of the nostrils. Nasal fluid is blown out through these tubes (Plate 2.1) and Schmidt-Nielsen (1959) has suggested that they may be an adaptation for eliminating nasal fluid. Continuous flight may hamper the flow of fluid from the nostrils because of the current of air over this region and the tubular extensions through which the fluid can be blown by a forced expiration could well act in the way Schmidt-Nielsen suggested. Some marine birds, like the Gannet (*Sula bassana*) and cormorants (*Phalacrocorax*), which dive into the sea from a great height in

Structure of salt glands

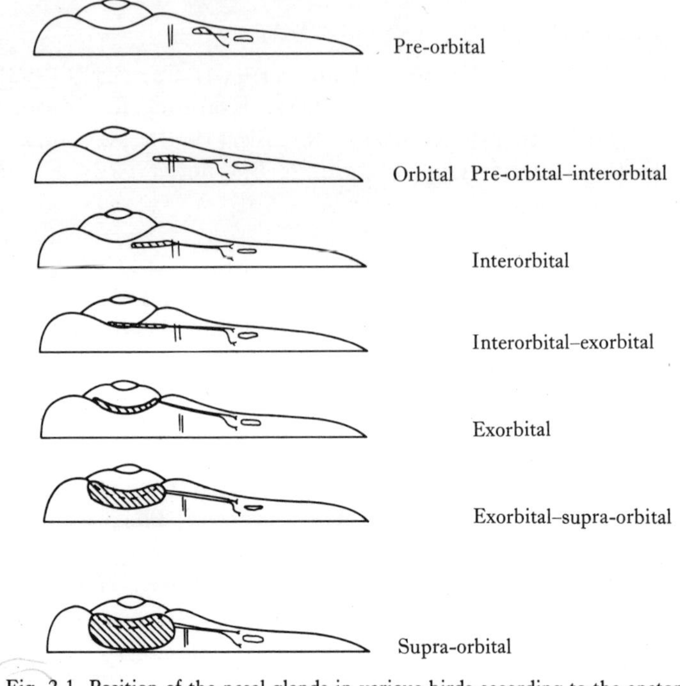

Fig. 2.1. Position of the nasal glands in various birds according to the anatomist G. Technau. Note the two ducts on each side (from Technau, 1936).

order to catch fish, have closed nostrils. In these forms secretion flows from the nasal cavity, through the internal nares and along the roof of the mouth to the tip of the beak.

General arrangement of secretory elements

The different parts of the duct system and secretory portions of this compound tubular gland (there are no acini) have been called by a variety of names. Different authors have used such terms as duct, ductule, lobe, lobule, tubule, lateral diverticulum, central canal etc., to describe what is essentially a simple system. In avian nasal salt glands, the *lobes* or *lobules* are surrounded by connective tissue. Radially-arranged *secretory tubules* open into the *central canal* of the lobe. In most species these central canals, which are ducts (i.e. non-secretory), join to form *secondary ducts* which in turn enter the *primary* or *main*

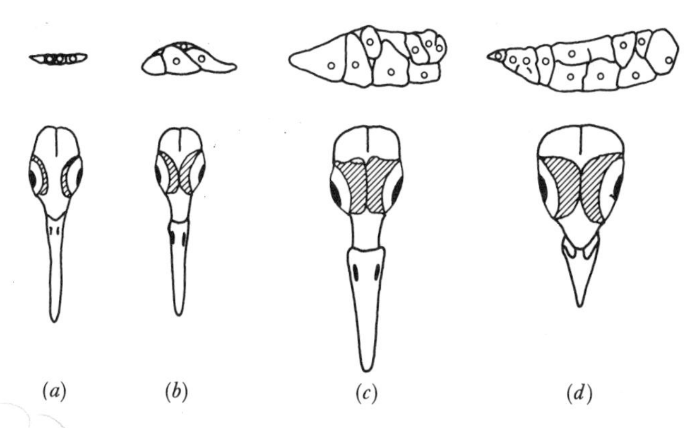

Fig. 2.2. The position and size of the salt glands in four species of wading-bird (all of which eat invertebrates) according to the time spent in a marine environment. Note the larger glands in the more marine species (towards the right). The cross-sections of the glands show the lobes and central canals: (*a*) Wood Sandpiper; (*b*) Common Sandpiper; (*c*) Knot; (*d*) Little Auk (from Staaland, 1967*b*).

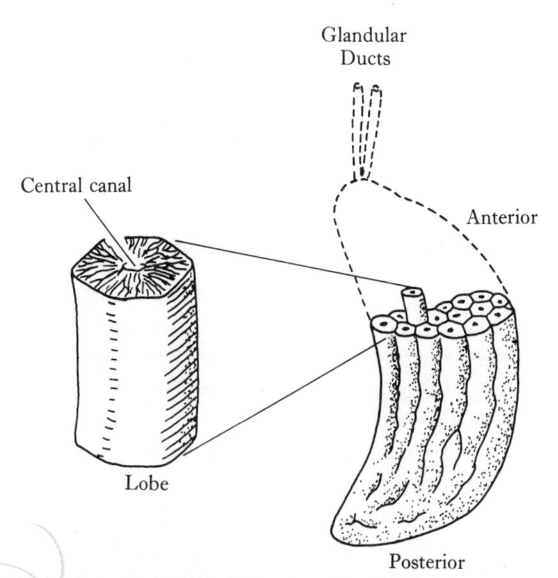

Fig. 2.3. The arrangement of lobes in the salt gland of the Herring Gull, *Larus argentatus* (from Fänge, Schmidt-Nielsen & Osaki, 1958).

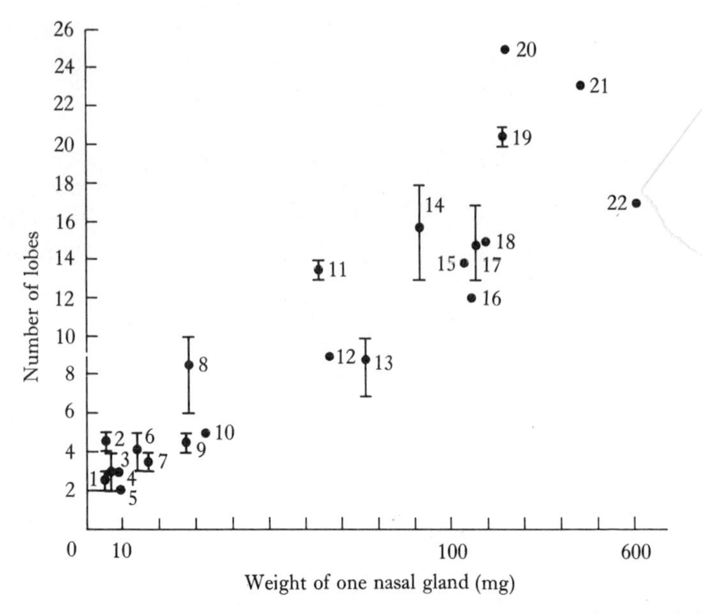

Fig. 2.4. The relation between the number of lobes and the weight of one nasal gland in various wading-birds (Charadriiformes). Vertical lines are the ranges of variation and the dots are the mean values. (1) Green Sandpiper, (2) Common Sandpiper chicks, (3) Wood Sandpiper, (4) Little Stint, (5) Snipe, (6) Common Sandpiper adults, (7) Ringed Plover, (8) Dunlin, (9) Golden Plover, (10) Sanderling, (11) Bar-tailed Godwit, (12) Knot, (13) Black-headed Gull, (14) Little Auk, (15) Kittiwake, (16) Razorbill, (17) Black Guillemot, (18) Puffin, (19) Guillemot, (20) Brünnich's Guillemot, (21) Herring Gull, (22) Glaucous Gull (from Staaland, 1967*b*).

ducts. It is the main ducts which leave the gland and run to the nasal cavity.

The general arrangement of the gland, although of this basic arrangement, varies between species. In many marine birds, for example the Herring Gull (*Larus argentatus*), the lobes run longitudinally for almost the entire length of the gland (Fig. 2.3). In others like the duck and goose the lobes are smaller and branches from the primary and secondary ducts arise all through the gland rather than at the anterior end. In these forms the lobes are smaller and are at an angle to, rather than parallel with, the main axis of the gland and a transverse section through the gland shows some lobes cut obliquely; both central canals and secretory tubules have been seen entering a main duct directly. For example, Marples (1932) found, in the duck, that the median duct continues into the gland and small lobes and tubules empty into it along its entire length. In contrast, the lateral duct also continues into the gland but has large branches which further divide into central canals.

The number of lobes in the gland of different species of one order of birds (Charadriiformes) was found by Staaland (1967*b*) to vary between two and 25, and was related to the absolute weight of the gland (Fig. 2.4); the minimum number of lobes presumably reflects the dual origin of the gland (see later).

Each lobe is surrounded by connective tissue which is continuous with the capsule surrounding the gland. In addition, the central canals through the lobes are also surrounded by connective tissue through which the secretory tubules pass. The tubules branch and radiate outwards, the blind ends reaching the perilobular connective tissue. The secretory tubules are tightly packed and run parallel to each other, separated by thin strands of connective tissue and blood capillaries.

Secretory tubules. The secretory tubules are branched and form two main regions which can be differentiated by staining and histochemical techniques. Towards the central canal the tubules consist of a single layer of columnar (principal) cells and the lumen is apparent. In the peripheral region, the diameter of the tubules is not so great, the lumen is less obvious and the cells are smaller. This difference is related to the volume of the cytoplasm rather than to that of the nuclei.

The peripheral cells, i.e. those towards the blind end of the tubules, can be distinguished from the central or principal cells by their positive reaction for the presence of alkaline phosphatase, far fewer mitochondria and less indication of enzymes associated with metabolic activity. By contrast the principal cells are columnar, rich in phospholipid, mitochondria, and such enzymes as succinic dehydrogenase and acid phosphatase. Their striated appearance reflects the basal infoldings seen in the electron microscope (p. 22).

In terms of function there is no doubt that the central principal cells are the ones primarily concerned with ion transport. However, Ellis (1965) showed clearly that the peripheral cells are the ones that divide, and therefore, are important during the growth of the gland which occurs when birds start to drink salt water (Chapter 9). He did this in two ways. Firstly by injecting [^{3}H]thymidine into ducklings given either fresh water or salt water to drink; the location of tritium in the tubules was then determined at intervals by autoradiography. Two days after the thymidine was given, label was seen only in the nuclei of the peripheral cells. After six weeks the cells mid-way between the central canal and the periphery were labelled, and after twelve weeks tritium was only found in a few cells near the central canal. Secondly, cell division was arrested by exposing the salt glands to a strong source of X-rays. The presence of alkaline phosphatase was then used as an indicator of the presence of peripheral cells. One week after irradiation no cells showing a positive reaction for alkaline phosphatase were seen. Three weeks later the glandular epithelium of the tubules was flattened towards the centre of the lobe and the tubules appeared dilated. These experiments indicate that the small peripheral cells are the site of cell division, that there is a slow turnover of cells in the tubule and that old cells are lost at the centre of the lobes.

Ducts. The duct system within the gland, including the central canals, are lined by a multi-layered epithelium two to four (usually two) cells thick; most authors have called it a pseudo-stratified epithelium. Towards the openings into the nasal cavity the epithelium of the main ducts is keratinized.

The cells of the duct system do not appear to be very metabolically active. There are few mitochondria, little phospholipid or other histochemical signs of high activity, although they do show a reaction for acid phosphatase.

Other histological features. In birds that have not been given salt water at all or for some time, material that looks like cellular debris can be seen in the lumen of the ducts and secretory tubules (in ducks and geese this is washed out when secretion is induced for the first time, the samples of nasal fluid being cloudy for up to ten minutes after the onset of secretion).

The gland virtually secretes only salt water to the exterior. There is no sign of mucus in the secretion, in the contents of the ducts or in the secretory cells (as shown by staining with periodic acid–Schiff reagent or toluidine blue). Scothorne (1959*a*) also noted that there is no histological evidence to suggest that it is a serous gland which secretes enzymes as well as salts and water; perhaps since the gland discharges its contents to the exterior this is not surprising.

Fänge, Schmidt-Nielsen & Osaki (1958) could not detect any smooth muscle in the gland except in the walls of arteries. In some birds cells containing a black pigment have been seen in the connective tissue. Zaks & Sokolova (1961) found a narrow band on the basal side of the secretory tubules which stained metachromatically with toluidine blue. This indicates the presence of a mucopolysaccharide which has also been observed in reptilian salt glands (p. 235).

Blood supply

There is considerable confusion concerning the nomenclature of blood vessels in the head of birds. We shall therefore follow most recent usage and ignore the problem of homology. In fact, from a functional point of view, the best and most useful account is also the most recent, that by Fänge, Schmidt-Nielsen & Osaki (1958) (Fig. 2.5). The large arteries supplying the salt gland arise from the internal carotid. The main artery enters the orbit through a foramen situated just above the optic nerve; this is usually called the internal ophthalmic artery and it follows the wall of the orbit and then divides. The anterior branch sends small vessels into the anterior part of the gland and surrounding structures and then continues into the beak. Jobert (1869) found that this branch anastomoses with another artery from below at the anterior edge of the orbit. The posterior branch of the internal ophthalmic artery supplies vessels to the rest of the gland, and may pass through foramina in the skull. Near, or on the undersurface of the gland this posterior branch anastomoses with another artery, the external ophthalmic. From this junction an anastomotic vessel runs forward along the base of the gland to join the anterior branch of the internal ophthalmic. Blood vessels from this anastomotic vessel pass into the gland. Clearly this is a complex system and Fänge, Schmidt-Nielsen & Osaki (1958) have suggested the

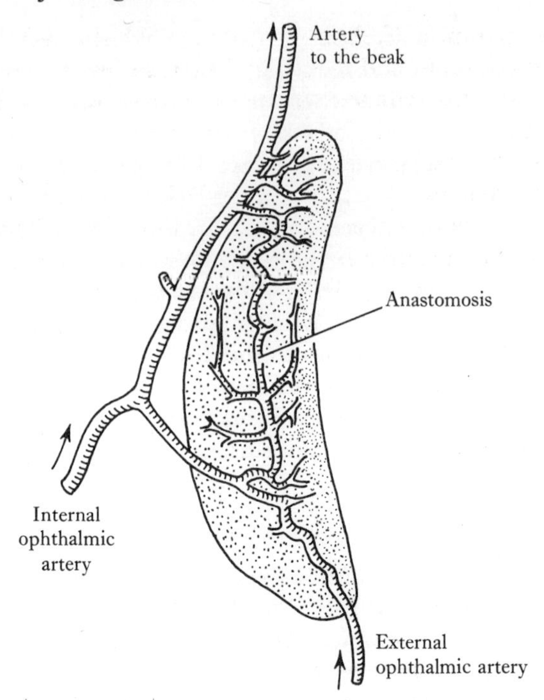

Fig. 2.5. The arterial supply of the salt gland in the Herring Gull (*Larus argentatus*). The sketch, drawn from a plastic cast of the vascular system, shows the left gland from below (from Fänge, Schmidt-Nielsen & Osaki, 1958).

arrangement may serve to ensure that other structures supplied by these arteries have an adequate blood supply whether the glands are secreting or not. They point out that the arteries to the salt gland of the Herring Gull are amongst the largest in the head because the gland needs a good blood supply to support the very high rate of blood flow that occurs during secretion (p. 70).

The veins in general follow the course of the arteries but nothing is known about the lymphatic drainage of the gland.

It may be mentioned that, as far as physiological experiments are concerned, the complexity and inaccessibility of the blood vessels of avian salt glands precludes the possibility of measuring blood flow by direct methods, perfusing the gland in isolation, transplanting it to a more convenient site, collecting samples of venous effluent or making close-arterial injections!

Fänge, Schmidt-Nielsen & Osaki (1958) studied the blood supply of the lobes and they found that the small arteries pass straight into the connective tissue surrounding the central canal from the perilobular connective tissue, without sending branches into the secretory tissue. In the central region the arterioles then break up into capillaries which radiate as a network between the secretory tubules and eventually reach the interlobular connective tissue. Here they join to form a venous plexus which in turn is drained by veins (Fig. 2.6, 2.7, Plate 2.6). This arrangement means that blood in the capillaries flows in a counter-current manner to the flow of secretion down the tubules – a point to which we shall return in other chapters. Further evidence in support of this postulated arrangement of the micro-circulation is that veins have not been seen in the central region of the lobes.

Innervation

The salt gland is innervated by nerve fibres which enter the anterior end of the gland from a ganglion situated on the Vth cranial nerve, as it runs through the orbit. A nerve from below, the 'secretory nerve', which probably carries fibres of the VIIth, IXth cranial nerves and sympathetic system, enters this ganglion (the ethmoidal or 'secretory nerve ganglion'). See also 'Nervous control of secretion' in Chapter 4.

Ultrastructure

Histological and cytological techniques have, to a great extent, been superseded by electron microscopy. It is not necessary to give all the older evidence which suggests, for example, that mitochondria are abundant, when later ultrastructural studies have provided unequivocal evidence that indeed they are.

Examination of the salt glands of marine birds in the electron microscope began almost as soon as their function was discovered. The paper by Doyle (1960) was the first and has been widely quoted but his observations have been superseded, as techniques improved. However he did show the extensive infoldings of the basal cell membrane, a characteristic feature. The salt gland of the Herring Gull (*Larus argentatus*)

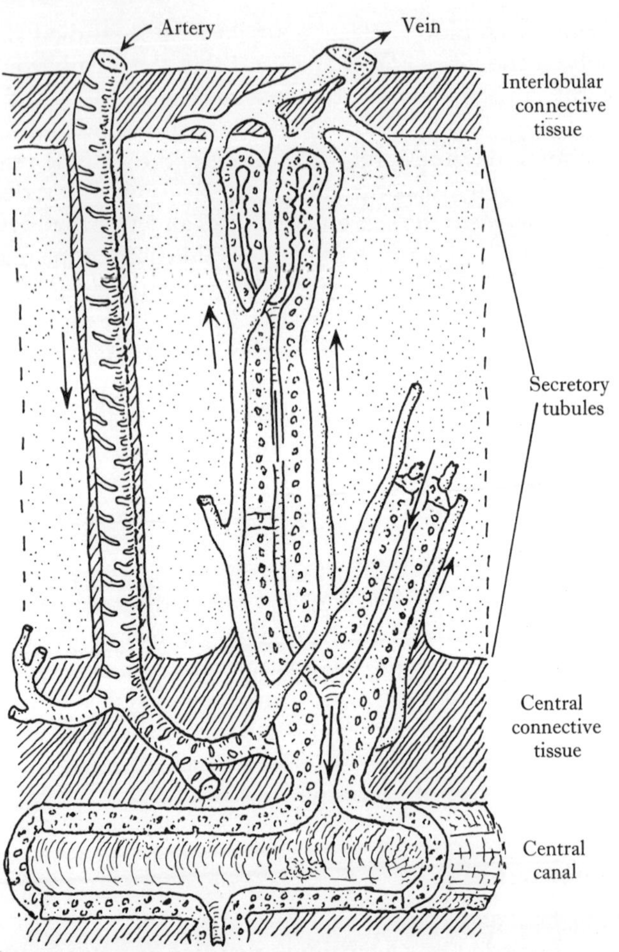

Fig. 2.6. Diagram of the micro-circulation showing the counter-current arrangement of blood flow relative to the flow of secretion along the tubules (from Fänge, Schmidt-Nielsen & Osaki, 1958).

was later fully, even exhaustively, studied by Komnick and his co-workers and their work alone accounts for 215 journal pages. Unfortunately it is difficult to separate observations from tendentious deductions on the secretory mechanism in their papers. In general these deductions cannot be supported by physiological evidence and we have therefore restricted our consideration to descriptive passages in his papers and to his

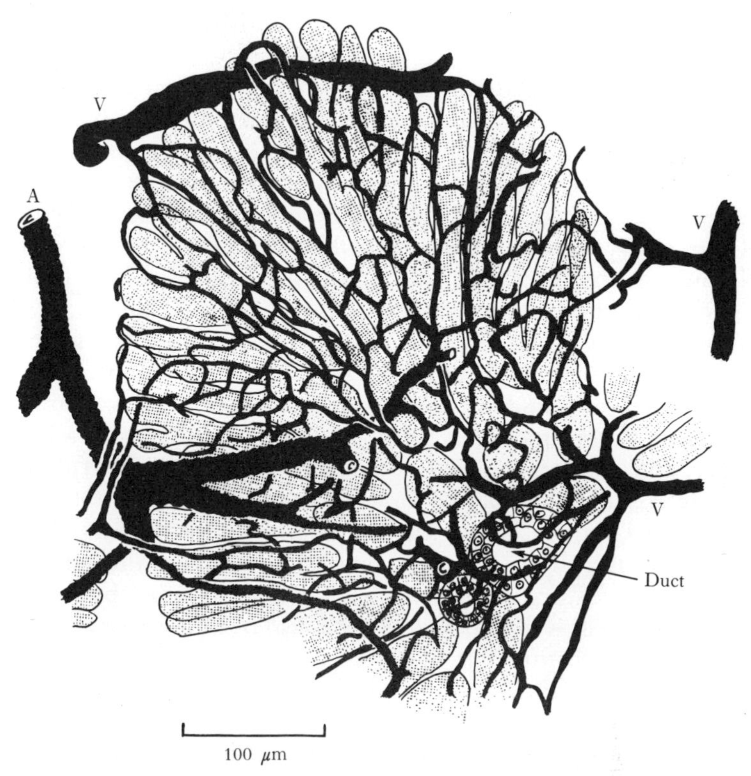

Fig. 2.7. Diagram of the complete blood supply to part of a lobe of the salt gland of a duck. At the height of secretion following a salt-load, the bird was killed by the intravenous injection of pentobarbitone. The head was immediately perfused via the carotids with a depolarising solution (high potassium, low sodium), to induce relaxation of vascular smooth muscle, and then with carmine gelatin. After fixation in formol saline 100 μm frozen sections were cut, cleared and then mounted unstained. The tissue appeared to have all vessels filled with the injection mass. The diagram is based on camera lucida drawings and photomicrographs. The area was also carefully examined by phase contrast, to identify arteries and veins, and with a binocular stereosopic microscope, to differentiate between superficial and deep vessels. V, vein; A, artery.

published electron micrographs which are of good quality. We have drawn on the findings of other workers (p. 11), but the structure of the cells is modified considerably by the salinity of the food and water consumed by such birds and these changes are also considered in Chapters 7 and 9.

Principal cells

The centrally-located principal cells of the secretory tubule can be distinguished from those at the extreme blind end by their appearance in the light microscope. However, any doubt that these principal cells are highly active is immediately dispelled by their appearance in the electron microscope. In salt glands from truly marine birds the principal cells show extreme folding of the basal and lateral cell membranes. These folds extend as far as, and even beyond the nucleus (Plate 2.2, 2.3, Fig. 2.8) and must present a very large surface area to the interstitial fluid. The cytoplasm of course extends into these finger-like folds and is packed with mitochondria. In contrast the luminal or apical cell membrane is almost flat with a few microvilli extending into the small lumen of the tubule (Plate 2.3). The Golgi apparatus, although usually not well developed, is present in the cytoplasm around the nucleus and although there is some endoplasmic reticulum it is not a prominent feature of these cells. Vesicles bounded by membranes as well as tonofilaments have also been observed in the cytoplasm. All the cells lie on a basement membrane which separates the epithelium from underlying connective tissue and blood capillaries.

The secretory cells, including those at the blind end of the tubules are joined to each other by typical junctional complexes consisting of zonulae occludentes (tight junctions), zonulae adhaerens and desmosomes. These junctional complexes are situated on the lateral cell membrane very close to the lumen of the tubule (Plate 2.4). Doyle (1960) failed to find them and he felt there might be some continuity between the interstitial space and the lumen. However Scothorne & Hally (1960) and all subsequent workers have reported their presence. Desmosomes also occur along much of the lateral membrane where folds from adjacent cells are in close apposition. These would appear to be 'spot welds' in contrast to the continuous tight junction which extends right round each cell and forms some sort of structural seal between the lumen and the interstitial space.

When birds with fully functional salt glands are kept on fresh water the amount of basal infolding is markedly reduced and

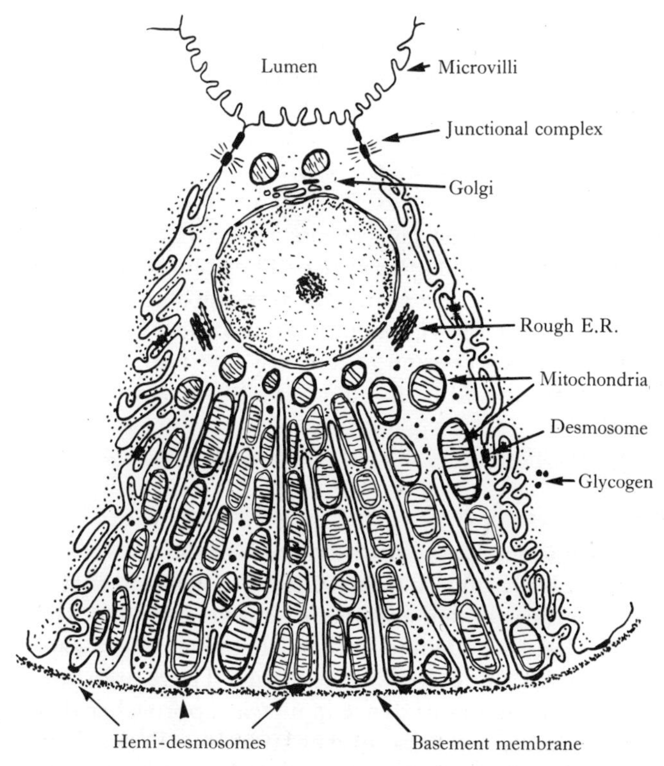

Fig. 2.8. Diagram showing the structure of a salt-gland secretory cell: E.R., endoplasmic reticulum.

there are fewer mitochondria. Although these changes are discussed more fully in Chapter 9, we should point out that Ernst & Ellis (1969) have classified the types of principal cell present. Adjacent to the peripheral cells (see below) the principal cells show some lateral infolding but the basal membrane is almost flat and there are relatively few mitochondria ('partially-specialized cell'). Nearer the centre of the lobe the 'fully-specialized' type of cell is present in two stages of development, showing progressively more basal and lateral folding and more mitochondria. In birds on salt water the highest degree of development is reached. Evidence has already been presented that the cell population is renewed by division of the peripheral cells and that development into progressively more specialized cell types occurs along the tubule, the final degree of development and total number of cells being dependent on osmotic conditions in the external environment (p. 15, see p. 163).

Peripheral cells

Ultrastructurally these cells do not appear to be very active and have an unspecialized appearance. The volume of cytoplasm is not large and there are abundant free ribosomes, some rough endoplasmic reticulum, smooth vesicles and a Golgi apparatus in the perinuclear region; there are few mitochondria. The basal membrane and lateral membranes are virtually unfolded (Plate 2.4).

Ducts

The only studies on the ultrastructure of the duct system have been made by Komnick (1963*a*; 1964). He found that the cells form a pseudostratified epithelium and are cuboidal, the nuclei ovoid or slightly indented and, in the lower layers especially, the volume of the cytoplasm is small. There are some mitochondria but their abundance in no way approaches that of the principal secretory cells of the tubules. From Komnick's micrographs it appears that the cells of the different layers are joined by desmosomes; there is a small amount of interdigitation between neighbouring cells, and the whole epithelium lies on a basement membrane.

The layer of cells lining the duct lumen has a somewhat different structure. The cells are more columnar but the cytoplasm nearest the luminal membrane (accounting for about a quarter of the cytoplasm) is a clear layer with inclusions of tonofilaments and some endoplasmic reticulum. Mitochondria are not found in this region but are present between this layer and the nucleus, as well as in the rest of the cytoplasm. It would appear that the cells of this upper layer are joined to each other by typical junctional complexes near the lumen and by desmosomes along the lateral membranes. Desmosomes also connect these cells to those lying below (Fig. 2.9).

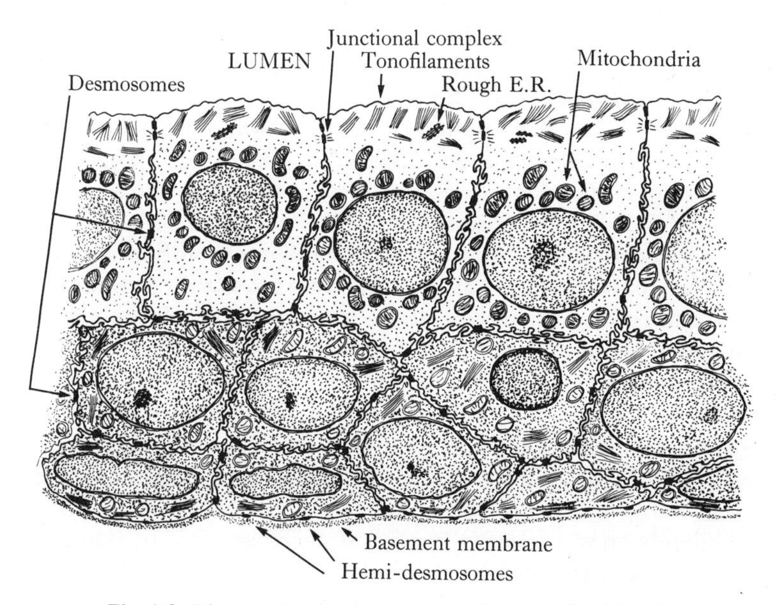

Fig. 2.9. Diagram showing the structure of an intraglandular duct: E.R., endoplasmic reticulum.

Blood capillaries

The capillaries supplying the secretory cells are fenestrated with pores 300–400 Å in diameter and vacuoles are present in the cells of the wall. There is only a short distance between a capillary and the basement membrane of the secretory epithelium.

Nerves

See Chapter 4.

Myoepithelial cells?

Although there is no doubt that secretion flows *vis à tergo* down the tubules and ducts to the exterior there is slight evidence for the presence of myoepithelial cells in the salt gland which might by their contraction aid this egress. Håkansson & Malcus (1970) found that, before the pressure in the ducts increased during secretory nerve stimulation, a small decrease in pressure occurred. This, they suggested, could be due to an increase in the diameter of the glandular lumen and might be due either to vasoconstriction or to the action of myoepithelial cells. We presume that in the latter case they envisage a shortening and widening of the secretory tubules or ducts, as in the mammary ducts (Linzell, 1955).

There is also some ultrastructural evidence that there may be some myoepithelial cells around the tubules. Cells of unknown nature have been observed in the secretory epithelium within the basement membrane and between the bases of the secretory cells (Ernst & Ellis, 1969). By light microscopy using conventional staining techniques for myoepithelial cells we could not convince ourselves of their presence although there remained the possibility that some stained structures could be such cells. We believe that if myoepithelial cells are present in salt glands they are very small and there are few of them in comparison with salivary glands.

Embryology

In the majority of birds each gland has two ducts and in the embryo these arise from each side of the rudiment of the nasal cavity and then grow backwards to a position where the gland will eventually be formed by branching. As in many branches of comparative anatomy there is some confusion about nomenclature. The two ducts are generally called the *inner* and the *outer*, taking as the point of reference their origin in the nasal cavity. The inner duct, which usually opens on the nasal septum, is therefore the duct of the medial nasal gland; the outer duct which opens on the lower median side of the vestibular concha is the duct of the lateral nasal gland. It is the latter which is probably homologous with Steno's gland in mammals.

As the ducts pass back from the nasal region they lie in close apposition but one above the other. The dorsal one is the outer duct (superior duct of Jobert, 1869); the ventral or inferior duct is the inner duct. However, serial sections of this region clearly show that the outer duct is in fact slightly medial to the inner duct – a point of considerable confusion when labels are added by authors to sections taken only from this region because then, of course, the more lateral duct is draining the medial gland and vice versa!

Some birds have only one duct on each side, for example, the galliform birds (which includes the domestic fowl), the heron *Ardea*, the stork *Ciconia*, the ostrich *Struthio camelus* and several more. Technau (1936) states that penguins have only one duct but the sections of Bang & Bang (1959) suggest that two are present at least in the Adélie Penguin (*Pygoscelis adeliae*). The question is whether both parts of the gland were present in these birds with one duct and the ducts have fused during development, or whether one duct system has entirely failed to develop. This cannot be answered because the ontogeny of the gland has not been studied in most species. However both Ganin (1890) and Marples (1932) clearly showed in the domestic fowl that the outer duct never arises from the nasal cavity, and therefore, that the medial nasal glands are the only ones present. The presence of one or two ducts on each side is not related to whether the nasal gland of a particular species acts as a salt gland since some with only one duct secrete salt while many with two do not.

Marples (1932) found that the gland arises by branching of the two main ducts. Both he and Zaks & Sokolova (1961) found development of the secretory tubules to be very rapid at about the time of hatching. Two to three days before hatching the outlines of the lobes can be seen with the central canal showing the rudiments of the secretory tubules. Growth of the tubules is then rapid and the day after hatching the structure of the gland is complete and further development is achieved by further elongation of the tubules which thus increases the size of the lobes (see Chapter 9) (Plate 2.5). In non-marine birds, the degree of development is so meagre that the nasal gland must be regarded in most cases as a vestigial organ; some species only have the two main ducts with a few lateral branches opening into them.

It is not possible to detect from the macroscopic appearance of the gland that it has a dual origin. Fänge, Schmidt-Nielsen & Osaki (1958) stated that, since the two parts of the gland are so similar in structure and are joined so closely together, 'they can be considered as one functional unit and may be regarded as one

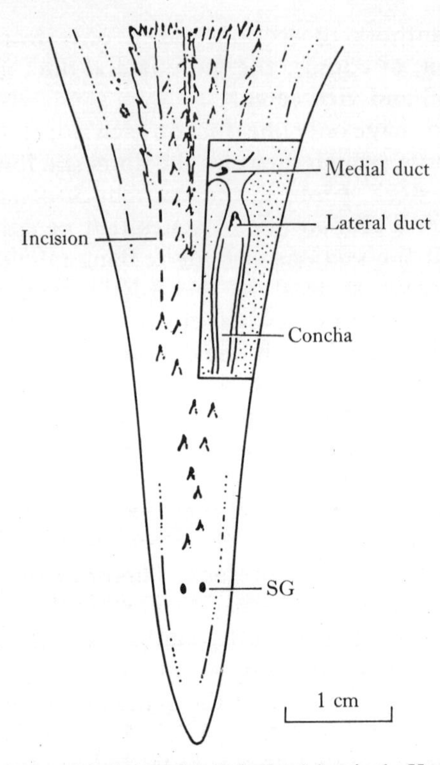

Fig. 2.10. Approach for cannulation of the salt-gland duct in the Herring Gull, *Larus argentatus*, showing the palate from below. The left half of the drawing shows the location of the incision. In the right half a part of the mouth roof is removed exposing the region of the vestibular concha with the two openings of the ducts from the salt gland. SG, openings from the supra-maxillary glands (from Fänge, Schmidt-Nielsen & Robinson, 1958).

gland'. This has been confirmed by Håkansson & Malcus (1969) in the Herring Gull. They cannulated the openings of the two ducts in the nasal cavity after making an incision in the palate (an approach originally adopted by Fänge, Schmidt-Nielsen & Robinson (1958) when they cannulated the outer duct) (Fig. 2.10). They found no difference in the rate of flow or in the composition of the secretion from the two parts of the gland when secretion was induced by stimulation of the secretory nerve.

Later, these same workers (Håkansson & Malcus, 1970) showed, by recording pressures in the two ducts and by taking radiographs of the gland after the retrograde injection of radio-opaque material into the lateral duct, that there are anastomoses within the gland between the lateral and medial duct systems. Therefore if one duct is blocked, the other may be able to carry secretion from both parts of the gland.

3 METHODS

The purpose of this chapter is not to provide a catalogue of techniques that have been used to study the salt gland but merely to consider the methods employed for the initiation of secretion in physiological experiments and for the collection of secretion in conscious ducks, geese and gulls – the three species most commonly used. These are important aspects of investigations in this field and the advantages and disadvantages should be realized, in order that the reader may assess any shortcomings or limitations of work reported in the literature. We shall restrict this discussion to short experiments because long-term adaptation to salt water is dealt with in other chapters.

For most studies hypertonic sodium chloride has been administered orally (i.e. via a tube passed into the proventriculus) or intravenously; sea water (or artificial sea water) has also been used for oral loading but stronger solutions can actually inhibit secretion (p. 135).

Intravenous salt-loads are usually injected fairly rapidly and induce secretion for many hours. The concentration and volume employed by Phillips & Bellamy (1962) is the most useful (0.5 M sodium chloride, 18 ml per kg body-weight) because renal excretion is often completely suppressed. The advantages of intravenous administration are that secretion can be studied without the additional variable of intestinal absorption and nasal fluid appears in the nostrils within two to three minutes. There are however very marked cardiovascular changes in that cardiac output, heart rate and stroke volume increase dramatically (Hanwell, Linzell & Peaker, 1971*b*) and such effects may be undesirable in some experiments. These marked changes do not occur with oral loading but the latency between administration and secretion is longer and more variable, and the results can be more difficult to interpret. For example Ensor, Thomas & Phillips (1970) found that thyroidectomy affected nasal secretion in response to an oral but not to an intravenous salt-load, which implies that a change in the absorption of salt from the gut was responsible.

Continuous slow infusions of hypertonic sodium chloride have also been used to elicit secretion. One of the most useful techniques in this respect is the 'minimal stimulatory salt-load', first used by McFarland (1964*b*; 1965). In such studies ten per cent sodium chloride is slowly infused intravenously until secretion is first observed. Because the response is not maximal the effect of agents in enhancing as well as inhibiting secretion can be studied (see, for example, Peaker, Phillips & Wright, 1970).

Whatever route of administration is employed there is the complication of renal excretion. Therefore, when factors that might affect secretion are being studied, care must be taken to ensure that an effect on the kidney is not responsible and that the salt gland is not secondarily influenced by changes in plasma composition. It is wise therefore to collect cloacal fluid as well as salt-gland secretion.

A major difficulty is that the nasal gland ducts are embedded in the beak and can only be cannulated after considerable dissection under general anaesthesia – a procedure which abolishes secretion in response to salt-loading. Therefore the collection of nasal fluid must be made from the nostrils or beak. Some workers have simply collected fluid dripping from the head or have placed a vial around the upper beak. However there is a serious risk of contamination from oral glands or from the regurgitation of sea water and great care must be taken in interpreting such data. Collections made directly from the nostrils are certainly more reliable. The simplest method is to collect the secretion in a Pasteur pipette as it appears (see Lanthier, Pépin & Sandor, 1965, for example) but the birds can be upset by the continual handling that this entails. There is the additional risk that fluid may be lost because, as it flows into the nostrils, some birds shake it out with a characteristic flick of the head.

In view of these difficulties Wright, Phillips & Huang (1966) fastened a plastic 'nose cone' to the beak of ducks by means of sticky tape (Fig. 3.1), and collected the secretion in washed,

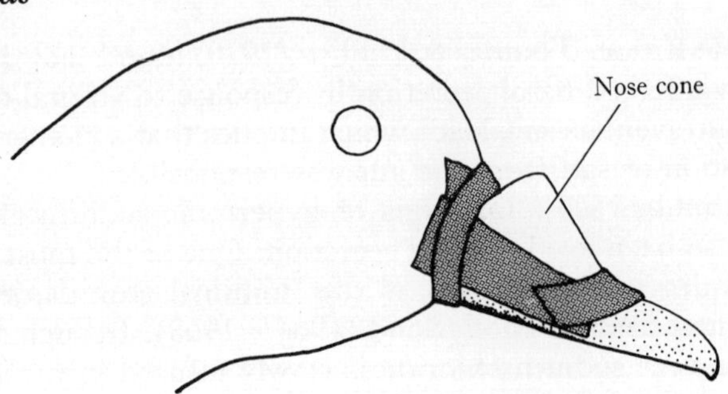

Fig. 3.1. Simple method, first used by Wright *et al.* (1966) for the collection of nasal fluid in ducks. Cotton-wool swabs are held beneath the nostrils by the plastic 'nose cone' which is attached to the beak with adhesive tape.

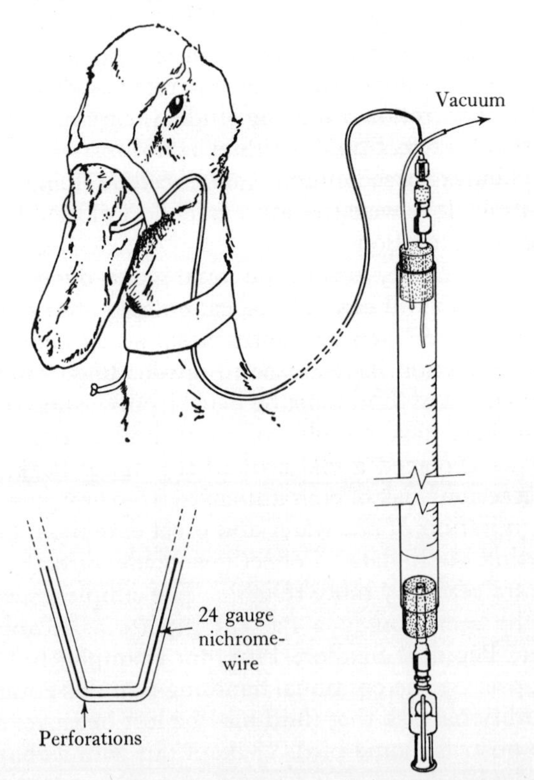

Fig. 3.2. Collection of nasal fluid using polyethylene tubing passed through the opening joining the two nostrils and held in position with adhesive tape. The section of tube which passes through the nares is perforated and has a length of wire inserted for rigidity. Any fluid appearing in the nostrils is thus aspirated and trapped in the burette (from Stewart, 1972).

dry, absorbent cotton wool, which was changed at intervals. However it is difficult to fit such a device to the beak of geese and moreover these birds resented its presence. Hanwell *et al.* (1970*b*) therefore fastened steel cannulae to the beak with dental cement so that the tips were situated on the floor of each nostril. The fluid was then continuously aspirated from each nostril through fine plastic tubing down the back of the neck, using a peristaltic pump and fraction collector (Plate 3.1). A similar method has been described for ducks by Stewart (1972) but he passed a thin plastic catheter, with holes in the side, through the nostrils as shown in Fig. 3.2. We have used both these methods in geese and although Stewart's method is easier to set up, adjustment is necessary when secretion has started in order to avoid loss of fluid. For experiments in which the rate of secretion must be determined as soon as secretion starts we therefore prefer to have the steel cannulae in position before the experiment. With both methods of course secretion can be collected automatically and continuously for long periods.

It is difficult to separate fluid from the left and right glands because the nostrils are connected by an orifice in the nasal septum, and all the fluid may be collected from one side or the other depending on how the bird holds its head. Separation is not therefore normally attempted but to study unilateral effects Gill & Burford (1968) closed the hole in the septum with paraffin wax.

4 NERVOUS CONTROL OF SECRETION

Following the demonstration by Schmidt-Nielsen and his co-workers of the true function of the nasal glands in marine birds, the same group applied themselves to a study of the way in which secretion is induced. Their studies and conclusions, which were acceptable to most physiologists, showed that secretion is induced by cholinergic nerves – as in many other cranial glands. It was surprising therefore to some workers in the field when several years later a group of comparative endocrinologists published a series of papers purporting to show that secretion is under hormonal control. These and other hormonal influences are considered in detail in Chapter 8 but no review has appeared to present the considerable body of evidence in support of the original conclusions of Fänge, Schmidt-Nielsen & Robinson (1958) that nerves do in fact control secretory activity and that hormones play a permissive or secondary role in the initiation and maintenance of secretion.

Innervation of the salt glands

Long before the function of the gland was discovered, several anatomists described nerves entering the anterior and posterior regions of the gland in a number of species. The origin of these nerves was not traced by every author – indeed this is a difficult task because the nerves are small and pass through dense bone where branches are easily broken or lost. According to Cords (1904), Ash *et al.* (1969) and our own dissections in geese and ducks, the main supply is from the VIIth cranial nerve, and Cords and ourselves have found connexions with the IXth and the sympathetic system. The physiologist will at once note that this type of innervation is similar to that of mammalian salivary glands and suggests both a parasympa-

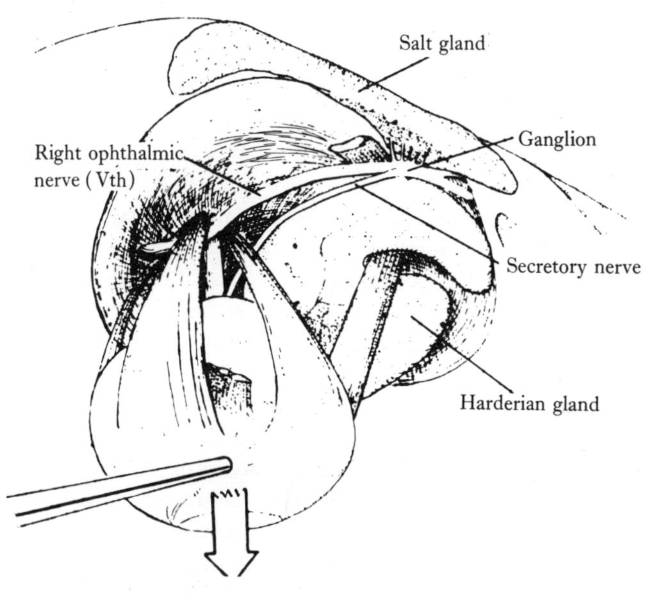

Fig. 4.1. Dissection of the right side of the head of a gull to show the gross innervation of the salt gland. Note the secretory nerve ganglion and the 'fan' of postganglionic nerves entering the anterior of the gland (from Fänge, Schmidt-Nielsen & Robinson, 1958).

thetic and a sympathetic innervation. The situation is complicated by the fact that, near the gland in the orbit, the main 'secretory nerve' is closely applied to the ophthalmic branch of the Vth, and some authors have wrongly concluded that the main innervation is from the Vth. Serial sections of this region show no apparent nervous connexions between the salt-gland nerve and the Vth but reveal that there are a number of ganglion cells (Ash *et al.*, 1969; Håkansson & Malcus, 1969; Cottle & Pearce, 1970). These authors produced evidence that this rather diffuse ganglion (which is in all probability the *ganglion ethmoidale* of early anatomists) is a typical peripheral parasympathetic ganglion where preganglionic fibres relay with numerous postganglionic fibres, which end in the gland.

Fig. 4.1 shows the gross innervation of the salt gland as found by the following authors: Fänge, Schmidt-Nielsen & Robinson (1958); Ash *et al.* (1969); Håkansson & Malcus (1969);

A. Hanwell, J. L. Linzell & M. Peaker (unpublished). This arrangement is in fact the same as that described by Cords (1904). Ash *et al.* (1969) have also described a nerve entering the posterior part of the gland in the duck; this they called the *posterior nerve* and found it to be a branch of the maxillary division of the Vth. However they considered that it passes through the gland and then ramifies in subcutaneous tissue above the gland. Hsieh (1951) in the domestic fowl has described two nerves entering the ganglion ethmoidale but there is no indication that this is the situation in birds with a functional nasal salt gland.

Studies on the innervation of the salt gland and the secretory nerve ganglion have been made using histochemical methods for demonstrating cholinergic and, to a lesser extent, adrenergic nerves. The method for adrenergic nerves detects the transmitter substance, i.e. the catecholamines, but for cholinergic nerves the method relies on the specific detection not of the actual transmitter itself but of acetylcholinesterase (AChE), which breaks down acetylcholine. While there seems little doubt that this latter technique is of great value in studying cholinergic innervation at the microscopical level, there is evidence that AChE is also associated, albeit to a lesser extent, with adrenergic and other nerve axons (see Furness & Iwayama, 1972 for references) so that in the absence of supporting physiological data the localization of AChE can be misleading; the possibility also exists that structures other than nerves may show AChE activity. The situation is complicated still further by the presence in many tissues of butyryl- or pseudocholinesterase (BuChE) which occurs in non-nervous structures. Both AChE and BuChE will catalyse the hydrolysis of acetylthiocholine, the substrate used in the histochemical method. However AChE does not catalyse the reaction with butyrylthiocholine, so that any activity seen with this substrate is due to BuChE. The identification of AChE in the presence of BuChE therefore requires the use of substances which inhibit BuChE to a much greater extent than AChE; the tissue must be incubated with them before, and also during, incubation with acetylthiocholine. There is the added complication when this method is used in birds that the substrate specificity may be different from that in mammals.

Ellis *et al.* (1963) were the first to apply a histochemical method for cholinesterase to the salt gland. Their micrographs suggested a rich cholinergic innervation in the duck. Later Fourman (1966), in an abstract, stated that studies with inhibitors showed the enzyme present to be BuChE, but subsequent work by Ash *et al.* (1966; 1969) clearly demonstrated the presence of both enzymes, with AChE predominating. Similar results were later obtained by Ballantyne & Fourman (1967) using a chemical technique.

At this stage we shall abandon further discussion of BuChE and raise the question again in Chapter 9, because it is clear that this enzyme is, under certain circumstances, associated with secretory cells. Therefore its presence is not entirely related to the question of innervation.

Cholinergic innervation

Ash *et al.* (1969) made a detailed study of the distribution of AChE in the salt gland of the duck. They found structures showing activity mainly concentrated around the centre of each lobe, between the secretory tubules and also along the walls of the ducts. A similar pattern was obtained with a silver staining procedure and it appeared that the AChE activity observed was present in nerves. This conclusion was strengthened when they were able to show apparent continuity between the structures showing AChE activity in the gland with fibres entering the gland from the secretory nerve ganglion. In addition 'looped structures' at the periphery of some lobes contained AChE. They considered it possible that these were blood vessels with activity in their walls. However by injecting carmine–gelatine into the blood vessels it could be seen that 'while stained fibres often accompanied the blood vessels, the vessel walls were free of activity'. The pattern of activity therefore suggested that the gland is very richly innervated by cholinergic fibres, to both secretory cells and blood vessels.

Fourman (1969) applied a similar technique, again in the duck, but suggested that structures other than nerves may show a positive reaction. The general pattern of staining was similar to that described by Ash *et al.* (1969) but she also injected the blood vessels and considered that some of the capillaries were outlined by the reaction product while others were not. Fourman also compared AChE activity with silver staining and concluded that, 'although the gland has a profuse innervation, this is not so extensive as the AChE-positive network might suggest'.

Neither Ash *et al.* (1969) nor Fourman (1969) found an increase in AChE activity or distribution in active salt glands. Similarly Ballantyne & Fourman (1967) using a chemical method found no increase per unit weight of tissue in birds given salt water for several days, although the total amount of AChE increased as the glands increased in weight.

A similar pattern of AChE activity has also been obtained in the goose by Dr Ann Silver (in Hanwell *et al.*, 1971*a*) and Dr J. R. McLean (unpublished). In addition, Dr McLean, together with Dr Catherine Hebb and S. P. Mann, have examined the larger blood vessels supplying the gland. Nerves containing AChE were seen in the walls and the possibility exists that some cholinergic nerves may enter the gland by this route as well as from the secretory nerve ganglion.

The richness of the cholinergic innervation has been confirmed by measurements of the choline acetylase activity in the salt gland. This enzyme is involved in the formation of acetylcholine and Dr Catherine Hebb (in Ash *et al.* 1966) obtained a figure of 400μg ACh formed/g tissue/hr in the duck salt gland – a very high activity for peripheral (i.e. non-nervous tissue); subsequent estimations in the goose are similar (J. R. McLean & C. O. Hebb, unpublished).

Adrenergic innervation

The adrenergic innervation of the salt gland has only recently been investigated by treating the tissue with formaldehyde vapour so that the catecholamines fluoresce (Falck technique). Haase & Fourman (1970) found many adrenergic fibres in the duck salt gland with a distribution similar to that of AChE-containing fibres; fluorescent fibres were not seen if the ducks were treated with reserpine. With Dr J. R. McLean and Dr Catherine Hebb we have found that, in the goose, most of the adrenergic fibres are present in the walls of larger blood vessels but that, in addition, small fibres with varicosities pass along the secretory tubules. This latter finding could indicate either an adrenergic innervation of the secretory cells (for which there is no physiological evidence), the capillary walls (an unorthodox suggestion) or other cells in close association with the secretory tubule. It is probable that adrenergic nerves enter the gland both in the walls of arteries and also along the 'secretory nerve route' through the secretory nerve ganglion because, after postganglionic denervation, many fluorescent fibres remained and a complete absence of fluorescence was only obtained after both denervation and 'freeze-clamping' the blood vessels to the gland. However it seems unlikely from our observations that those fibres entering from the secretory nerve only innervate the tubules because small fibres could be seen leaving the rich perivascular network and running between the secretory tubules.

Photomicrographs showing the innervation of the salt gland are shown in Plates 4.1 and 4.2.

Ultrastructure

Several workers have studied the innervation of the salt gland using the electron microscope, but only one paper has appeared in anything but abstract form. Fawcett (1962) noted nerve endings in close association with the basal membrane of the secretory cell. He also sent us micrographs showing a number of unmyelinated nerve fibres running in a Schwann cell outside the basement membrane (Plate 4.3). Kühnel (1972) has also observed axons, surrounded by Schwann cells, running outside the basement membrane of the tubules and central canals in the perivascular region, as well as independently of blood vessels. Other axons were seen penetrating the basement membrane, ramifying and ending between the basal membranes of secretory cells and the basement membrane; some endings extended into the space between adjacent cells (Plate 4.4). The endings were seen to be terminal swellings of the axon separated from the basal membrane by about 200 Å. They were found to contain agranular vesicles (approximately 500 Å in diameter), large granular vesicles (1000–2000 Å) and mitochondria. It is generally agreed that agranular vesicles are characteristic of cholinergic endings and those Kühnel described are similar to those observed in mammalian salivary glands (see Hand, 1972). Kühnel made no mention of any small granular vesicles which are generally thought to contain stored catecholamines and, therefore, to be characteristic of adrenergic endings. However in order to see the granular cores fixation in permanganate is usually required (Richardson, 1966) and the possibility exists that at least some of the vesicles seen by Kühnel might have been granular and therefore adrenergic in type. Haase & Fourman (1970) noted the presence of granular vesicles but not their location in relation to the secretory cell. They were not present in salt glands of birds given reserpine so it would appear that they do represent adrenergic terminals. These authors stated that cholinergic and adrenergic nerves run in close proximity and discussed the possibility that AChE and catecholamines occur in the same axon terminal.

The secretory nerve, secretory nerve ganglion and the 'fan' of postganglionic nerves which enter the anterior of the gland have also been examined using histochemical techniques. Ash *et al.* (1969) found the secretory nerve in the duck to be almost free of AChE activity until it approached the ganglion. However cells with high activity were evident there and were apparently the source of AChE-containing fibres which could be traced to the gland. Like the preganglionic secretory nerve, the posterior nerve and ophthalmic branch of the Vth were virtually free of AChE. As well as being present in the ganglion itself, ganglion cells have been found in the 'fan' and, in some cases, even within the gland (Ann Silver, personal communication; Cottle & Pearce, 1970). The distribution of ganglion cells in the region of the secretory nerve ganglion has been studied by Håkansson & Malcus (1969).

Cottle & Pearce (1970) have examined the secretory nerve pathway in some detail. They prepared serial sections of the region in the duck and made a reconstruction to demonstrate the innervation; this is shown in Fig. 4.2. It appeared that although the secretory nerve and ganglion are closely applied to the Vth, they are independent. Evidence from their light and electron microscopical studies suggested that the ganglion is a synaptic site. Processes containing synaptic vesicles were seen ending on nerve cells. Agranular vesicles were predominant but

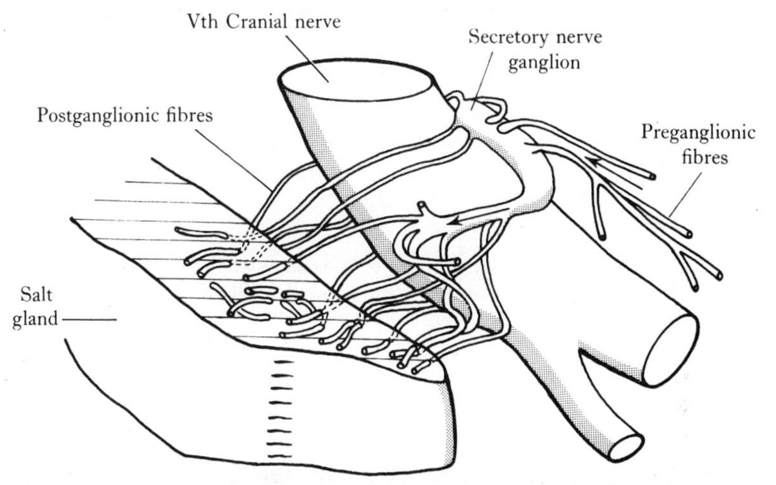

Fig. 4.2. Diagram showing the secretory nerve, secretory nerve ganglion and ophthalmic branch of the Vth cranial nerve in relation to the salt gland (reconstructed from serial sections) (from Cottle & Pearce, 1970).

granular vesicles were also seen in the same axon. While this evidence would indicate a cholinergic innervation of the ganglion cells, the granular vesicles would suggest the presence of catecholamines in the same ending, as some workers have suggested in other tissues. Cottle & Pearce also found large numbers of mitochondria and membrane thickenings characteristic of synaptic sites.

The presence of adrenergic endings suggested by the granular vesicles, has been confirmed by J. R. McLean (unpublished); he has found fluorescent endings around the ganglion cells (Plate 4.5), but so far their function is unknown.

Control of secretion – early work

In their first full-length paper on the discovery of salt glands in marine birds, Schmidt-Nielsen *et al.* (1958) established that secretion could be initiated by giving cormorants (*Phalacrocorax auritus*) sea water by stomach tube or hypertonic sodium chloride intravenously. In addition they found that hypertonic sucrose given intravenously also induced secretion, and concluded that 'the nasal secretory mechanism seems to respond to an osmotic load rather than specifically to the plasma

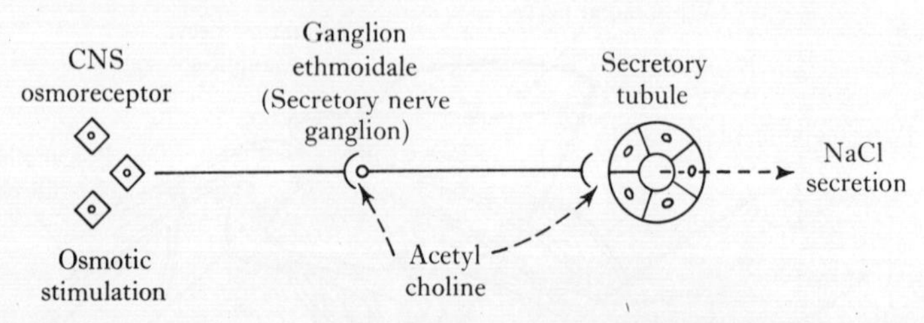

Fig. 4.3. Sequence of events suggested by K. Schmidt-Nielsen (1960) for the osmotic stimulation of salt-gland secretion (by permission of The American Heart Association).

concentrations of sodium and chloride'. Fänge, Schmidt-Nielsen & Robinson (1958) then went on to study the mechanism by which secretion is induced. They found that in the Herring Gull (*Larus argentatus*) secretion appeared one to five minutes after an intravenous salt load but that secretion is blocked by general anaesthetics (pentobarbitone or chloral hydrate). They then discovered that stimulation of a nerve in the orbit (the 'secretory nerve', see above) evoked secretion. Since acetylcholine injected into a carotid artery and methacholine (mecholyl or acetyl-β-methylcholine) given intravenously were also effective, and atropine blocked secretion in response to a salt-load, they concluded that salt-gland secretion is initiated by cholinergic nerves. Stimulation of the cervical sympathetic chain did not induce secretion and there is still no evidence for a functional innervation of the secretory cell by adrenergic fibres. Large doses of adrenaline caused a transient block of secretion but this could have been due to a vasoconstrictor effect on the blood vessels (see Chapter 5).

Thus Schmidt-Nielsen and his co-workers considered that secretion is induced by a reflex triggered by an increase in the osmotic concentration of the plasma. Later Schmidt-Nielsen (1960) suggested that the osmoreceptors may well be situated in the CNS; his very reasonable scheme for the secretory reflex is shown in Fig. 4.3. This is how knowledge of the nervous control of the glands stood within a few years of the discovery of their function. Later modifications and challenges to this hypothesis must now be considered.

Nature of the receptors

The view that osmoreceptors are responsible for initiating secretion was questioned by Holmes (1965) and later by Burford & Bond (1968). The increase in blood volume after the administration of hypertonic sodium chloride is well known (see McFarland, 1963*b*) and, apart from the volume of the load itself being responsible, water will be drawn into the extracellular compartment and result in further expansion of blood volume and in haemodilution. Holmes & Donaldson (in Holmes, 1965) therefore suggested that volume or stretch receptors might trigger secretion as the blood volume increases, presumably because atrial stretch receptors play an important part in the control of fluid balance in mammals. In an attempt to study this further they injected a sufficient volume of so-called 'isotonic saline' (i.e. 0.154 M sodium chloride) containing gum arabic to raise the blood volume to the same extent as that observed after a hypertonic salt-load and this they found induced secretion in ducks. However, their conclusion, that the increase in blood volume is the primary stimulus, depends on the assumption that in their experiments plasma tonicity was not affected; we shall raise this point again later (p. 44).

Burford & Bond (1968) also reported an increase in blood volume in anaesthetized geese given hypertonic sodium chloride or sucrose solutions intravenously. They suggested that the expansion might occur before a change in plasma composition and could therefore be the stimulus for secretion. However, because their geese were anaesthetized, the latency of the secretory response could not be determined simultaneously with the change in blood volume and composition. By contrast Hanwell *et al.* (1971*b*) found, in conscious geese, that the administration of hypertonic sodium chloride intravenously, did indeed increase blood volume as well as heart rate, cardiac output and stroke volume but had no consistent effect on blood pressure. Nevertheless there were also early changes in plasma composition. We continued...

it is obviously difficult to identify which of the many changes in the cardiovascular system, occurring when secretion starts after intravenous loading, might be the stimulus or stimuli for the receptors which initiate the cholinergic secretory mechanism. We considered that to investigate the stimulus for secretion the bird should be given sodium chloride or sea-water at concentrations it is likely to encounter in the wild and that these solutions should be administered by the natural route into the alimentary canal in quantities the bird might drink. Using this technique we found that at the time secretion started [5–14 minutes after oral loading] there were no changes in heart rate, cardiac output or arterial blood pressure and in three out of four birds, no decrease in haematocrit, all of which suggests that an increased plasma volume is not the normal stimulus for secretion in the goose. A slight increase in both plasma volume and cardiac output eventually occurred, but only some time after secretion had started... The only changes we could detect when secretion started after sea-water was given orally were small increases in plasma osmolality and sodium and chloride concentrations. These results thus support the hypothesis that the receptors respond to an osmotic stimulus.

Later work has suggested that in some cases, after an oral salt-load, blood volume may be actually *decreased* when secretion starts. This may be due to the passage of water into the lumen of the gut resulting in a raised plasma sodium chloride concentration before absorption of salt begins, but whatever the mechanism these findings argue against the raised blood volume starting secretion, (Hanwell *et al.*, 1972). Direct evidence was also obtained against this hypothesis by rapidly injecting homologous blood. Although amounts sufficient to raise the total blood volume by 9–16 per cent were injected rapidly into the right atrium secretion was not induced and, of course, the right atrium must have been grossly distended for a short time which would have energetically stimulated stretch receptors. Nevertheless, the injection of hypertonic sodium chloride soon after stimulated secretion within two minutes (Fig. 4.4).

Stewart (1972) devised an elegant method of distinguishing between volume and composition effects. He argued that, if water is removed from the extracellular compartment, blood volume will fall while the plasma solute composition will rise. If secretion then occurs it indicates that a raised blood volume is not the stimulus. To test this hypothesis he simply kept ducks without water for 33 hours during which time blood volume fell but plasma osmolality and sodium chloride concentrations increased. Salt-gland secretion started, thus providing additional evidence against the hypothesis that the initiation of secretion is due to an increase in blood or plasma volume.

We must now consider the problem of why 0.154 M sodium chloride containing gum arabic stimulated secretion in the

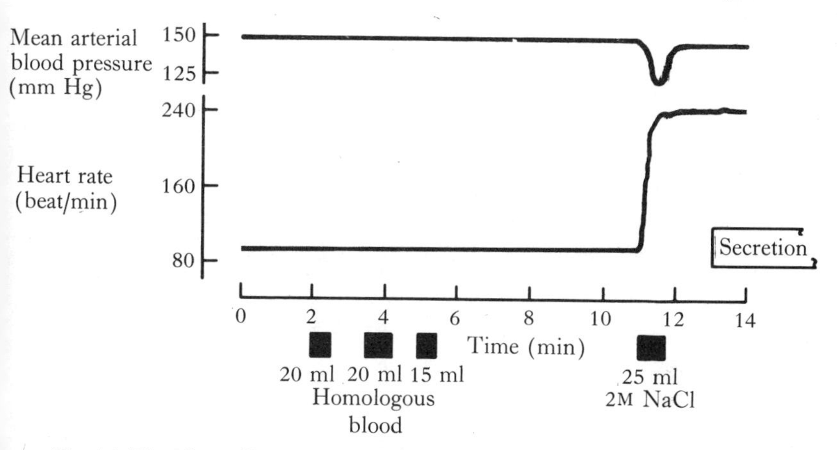

Fig. 4.4. The effects of injecting homologous blood or hypertonic sodium chloride into the right atrium on blood pressure, heart rate and salt-gland secretion in a conscious goose (from Hanwell *et al.*, 1972).

ducks studied by Holmes (1965). Later work in the duck confirmed that some birds do respond to large volumes of saline of this strength (S. J. Peaker, unpublished observations) or to 0.154 M sodium chloride with added dextran (Lanthier & Sandor, 1967). Furthermore, all geese tested were found to secrete in response to large volumes of saline of this strength given intravenously (Hanwell *et al.*, 1971*a*, *b*), whereas Hajjar, Sattler, Anderson & Gwinup (1970) found no stimulation in gulls (*Larus occidentalis*) with or without dextran in the solution (Table 4.1) although Hughes (1972*a*) noted a response in gulls to oral loading with 0.154 M sodium chloride.

At first sight these results would indicate a response to plasma volume expansion in the goose and in some ducks. However this is unlikely to be the explanation. Hanwell *et al.* (1971*b*) found that plasma concentrations of sodium and chloride did increase after loading with 0.154 M sodium chloride in geese and there seems little doubt that 'isosmotic saline' is in fact hypertonic as far as the cell membrane of the receptor is concerned. It is for this reason that we went on to use homologous blood rather than a salt solution containing a colloid to test for volume receptors. It is worth emphasizing that some confusion still exists in the literature about the difference between hyperosmotic and hypertonic. Even sixty

TABLE 4.1. *Response of avian salt glands to intravenous or oral administration of various solutions. Results of oral administration are included only when it was checked that plasma osmolality was increased. Negative effects for either route are excluded if the possibility that secretion was inhibited by the substances was not checked* (modified from Hanwell *et al.*, 1972)

Solution	Route	Response	Species*	References
Hyperosmotic NaCl	i.v. and oral	+	Many	Many authors
Hyperosmotic sucrose	i.v.	+	Cormorant (*Phalacrocorax*)	Schmidt-Nielsen, Jörgensen & Osaki (1958)
	i.v.	+	Duck	Lanthier & Sandor (1967)
	i.v.	+	Duck	Ash (1969)
	i.v.	+	Gull (*Larus occidentalis*)	Hajjar, Sattler, Anderson & Gwinup (1970)
	i.v.	+	Goose	Hanwell *et al.* (1972)
Hyperosmotic mannitol	i.v.	+	Duck	Lanthier & Sandor (1967)
	i.v.	+	Duck	Ash (1969)
	i.v.	+	Gull (*Larus occidentalis*)	Hajjar *et al.* (1970)
Hyperosmotic Na_2SO_4	i.v.	+	Goose	Hanwell *et al.* (1972)
Hyperosmotic NH_4Cl	i.v.	+	Duck	Lanthier & Sandor (1967)
Hyperosmotic LiCl	i.v.	+	Goose	Hanwell *et al.* (1972)
Hyperosmotic KCl	oral	−	Duck	Ash (1969)
Hyperosmotic urea	i.v.	−	Duck	Lanthier & Sandor (1967)
	i.v.	−	Duck	Ash (1969)
	oral	−	Coot (*Fulica americana*)	Carpenter & Stafford (1970)
	oral	−	Rail (*Rallus owstoni*)	Carpenter & Stafford (1970)
Hyperosmotic glucose	i.v.	− or slight +	Duck	Ash (1969)
	i.v.	−	Duck	M. Peaker & S. J. Peaker (unpublished)
0.154 M NaCl	i.v.	− or +	Duck	S. J. Peaker (unpublished)
	i.v.	−	Duck	Smith, Fourman & Haase (1971 *a*)
	i.v.	−	Gull (*Larus occidentalis*)	Hajjar *et al.* (1970)
	i.v.	+	Goose	Hanwell *et al.* (1971 *a*)
	oral	+	Gull (*Larus glaucescens*)	Hughes (1972 *a*)
0.154 M NaCl+dextran	i.v.	slight +	Duck	Lanthier & Sandor (1967)
	i.v.	−	Gull (*Larus occidentalis*)	Hajjar *et al.* (1970)
0.154 M NaCl+gum arabic	i.v.	+	Duck	Holmes (1965)
Homologous blood	i.v.	−	Goose	Hanwell *et al.* (1972)

years ago Bayliss in his classic text book *Principles of General Physiology* felt constrained to devote a page to this topic and this is still worth reading. He points out that tonicity must always be qualified by reference to some semipermeable membrane (*in vivo*, usually cell membranes). The reason why species differ in their response to 0.154 M sodium chloride is apparently related to the sensitivity of the receptors (see below). In descending order of sensitivity the species can be arranged as follows: goose > duck > gulls, and this is the same order as the response to 'isosmotic saline'. Thus it would appear that 'isosmotic saline' is not sufficiently hypertonic invariably to reach the threshold for stimulation of the receptors in the less sensitive species.

It is now clear that an increase in blood volume is not the primary stimulus for secretion but it is not known whether, under normal circumstances, the amount of secretion is in any way influenced by blood volume as well as by plasma composition. Long-term water deprivation does inhibit secretion but this may be mediated by the sympathetic nervous system (because it is relieved by the administration of an α-blocker) acting to constrict the blood vessels of the gland rather than by a complex interaction of volume and osmoreceptors (see Chapters 5 and 11).

The next problem to consider is whether the receptors are in fact osmoreceptors as Schmidt-Nielsen, Jörgensen & Osaki (1958) concluded or whether they respond specifically to sodium or chloride ions. Lanthier & Sandor (1967) found that hyperosmotic sucrose and mannitol as well as sodium chloride and sodium bicarbonate given intravenously evoked secretion whereas hyperosmotic urea did not. Ash (1969) also working with ducks obtained similar results and further showed that glucose and potassium chloride were ineffective. These results (Table 4.1) were interpreted as showing that the receptors respond to an osmotic stimulus rather than specifically to sodium or chloride concentrations (plasma sodium concentrations in fact fall after the administration of sucrose or mannitol).

The similarity between the receptors for salt-gland secretion and the ones in the brain discovered by Verney (1947), which cause antidiuretic hormone to be released in mammals, is evident (Table 4.1). The requirement for stimulation is therefore not simply an increase in plasma osmolality (i.e. the solute concentration determined by depression of freezing point) but in tonicity, which involves net water movement out of a cell in response to an osmotic gradient. Therefore for plasma tonicity to be increased the cell membrane in question must be relatively impermeable to the solute. Thus sodium chloride, sucrose, mannitol, etc. raise plasma tonicity which in turn presumably activates the receptor responding to cell shrinkage as water leaves the cells. On the other hand an increase in osmolality as a result of injecting hyperosmotic urea or glucose does not induce secretion. This is usually interpreted as being due to the rapid passage of these substances across the cell membrane so that osmotic equilibrium is achieved without the movement of water out of the cells; therefore cell shrinkage does not occur.

While the evidence indicates that osmoreceptors are responsible for initiating salt-gland secretion, Ash (1969) found that sucrose and mannitol induced a more transient secretory response than did sodium chloride. This may suggest that the receptors are more permeable to sucrose and mannitol than to sodium, or that the cells can maintain, as Ash put it, 'a critical difference in concentration of Na^+ across the cell membrane' by active transport.

Carpenter & Stafford (1970) have claimed that the receptors may be specifically sensitive to sodium concentration in coots (*Fulica americana*) and rails (*Rallus owstoni*). However their results are similar to those of Lanthier & Sandor (1967) and Ash (1969) in ducks on the effect of raising the osmolality with urea and their conclusion apparently neglects the properties of osmoreceptors discussed above; they did not investigate the effects of hyperosmotic sucrose or mannitol solutions on secretion in their birds. Additional evidence that specific sodium or chloride receptors are not responsible is that hyperosmotic sodium bicarbonate, sodium sulphate and lithium chloride all induce secretion (Table 4.1).

In conclusion therefore, current evidence favours the view that osmoreceptors as Verney described them in mammals are responsible for initiating salt-gland secretion in marine birds. In view of the apparent confusion that exists between the terms osmolality (or osmolarity) and tonicity, we have previously suggested (Hanwell *et al.* 1972) that they may be better described as 'tonicity receptors'. Under normal circumstances of course the increase in tonicity is brought about by increases in plasma sodium and chloride concentrations.

Sensitivity of the receptors

The intravenous administration of large quantities of hypertonic sodium chloride provides no information on the sensitivity of the receptors. McFarland (1964*b*) therefore infused ten per cent sodium chloride intravenously into gulls until secretion started and calculated, from plasma composition before and after infusion, the increase in sodium concentration required to stimulate secretion. An increase of approximately 21 mmol per litre (i.e. about ten per cent) was found to be needed which, McFarland noted, is about five times that required to induce drinking in mammals. Later we found that the duck and particularly the goose require less of an increase in plasma sodium and chloride concentrations on a body-weight basis than the gulls studied by McFarland (1964*b*) (Hanwell *et al.*, 1972; Peaker, Peaker, Hanwell & Linzell, 1973). When these data together with others from the literature were examined it was evident that marked differences exist between species (Table 4.2). McFarland had previously noted that in his investigations the larger Glaucous-winged Gull (*Larus glaucescens*) showed a tendency to require less salt than the smaller California Gull (*Larus californicus*). When the percentage increase in plasma sodium required to initiate secretion was plotted against body-weight a clear inverse exponential relationship became apparent for five species. On a log–log basis a significant, negative correlation coefficient ($r = -0.932$) was obtained and the slope of the line (i.e. the exponent of body-weight) was -0.76 (Peaker *et al.*, 1973) (Fig. 4.5). Thus it is clear that the receptors are more sensitive in the larger species but the reasons why this should be so are unknown. There appears to be no less salt-gland tissue, when related to body-weight, in small birds; indeed the data shown in Fig. 11.1 imply that they have a little more. So until we have more information on renal and extra-renal excretory capabilities in

TABLE 4.2. *Variables that can be related to the sensitivity of the receptors for salt-gland secretion in different species* (from Peaker *et al.*, 1973)

Species	Maintained on fresh water (FW) or salt water (SW)	mmoles NaCl (as hypertonic solution i.v.)/kg body-weight required to induce secretion	Percentage increase in plasma concentrations when secretion started: After oral salt load			During i.v. infusion of hypertonic NaCl			Reference
			Na	Cl	Na+Cl	Na	Cl	Na+Cl	
Goose (*A. anser*)	FW	1.9–2.1	0.3–6	1.3–7.1	0.7–6.3	0–2.7	2.5–5.4	1.4–3.9	1
Duck (*Anas platyrhynchos*)	FW	8–29	—	—	—	6.9–8.6	—	—	2
	FW	—	4.8–8.8	—	—	—	—	—	3
	FW	—	11.1	—	—	—	—	—	4
						8.4	—	—	5
Gulls (see below)	FW	6.1–10.6	—	—	—	12.4	24.2	17.4	6
	FW or SW	—	—	—	—	6.7–13.3	8.7–14.4	7.6–15.1	7
American coot (*Fulica americana*)	FW*	—	40	27	38.8	—	—	—	8
Guam rail (*Rallus owstoni*)	FW+SW		21.4	23.8	22.3	—	—	—	8

* On fresh water for only a few days after capture from salt water.

Data calculated from (1) Hanwell *et al.* (1971*b*) and present data; (2) Present data; (3) Ash (1969); (4) Holmes, Phillips & Butler (1961); (5) Lanthier & Sandor (1967); (6) McFarland (1964*b*) in *Larus glaucescens, L. occidentalis* and *L. californicus*; (7) Nechay, Larimer & Maren (1960) in *L. delawarensis* and *L. argentatus*; (8) Carpenter & Stafford (1970).

Mean or range of values given.

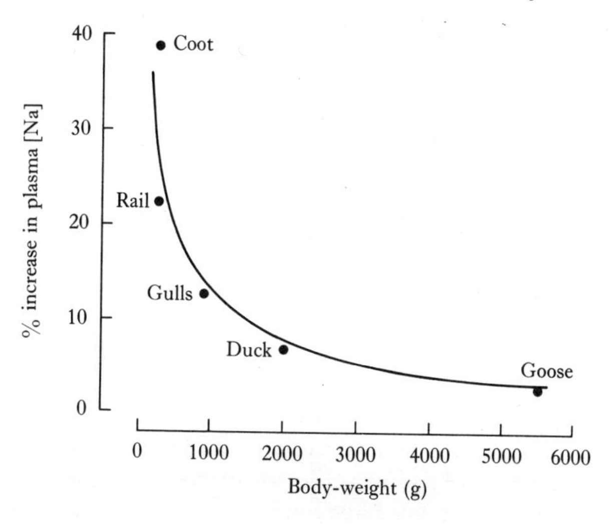

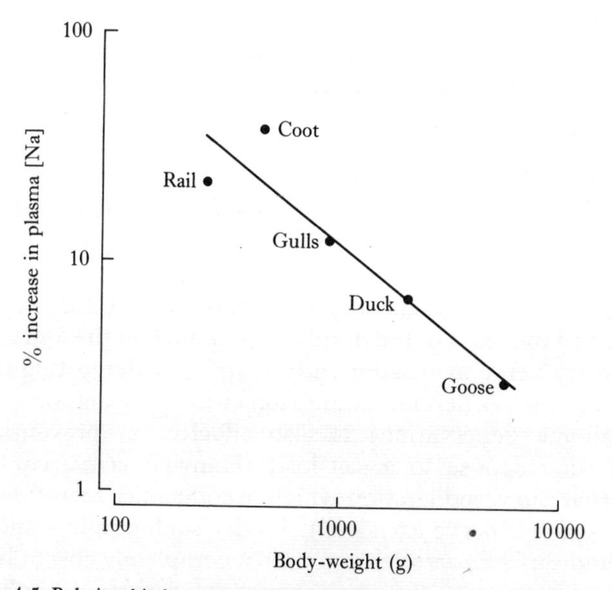

Fig. 4.5. Relationship between body-weight and percentage increase in plasma sodium concentration required to initiate salt-gland secretion. The slope of the log-log line (bottom) is −0.762 and the correlation coefficient is −0.93 ($P < 0.05$) (from Peaker *et al.*, 1973).

birds of different size we cannot attempt to offer an explanation of the interesting relationship between receptor sensitivity and body-weight.

Taking data from the most sensitive species so far studied, the goose (Table 4.2), it is clear that an increase in plasma tonicity of approximately two per cent is required to initiate secretion. The receptors in this species are therefore similar in sensitivity to those which induce drinking (Fitzsimons, 1963) and antidiuretic hormone release (Verney, 1947) in mammals.

Location of the receptors

Early experiments done *in vivo* and *in vitro* established that the cells of the salt gland are not directly stimulated by an increase in the concentration of sodium chloride in the extracellular fluid but that cholinergic nerves have both a secretomotor and, in all probability, a vasodilator action. We shall return to the local action of acetylcholine later but as far as this discussion is concerned the very reasonable suggestion made by Schmidt-Nielsen (1960), that the receptors are probably situated in the central nervous system (CNS), was not examined for a number of years.

While studying the innervation of the salt gland in the duck, Ash *et al.* (1969) were led to the conclusion that the receptors are located elsewhere. They confirmed that atropine blocks secretion in response to salt-loading and also showed that denervation by 'nerve excision' (removal of the secretory nerve ganglion and the parts of the ophthalmic branch of the Vth and the secretory nerve associated with it) or by undercutting the gland so as to sever nervous connexions with the ganglion (i.e. postganglionic denervation) is also effective in preventing secretion in response to a salt-load (Hanwell *et al.* 1971*a*, showed that the vasodilatation which accompanies secretion is also blocked by 'nerve excision'). Under such circumstances AChE (and BuChE) activity was almost completely absent but the administration of the cholinomimetic carbamyl choline (carbachol) evoked secretion. However when Ash *et al.* (1969) cut the secretory nerve and the ophthalmic branch of the Vth preganglionically (as well as the posterior nerve), secretion still occurred although the response to salt-loading in terms of the

Plate 2.1
Salt-gland secretion being ejected from the tubular nostrils of a petrel (from K. Schmidt-Nielsen, *Scientific American* (1959) **200**, 109).

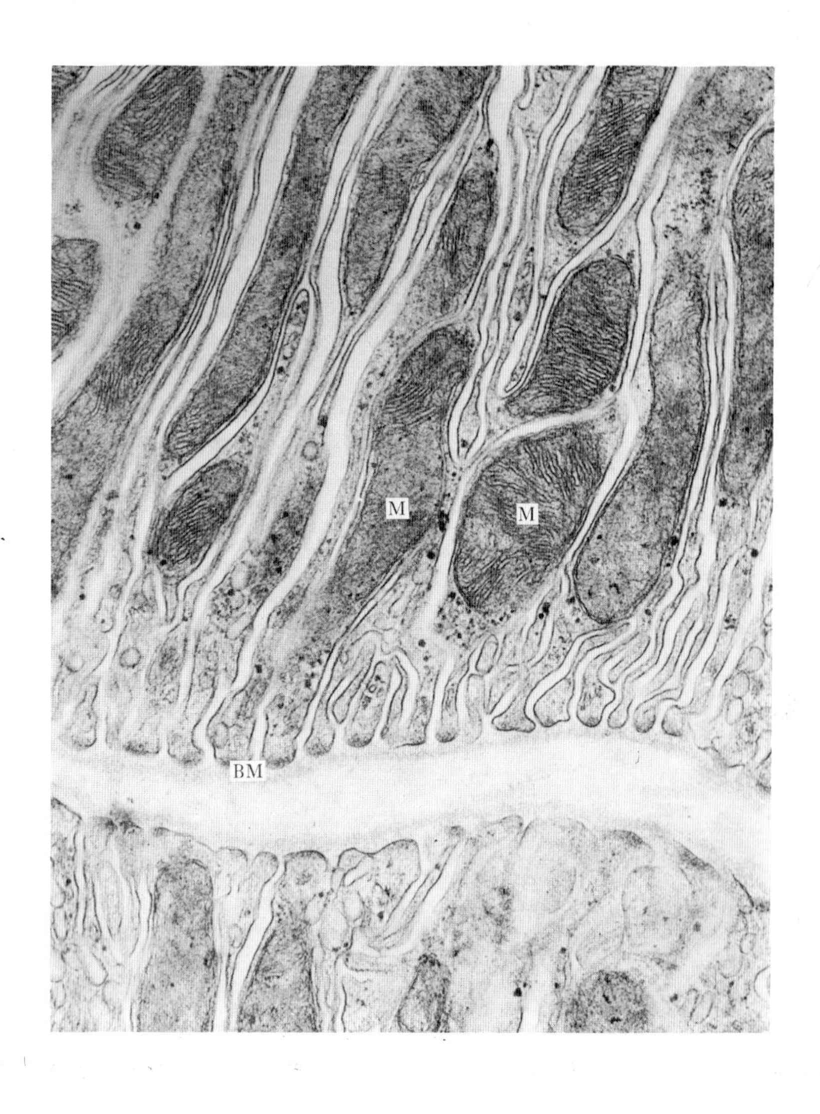

Plate 2.2
Electron micrograph showing the greatly infolded basal membrane of a secretory cell. Note the basement membrane (BM) of the epithelium, and the mitochondria (M) (by courtesy of Professor D. W. Fawcett).

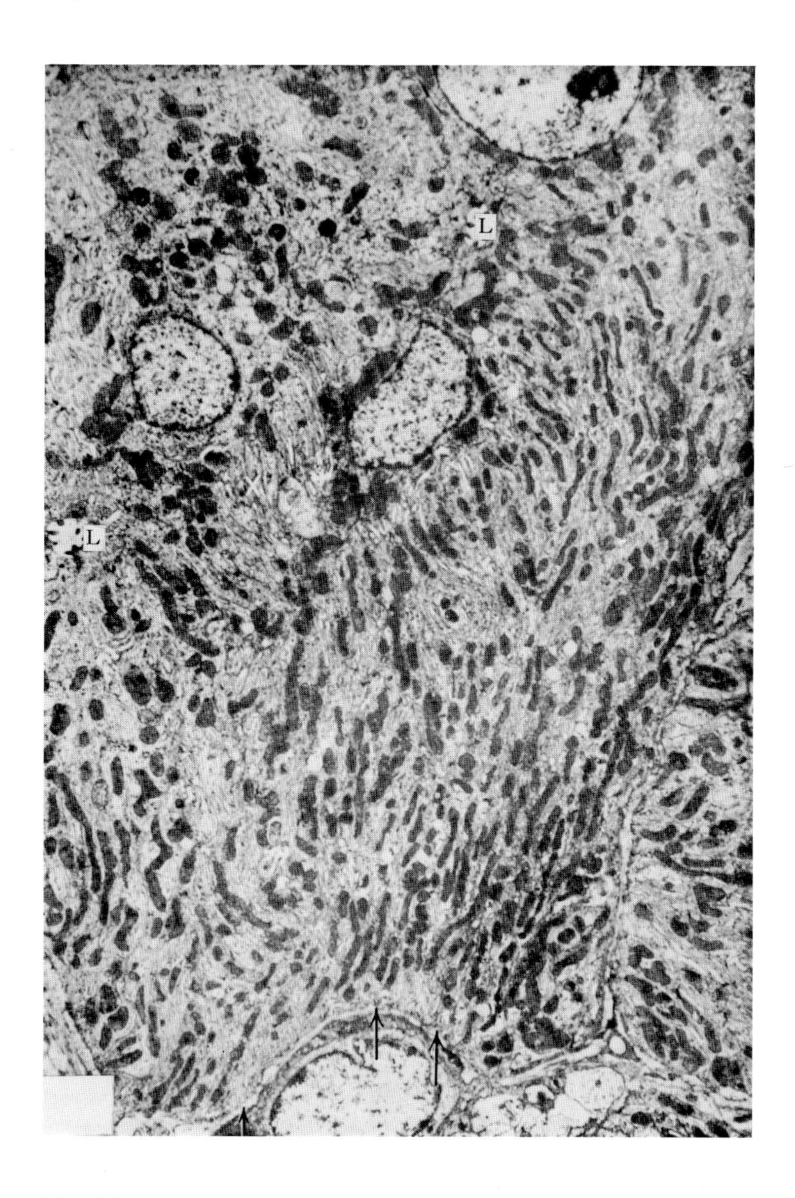

Plate 2.3
Fully-specialized secretory cells from the salt gland of a duckling given salt water to drink for eleven days. Note the complex folding of the lateral (X) and basal (arrows) cell membranes and, in contrast, the area of the luminal membrane with a few microvilli protruding into the lumen (L) (from S. A. Ernst & R. A. Ellis, *Journal of Cell Biology* (1969) **40**, 305).

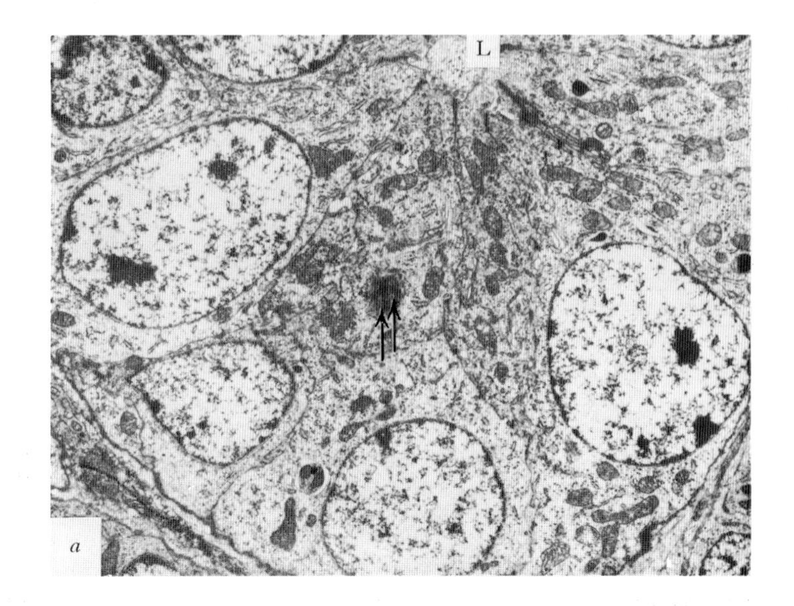

Plate 2.4
(*a*) Peripheral cells in the tubule of a salt gland from a duckling. The lateral and basal surfaces are nearly flat with only a few folds. Some cells contain more mitochondria than others. Lumen (L), nuclear pores (arrows).

(*b*) The apical surfaces of the secretory cells from the salt gland of a two-day old duckling have short microvilli (Mv) extending into the lumen (L) of the tubule. The cytoplasm contains abundant free ribosomes and a few mitochondria (M). The opposed lateral surfaces of the cells are usually flat, but some bear short folds; intermittently, they are joined by desmosomes (arrows). The junctional complexes (J) presumably block direct passage from the intercellular space to the lumen. Multivesicular bodies (MB) are found commonly near the apical surface and profiles of rough-surfaced endoplasmic reticulum (RER) and smooth vesicles (V) are scattered randomly throughout the cytoplasm (from S. A. Ernst & R. A. Ellis, *Journal of Cell Biology* (1969) **40**, 305).

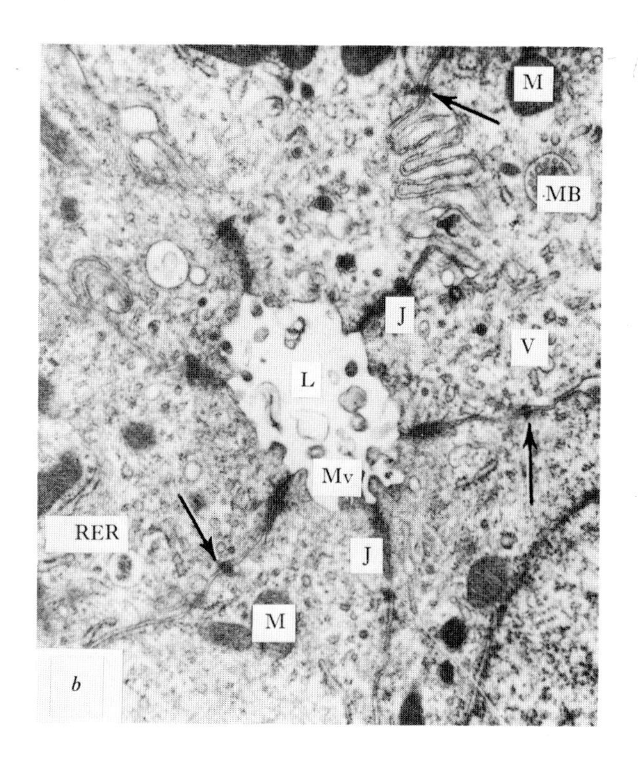
M
MB
J
V
L
Mv
RER
J
M
b

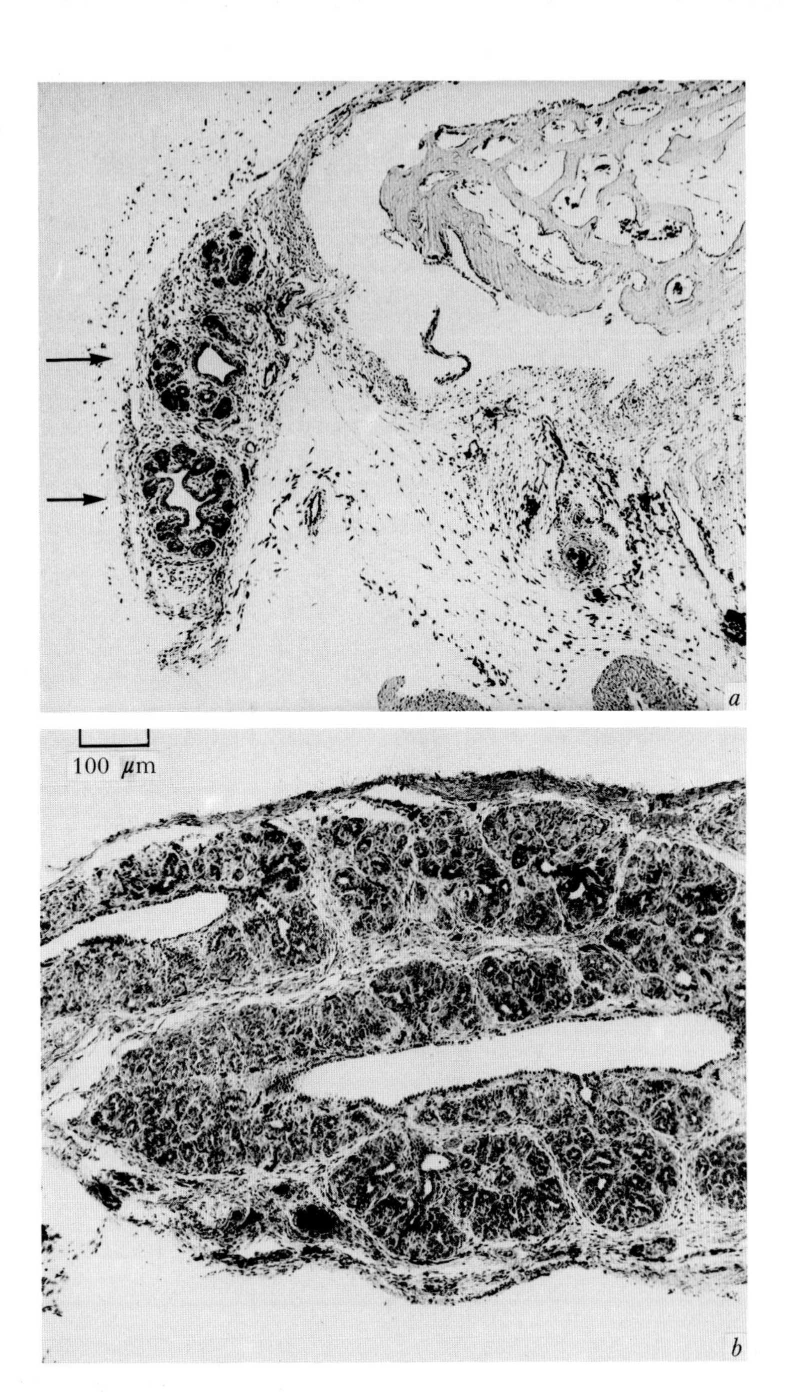

Plate 2.5
Photomicrographs, at the same magnification, of sections of the salt glands of ducklings about 2 days before (*a*) and the day after (*b*) hatching. In the younger bird the gland is minute and consists of the 2 main ducts (arrows) and a few vestigial tubules. By the day of hatching the gland has grown to about 75 mg and is capable of functioning. Masson stain. Sections prepared by Mr J. G. Jarvis. Photography, Mr A. L. Gallup.

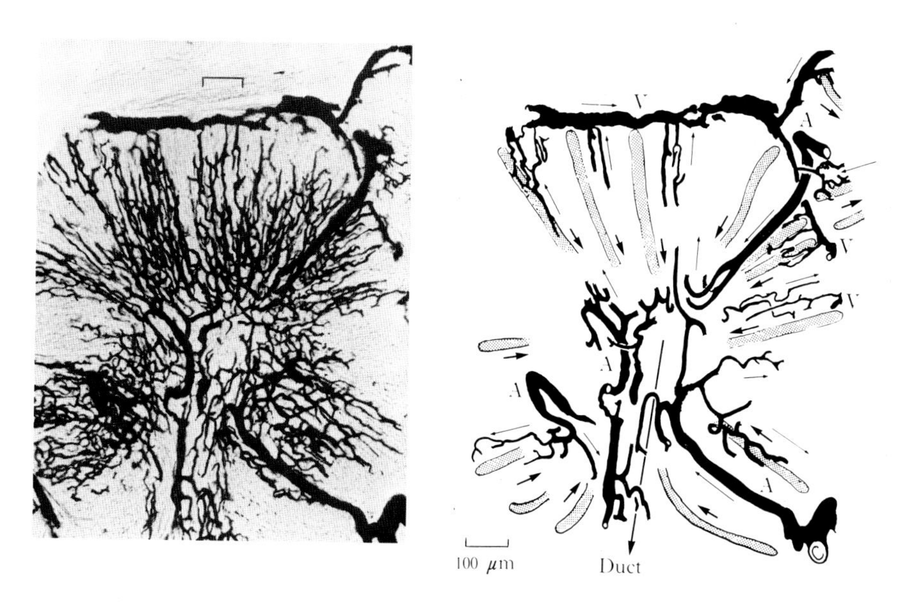

Plate 2.6
Blood vessels to part of a lobe of the salt gland of a duck, injected with Indian ink; unstained frozen section 100 μm thick. Not all the capillaries are filled but this makes it easier to see that they run parallel to the secretory tubules; these empty into a duct. The lobe was examined with a stereoscopic binocular microscope to prepare the drawing, which shows regions of vascular continuity; in the left photograph these are obscured by other vessels. Arteries were distinguished by their relatively straight course, uniform diameter and, in parts, wrinkles; veins were more irregular in diameter and course. Large arrows, direction of flow of secretion; small arrows, flow of blood.

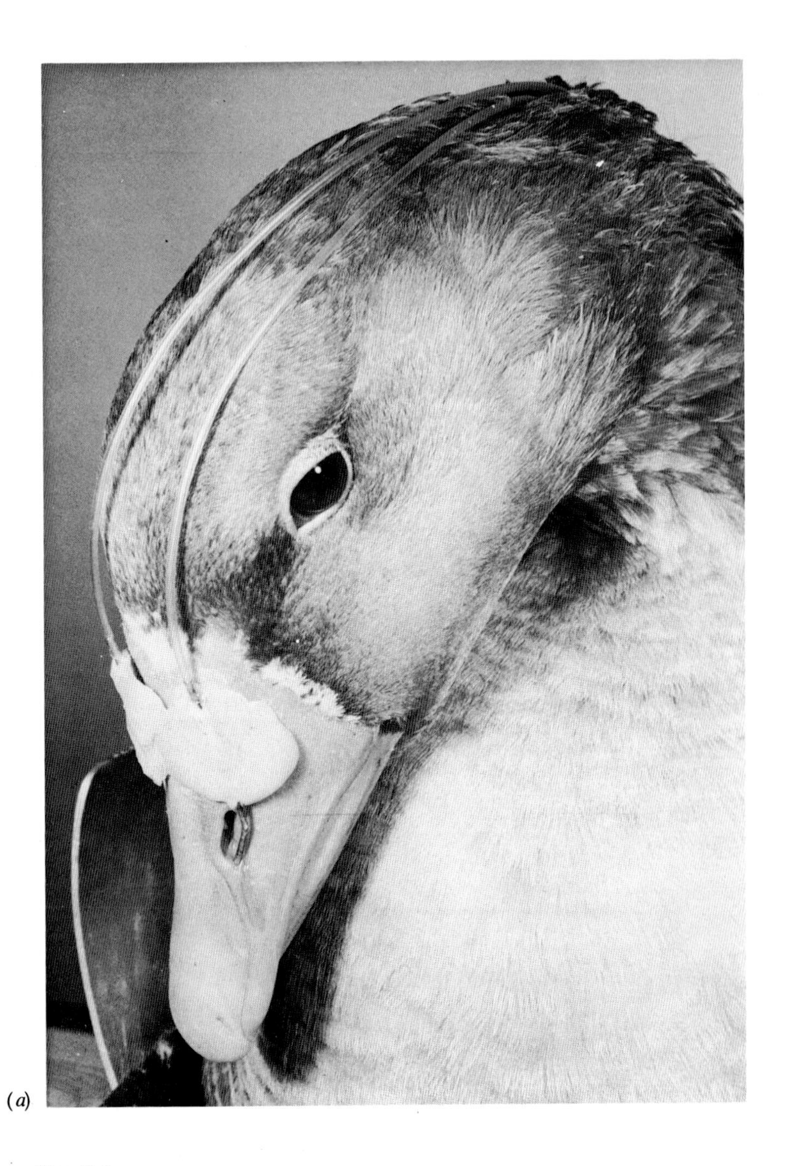

(*a*)

Plate 3.1
(*a*) Steel cannulae fastened to the beak of a goose for collection of salt-gland secretion by continuous aspiration. The tips of the cannulae lie on the floor of each nostril.
(*b*) Set-up for collecting salt-gland secretion. Note the peristaltic pump and fraction collector.

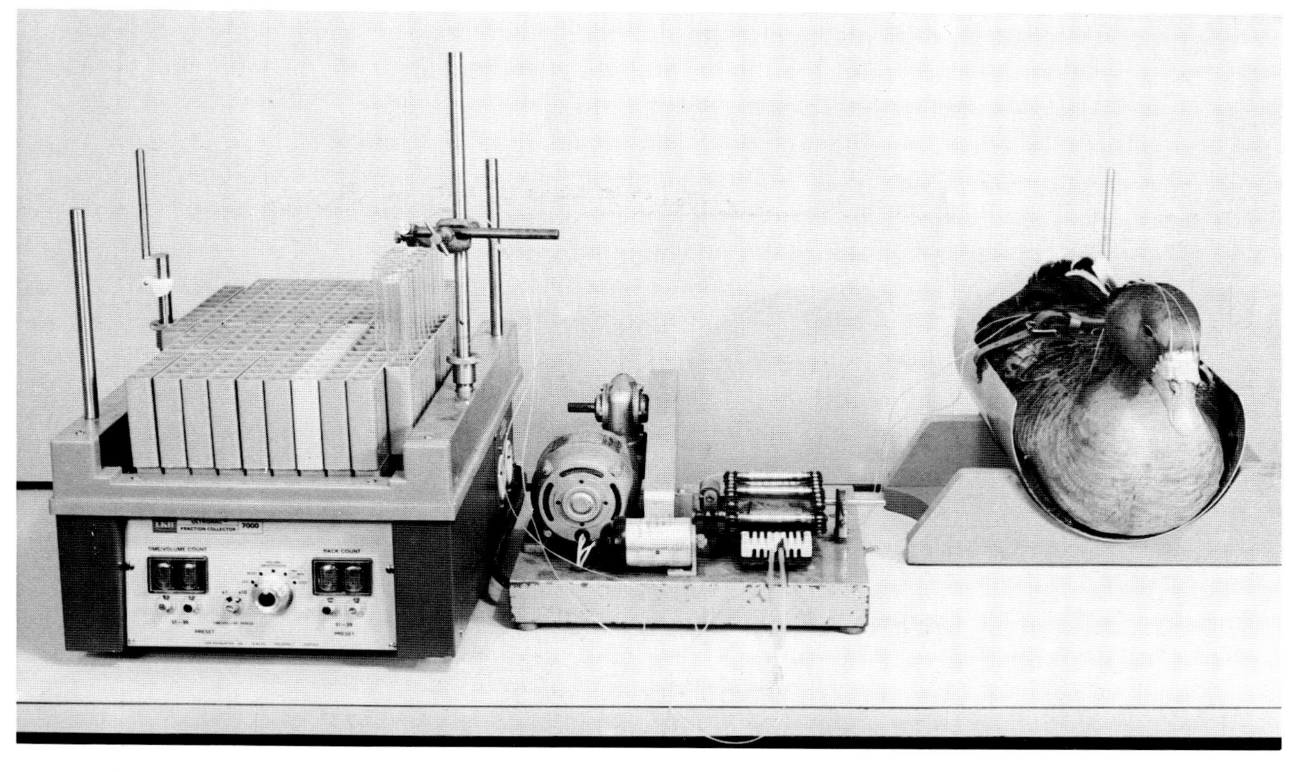
FRACTION COLLECTOR 7000
TIME/VOLUME COUNT
RACK COUNT
PRESET
PRESET

(*b*)

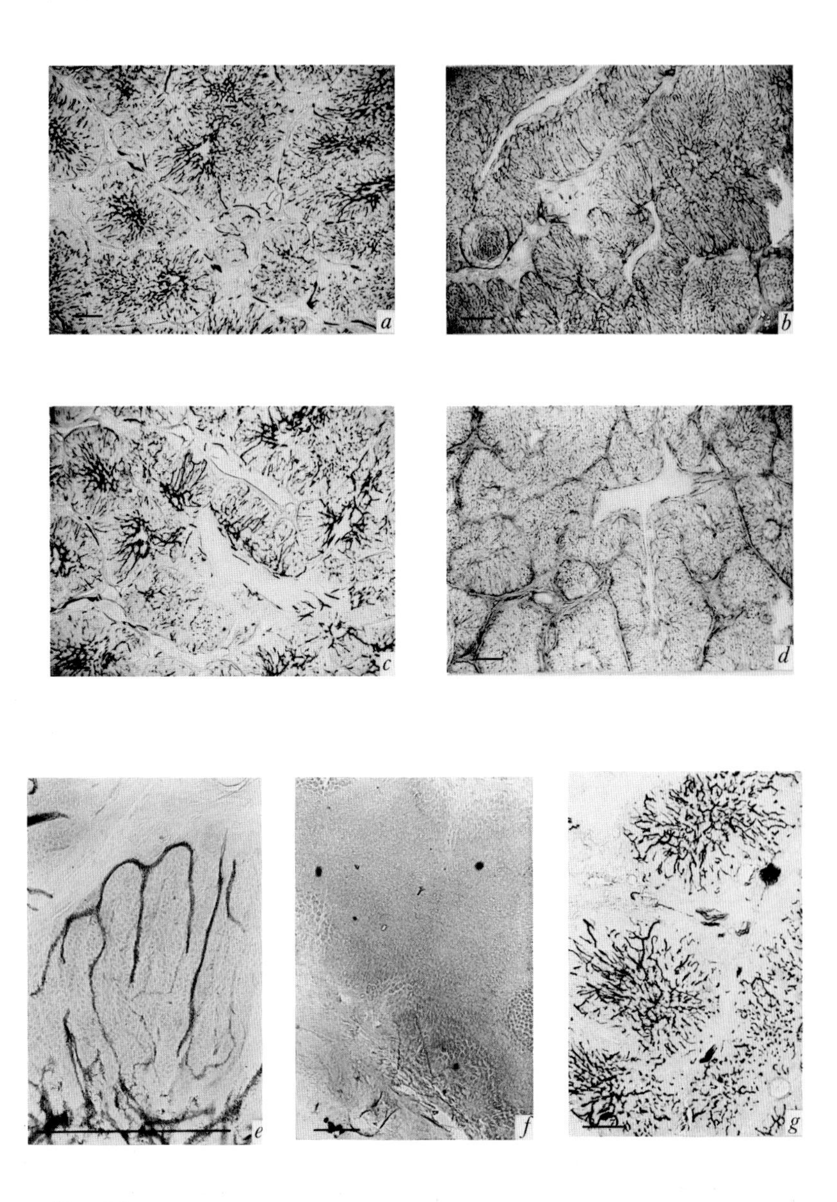

Plate 4.1
(*a*) Acetylcholinesterase (AChE) distribution in resting salt gland of the duck; (*b*) Butyrylcholinesterase (BuChE) distribution in resting salt gland; (*c*) AChE distribution in secreting salt gland; (*d*) BuChE distribution in secreting salt gland; (*e*) looped appearance of AChE distribution near periphery of salt gland lobule under higher magnification; (*f*) Absence of AChE staining in gland previously denervated by nerve excision (removal of secretory nerve ganglion); (*g*) Persistence of AChE staining in gland following section of its secretory nerve. Scale bars: 200 μm (from R. W. Ash, J. W. Pearce & A. Silver, *Quarterly Journal of Experimental Physiology* (1969) **54**, 281).

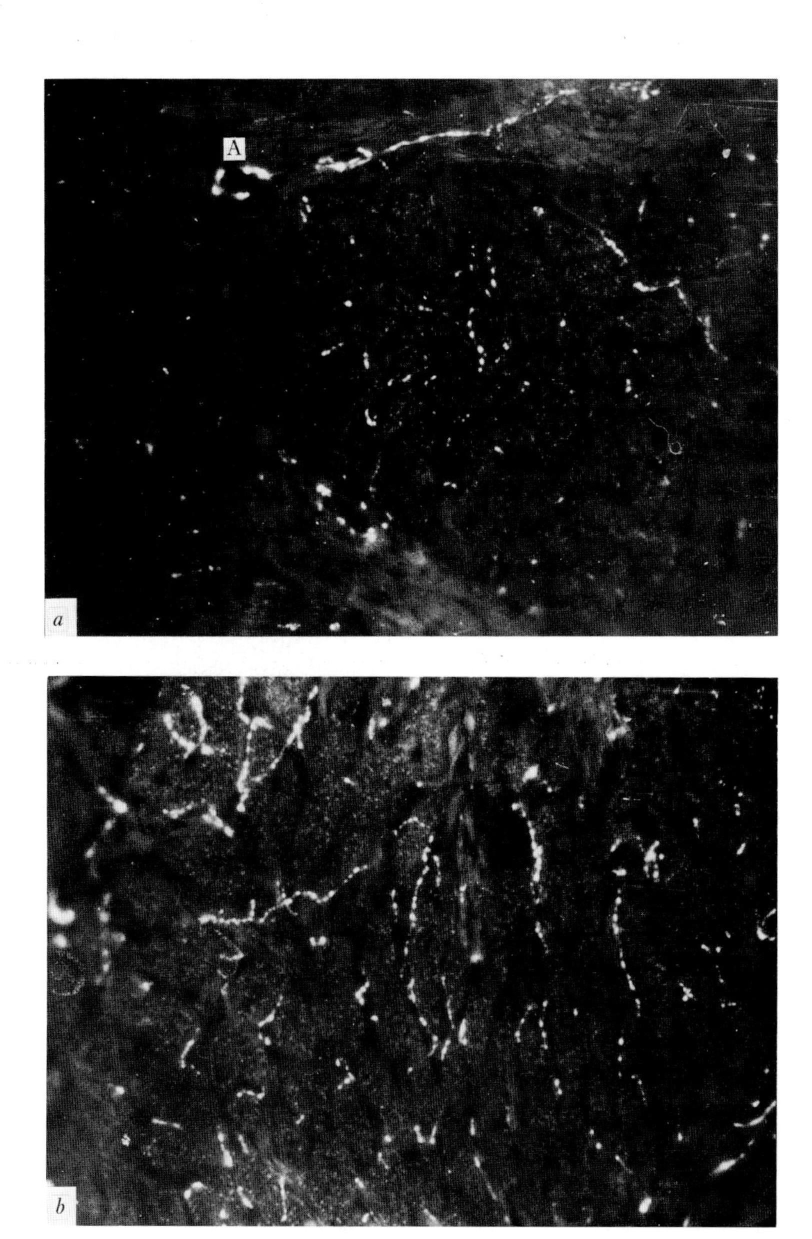

Plate 4.2
Photomicrographs showing fluorescent adrenergic fibres in the salt gland following treatment of sections with formaldehyde vapour: (*a*) Fibres running between secretory tubules; note also the fibres around an arteriole (A); (*b*) longitudinal section through a number of secretory tubules showing adrenergic fibres with varicosities (sections prepared by Dr J. R. McLean).

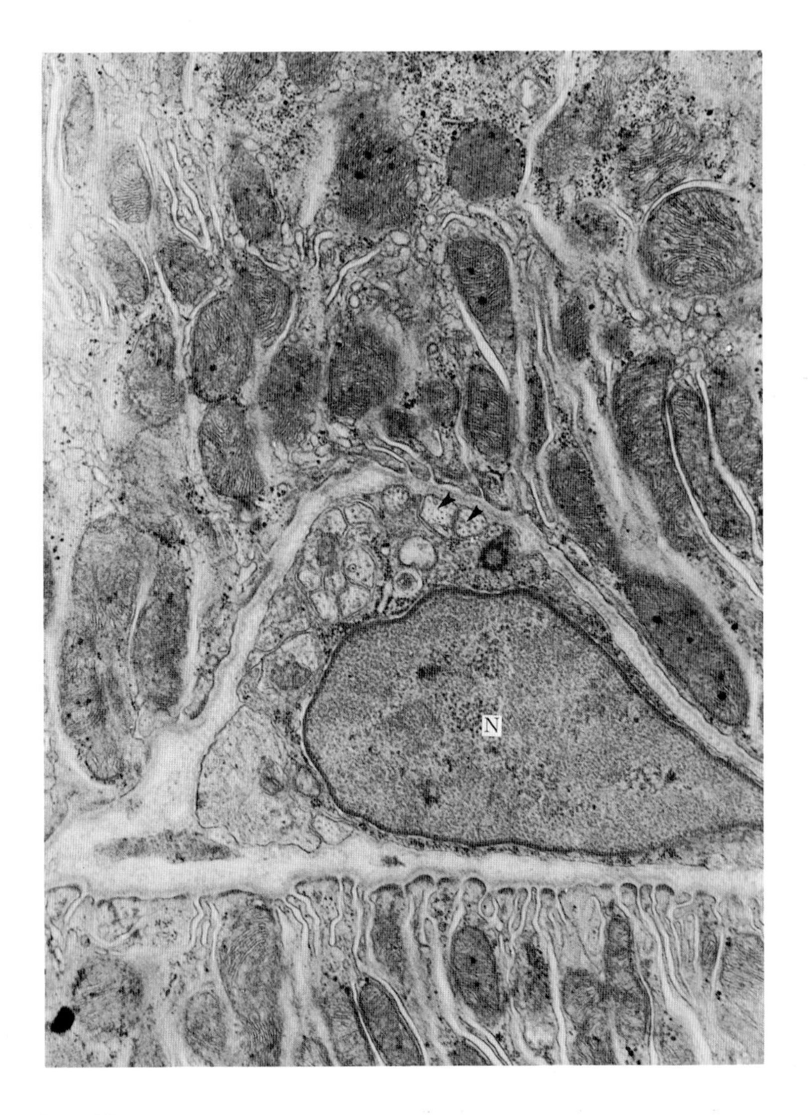

Plate 4.3
Schwann cell containing nerve axons (some of which are arrowed) running between the bases of two secretory tubules; N = Schwann cell nucleus (by courtesy of Professor D. W. Fawcett).

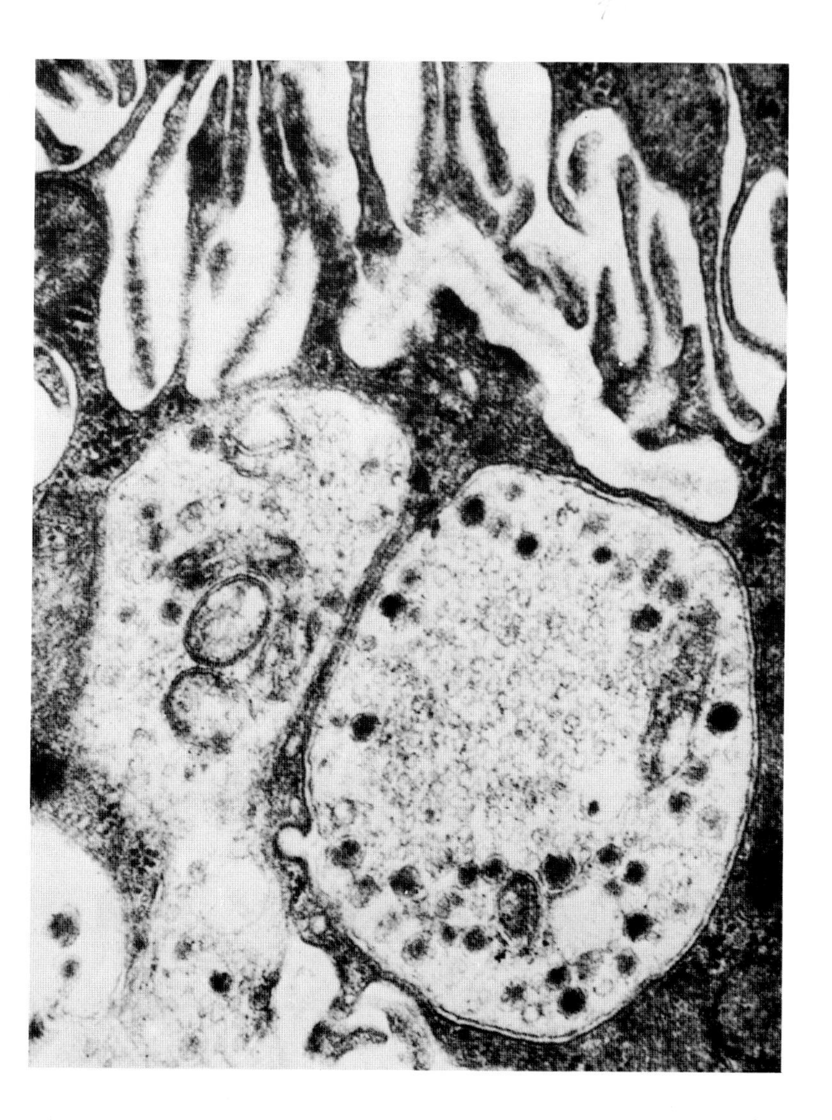

Plate 4.4
Two nerve terminals situated in furrows in the base of a secretory cell (from W. Kühnel, *Zeitschrift für Zellforschung und mikroskopische Anatomie* (1972) **134**, 435).

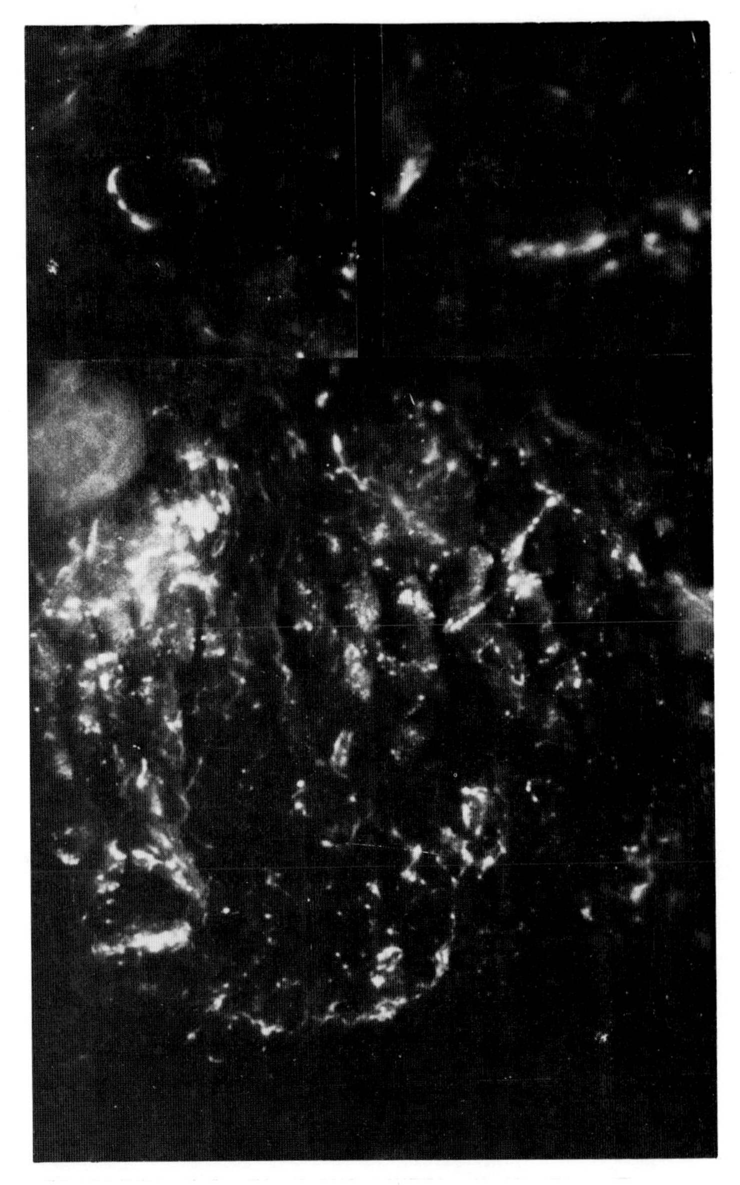

Plate 4.5
Photomicrographs showing fluorescent adrenergic fibres in sections of the secretory nerve ganglion following treatment with formaldehyde vapour. Fine fluorescent terminals are evident surrounding non-fluorescent nerve cell bodies; the insets show these varicosities in more detail (sections prepared by Dr J. R. McLean).

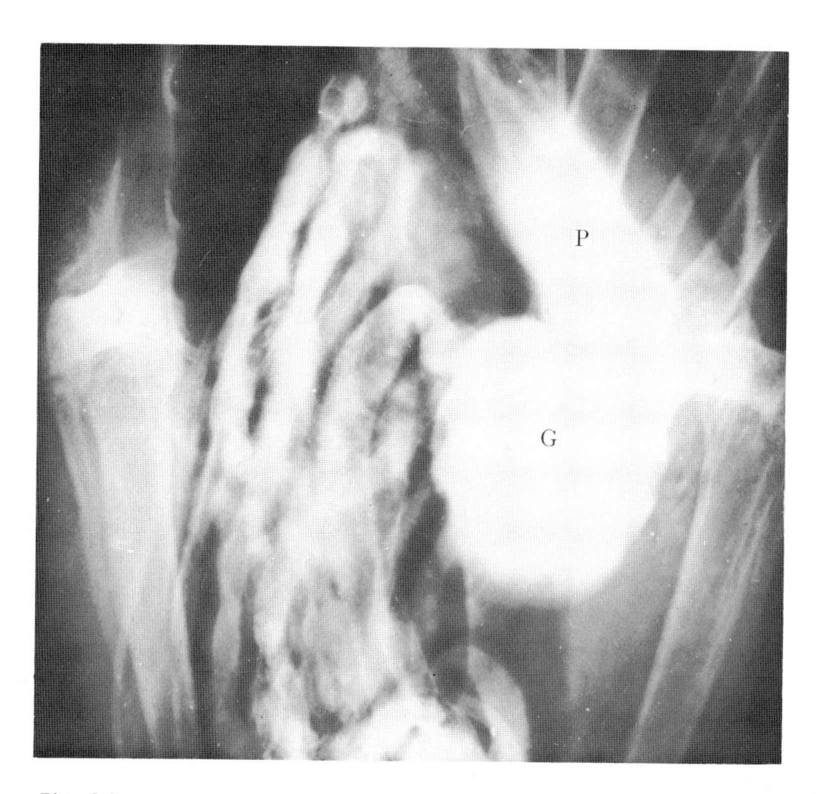

Plate 8.1
Radiograph (dorso-ventral) of a conscious goose taken two minutes after the administration of 100 ml sea water containing barium sulphate into the proventriculus, showing that the whole alimentary canal contains radio-opaque material; P = proventriculus, G = gizzard.

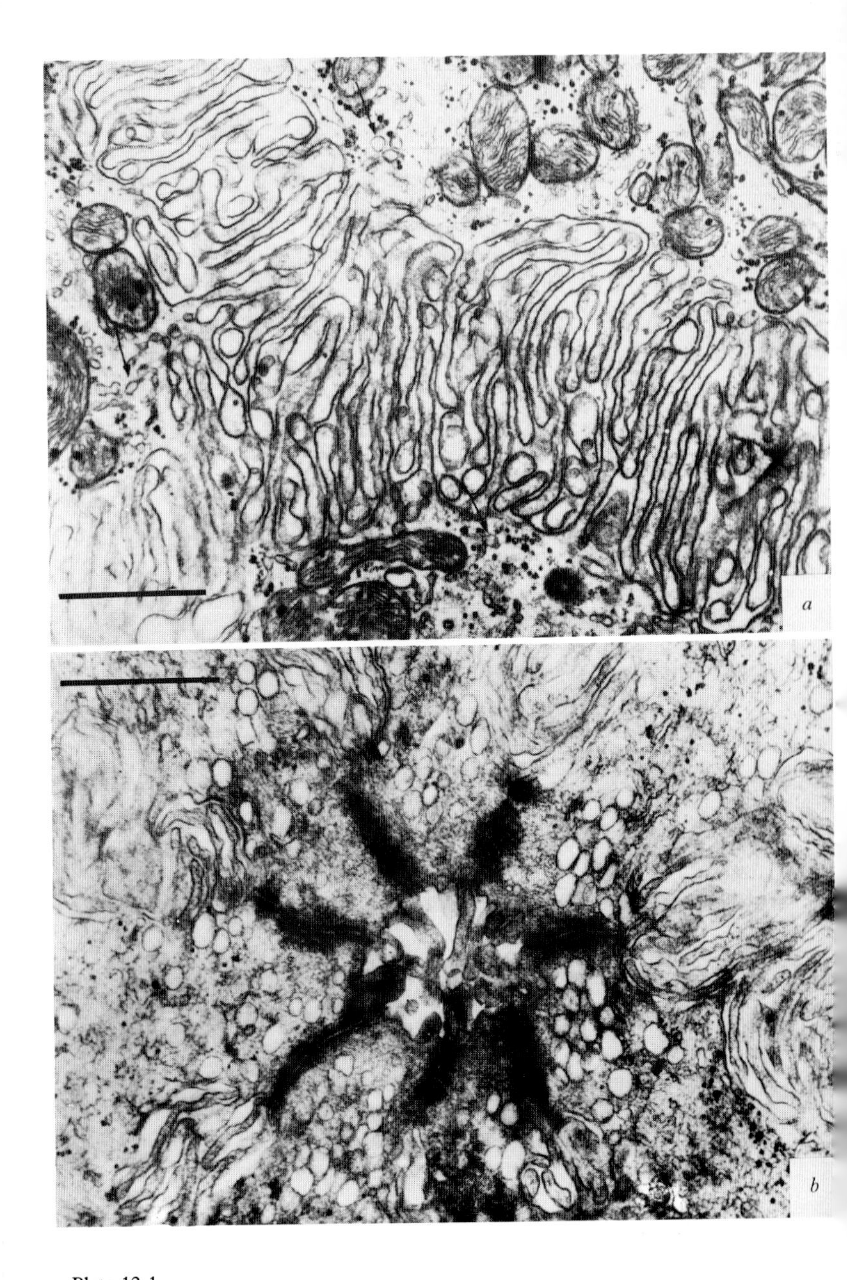

Plate 12.1
(*a*) Portions of the intercellular channels surrounding three cells of a turtle salt gland. Mitochondria and small vesicles (arrows) are evident near the bases of the microvilli. Bar = 1 μm.

(*b*) Seven cells of a turtle salt gland sectioned near their apices. Note the junctional complexes, lateral interdigitating microvilli and projections from the luminal membrane. Bar = 1 μm (from J. H. Abel & R. A. Ellis, *American Journal of Anatomy* (1966) **118**, 337).

volume secreted was less in ducks tested within twenty-four hours of the operation. In view of these results – abolition of secretion by postganglionic but not by preganglionic denervation – Ash *et al.* (1969) suggested that the receptors may in fact be situated in the secretory nerve ganglion. In support of this suggestion they considered that the ganglion may not be a normal synaptic site since the administration of hexamethonium (an agent which blocks ganglionic transmission) failed to abolish secretion, and the secretory nerve was found to contain little AChE except near the ganglion.

Whilst the evidence from denervation studies does suggest the possibility of receptors in the ganglion a number of important points are not covered by this hypothesis. Firstly, general anaesthesia blocks secretion, which would suggest some action in the CNS. Secondly, decerebration often results in inhibition of secretion in response to a subsequent salt-load (Ash *et al.*, 1969; Hanwell *et al.*, 1972). And thirdly, stimulation of the preganglionic (secretory) nerve initiates secretion and this, as Fänge, Schmidt-Nielsen & Robinson (1958) concluded, is the obvious route of excitatory fibres from the brain. The possibility that the preganglionic innervation might be inhibitory, as suggested by Cottle & Pearce (1970) to explain the results of the denervation experiments, therefore seems unlikely.

We were surprised that in marine birds nobody had repeated Verney's classical experiments in the conscious dog of inducing antidiuresis by infusing hypertonic sodium chloride into an exteriorized carotid artery, which led to his discovery of osmoreceptors in the hypothalamus. When we did this in the conscious goose we were, to our surprise, unable to induce salt-gland secretion. During infusions secretion eventually started but in all cases 2–2.3 mmoles per kg body-weight were required to initiate secretion. Since this amount was identical to that necessary when given by intravenous infusions it appeared that the receptors do not lie in the head. A cross-circulation preparation was therefore used to study this further; the arrangement is shown in Fig. 4.6. Blood from the donor's brachial artery was used to perfuse the head of the recipient via the carotid arteries; the venous outflow from the head was then returned to the brachial vein of the donor. In fact

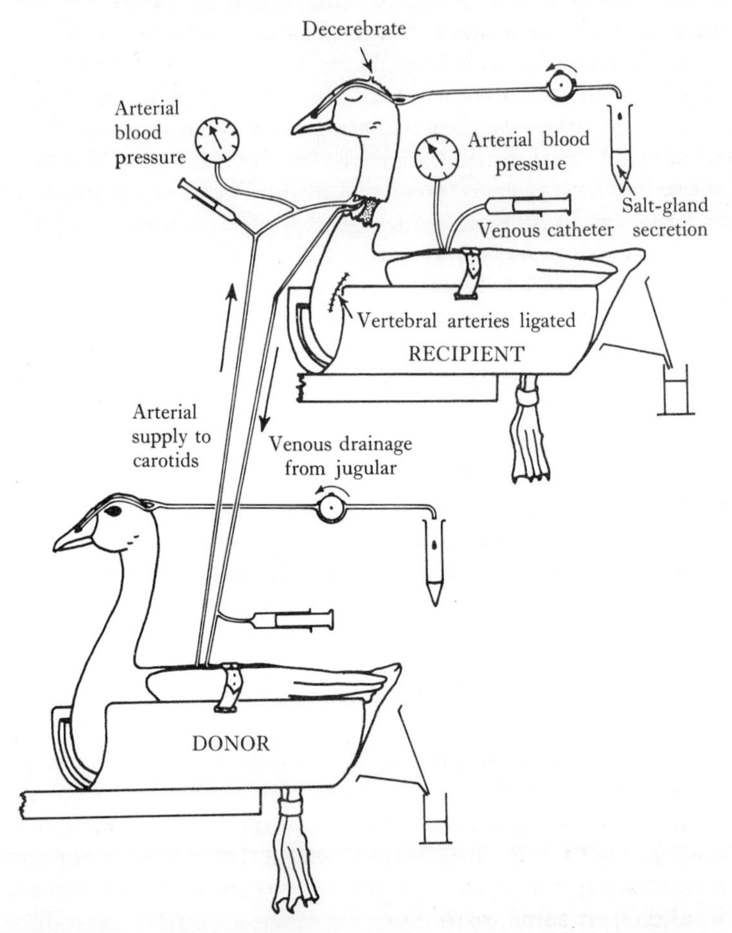

Fig. 4.6. The set-up for cross-circulation experiments used to study the location of the receptors in geese. Blood from the donor's brachial artery was used to perfuse the head of the recipient via the carotid arteries. Venous blood from the recipient's jugular was returned to the brachial vein of the donor. The vertebral arteries of the recipient were ligated low in the neck (from Hanwell *et al.*, 1972).

although the preparation sounds a simple one, difficulty was experienced in achieving a complete separation of donor and recipient blood in the head and neck. Furthermore for humanitarian reasons and because general anaesthesia blocks secretion, the recipient was decerebrated and this procedure itself was often found to inhibit secretion. However in one experiment in which we did separate the two circulations (confirmed by the failure of Evans Blue dye to pass from one bird to the other) and in which we were able to complete the whole series of tests, a very clear result was obtained. In this bird the vertebral arteries of the recipient were ligated low in the neck before they enter the vertebral column and all other tissue in the neck divided except for the spine and vagus nerves. When the donor was given hypertonic sodium chloride intravenously its salt glands started to secrete after the normal latency of approximately two minutes. However no secretion was observed from the glands of the recipient whose head was perfused with the same blood (the sodium concentration of which was raised by 15 mmoles per litre an increase of 9.7 per cent); no change in plasma composition was apparent in the body of the recipient. Thirty minutes after the donor was loaded, the body of the recipient was given hypertonic sodium chloride intravenously. The salt glands of the recipient began to secrete 2.5 minutes later and continued in a normal manner (Fig. 4.7). Thus we were able to conclude from the two types of experiment that nervous impulses from a part of the body, as opposed to the head, are necessary for salt-gland secretion. Incidentally the cross-circulation studies also showed clearly that hormones or other humoral substances are not responsible for the initiation of secretion, as has been proposed by some workers (Chapter 8).

Our conclusion, that the receptors are not located in the head, was not totally unexpected since at the time evidence was accumulating for peripheral osmoreceptors in the hepatic circulation in some mammals. The next approach was therefore to try to locate them. Catheters, through which hypertonic sodium chloride solutions were infused, were implanted in a number of blood vessels in the splanchnic area. Sites examined were the hepatic portal circulation, in case the receptors monitored blood arriving from the alimentary canal (catheters in the venous drainage of the gizzard, proventriculus and small intestine); the wall of the alimentary canal (catheters in the arterial supply of the same regions) and the renal portal system, since hepatic portal blood can be diverted to the renal portal system along the coccygeomesenteric vein

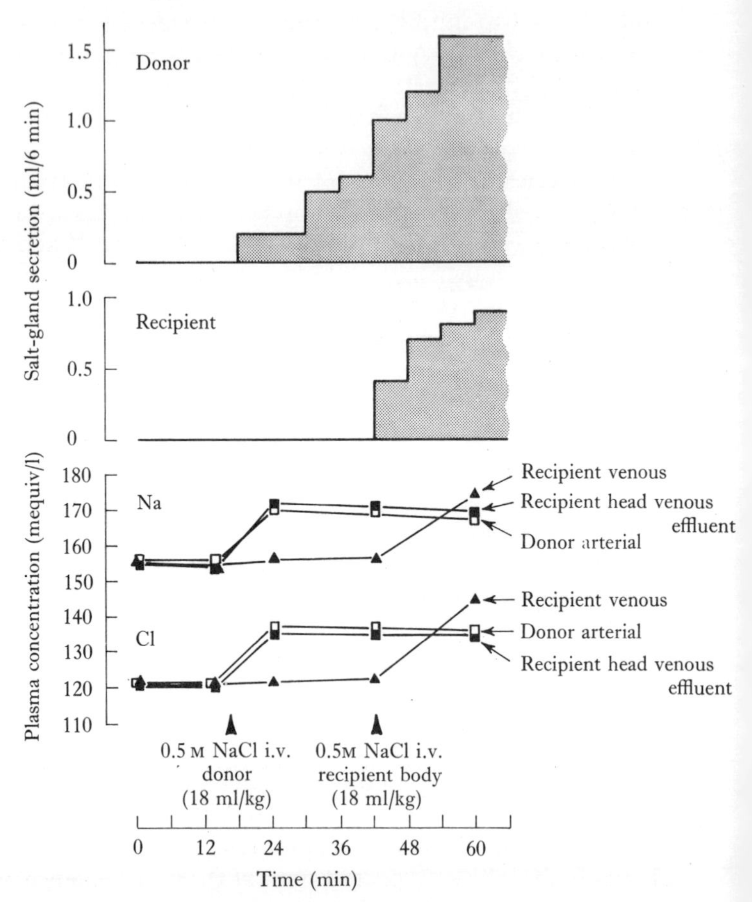

Fig. 4.7. Effect of salt-loading in a cross-circulation experiment (see Fig. 4.6). Time 0 is the time the preparation was finally set-up (from Hanwell *et al.* 1972).

(see Akester, 1967) (catheters in a renal portal vessel). However as with infusions in the carotid artery, secretion did not begin until the same amount of sodium chloride had been given as is required by an intravenous route (brachial or metatarsal veins) – the 'minimal stimulatory load'. At this time Dr Ann Hanwell argued that if impulses do arise from the body then the vagus nerves would be the most likely route to the CNS. When the vagi in the neck were cut (they are superficial and can easily be exposed under local anaesthesia) in geese secreting in

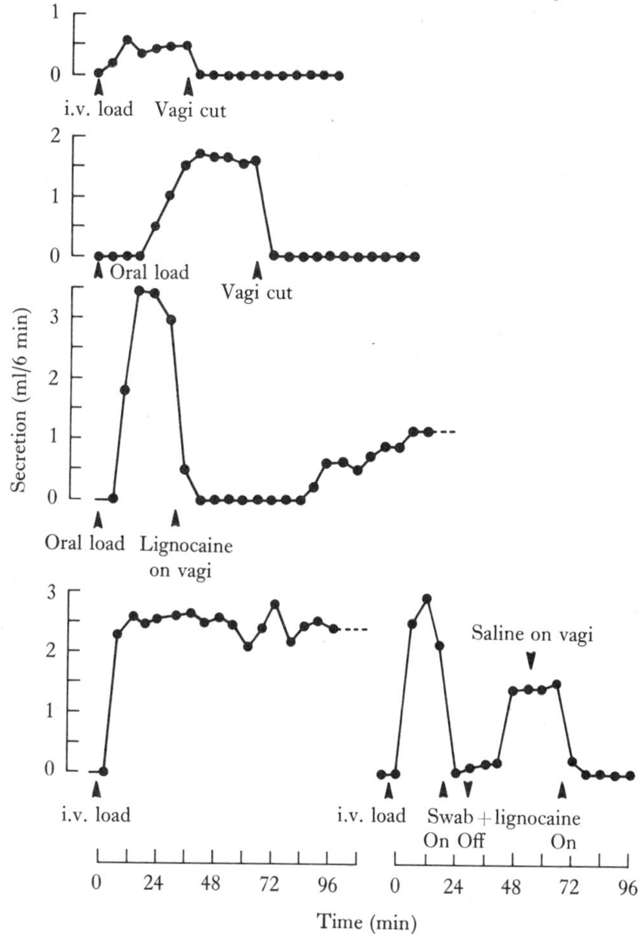

Fig. 4.8. Effect of cutting or blocking the vagi on salt-gland secretion in response to oral or intravenous salt-loading in four geese. In one experiment the control response to salt-loading was studied two days before (from Hanwell *et al.*, 1972).

response to a salt-load, secretion stopped within seconds. Furthermore when the local anaesthetic lignocaine was applied to intact vagi, secretion again ceased and then recovered after some time, presumably as the effect wore off (Fig. 4.8). Thus it seemed that the afferent fibres do run in the vagi. When the vagi were cut below the heart and a long-acting local anaesthetic was injected into the wall of the proventriculus in order to block any fibres that may have remained intact, the administration of a salt-load (as soon as the birds had recovered from halothane anaesthesia) induced secretion after the normal latency.

Dissections confirmed that between the cuts made in the vagi in the neck (which abolished secretion) and caudal to the heart (which did not) branches of the vagi pass to the heart, great vessels (including the carotid body), lungs and part of the oesophagus, so that any of these sites could have been the location of the receptors. The injection of large amounts of long-acting local anaesthetic into the crop region of the oesophagus failed to influence secretion so a site in the heart, lungs or great vessels became a distinct possibility. However cardiac nerves are extremely inaccessible to surgery from which the animal must be allowed to recover, so we used the technique of Ledsome & Linden (1968) of blocking cardiac nerves by injecting local anaesthetic into the pericardial sac. As in the cervical vagotomy experiments, this procedure abolished secretion in response to a salt-load (Fig. 4.9). We therefore concluded that the receptors are probably located in the heart or great vessels and that the axons pass through or near the pericardial sac. It seems unlikely that the endings are in the pericardial region because hypertonic sodium chloride injected into the sac failed to induce secretion; a site in the luminal wall of the chambers or vessels would appear likely.

It could be argued that, in the experiments in which the vagi or cardiac nerves were blocked, there may have been an adrenosympathetic discharge due to stress and hence secretion may have been inhibited by vasoconstriction in the gland. However the rapid and complete cessation after vagotomy under local anaesthesia via a small incision in the neck argues against this interpretation since in no experiments have catecholamine infusions completely inhibited secretion in geese. Furthermore carbachol induced secretion in vagotomized birds showing that the glands could still respond to a blood-borne cholinergic stimulation. Direct evidence for the vagi being the afferent pathway of a secretory reflex was also obtained. It was found that stimulation of the vagi in conscious birds evoked signs of discomfort. We therefore used a perfused decerebrate head preparation, completely isolated from the body, for further studies in order to overcome this problem as well as any

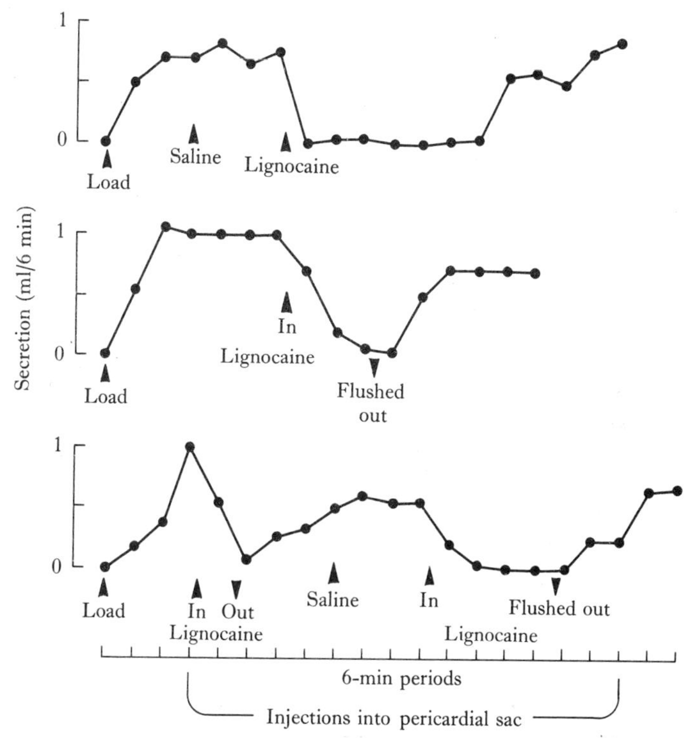

Fig. 4.9. Effects of injecting local anaesthetic (1–2 ml 2% lignocaine) or 0.154 M-NaCl into the pericardial sac, on secretion in response to an intravenous salt-load in geese (from Hanwell *et al.*, 1972).

cardiovascular effects. In these experiments stimulation of the right or left vagus evoked secretion at a high rate which quickly ceased when stimulation was discontinued (Fig. 4.10). As in the cross-circulation experiments a raised sodium chloride concentration in the homologous blood used as perfusate failed to induce secretion.

The conclusion from these experiments is clear (Hanwell *et al.*, 1971*c*, 1972; Peaker, Hanwell & Linzell, 1971). Osmoreceptors situated either in the heart or great vessels are responsible for initiating secretion and impulses travel in the vagus nerves to the brain. Arguing teleologically, our earlier thoughts that osmoreceptors might monitor sites of entry of salt into the body were wrong because in a sodium-deficient state it is disadvantageous to secrete and thus lose more salt.

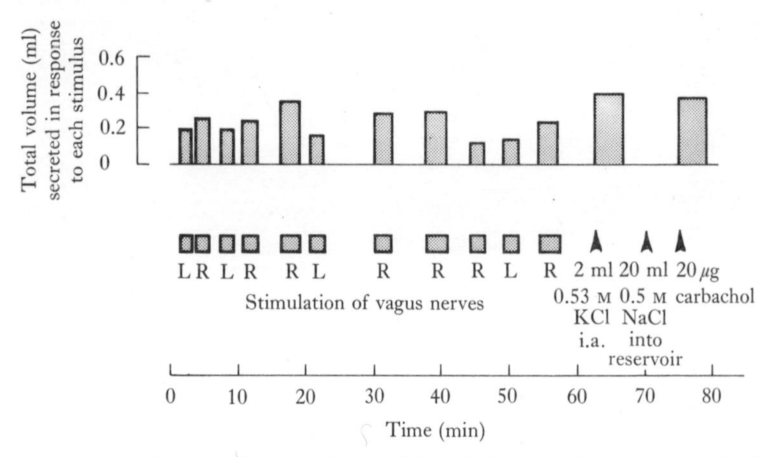

Fig. 4.10. Effects of vagal stimulation, KCl, NaCl and carbachol on salt-gland secretion in the isolated, perfused, decerebrate head of a goose. Time 0 is 30 min after the head was completely severed from the body. The composition of the fluid collected was approximately 340 mM for sodium, 15 for potassium and 346 for chloride. L, left; R, right (from Hanwell *et al.*, 1972).

The bird needs to know the overall state of its osmotic balance and of course the composition of blood, mixed from all parts of the body, in the heart reflects the osmotic status of the animal as a whole.

We must make it clear that we have no further information on the exact location of the receptors. This is a field for further discoveries but in view of the complexity of vagal branches in birds and the fact that general anaesthesia inhibits secretion, their exact location will probably prove difficult to determine.

Efferent pathway

The evidence on the site of the receptors and the afferent nervous pathway clearly provides no support for the suggestion that they are located in the secretory nerve ganglion. Rather it supports the original hypothesis of Schmidt-Nielsen and his group that the efferent pathway is the secretory nerve which synapses in the secretory nerve ganglion. In terms of interpreting denervation experiments we can do little more than to quote Burgen & Emmelin (1961) on the innervation of mammalian salivary glands. 'There is some uncertainty as to the route by which the nerves reach some of the glands. Evidence suggests that secretory fibres may arrive at some glands through unknown channels. This may invalidate inferences drawn from denervation experiments.' Later evidence from, for example, the dog parotid gland supports this view (Holmberg, 1971; Ekström & Holmberg, 1972). In the case of the salt gland the findings could be explained if another nerve containing afferent fibres enters the secretory nerve ganglion. Although this might be the case in the domestic fowl (Hsieh, 1951), there is no such indication from anatomical work in other birds that have been studied.

Our own investigations in the goose add little except possibly to suggest an alternative explanation for preganglionic denervation. In contrast to Ash *et al.* (1969) with the duck, we found, with the goose, that a somewhat higher dose of hexamethonium does block secretion in response to a salt-load, and when applied topically to the secretory nerve ganglion it also blocks secretion in response to preganglionic stimulation of the secretory nerve (Hanwell *et al.*, 1972). This, together with the anatomical findings, suggests that the secretory nerve ganglion is a normal synaptic site. Furthermore in two birds we found that six hours after bilateral preganglionic denervation salt-loading failed to induce secretion. After twenty-four hours some secretion appeared but the pre-operative response was not obtained until four days after the operation. Even one hour after extensive surgery unconnected with the secretory nerve, we have found secretion is produced normally, so that the response to preganglionic denervation at this time would appear to be a genuine effect. The slowly-returning response might suggest a phenomenon of altered- or supersensitivity of the ganglion cells to some agent.

The suggestion of a supersensitivity phenomenon in the salt gland is not new. Gill & Burford (1968) reported a unilateral increase in the secretory response to salt-loading twelve to fourteen days after ipsilateral denervation. Unfortunately they only stated that they removed 'five or six secretory nerves leading to the anterior base of the salt gland'. This might suggest an incomplete postganglionic denervation and supersensitivity of the secretory cells developing in response to the reduced levels of acetylcholine (some birds do in fact have ganglion cells actually in the gland, see above). We are still left with the problem of what might stimulate the denervated secretory nerve ganglion. A possible candidate is plasma potassium, the concentration of which increases after salt-loading. Potassium chloride infused into the carotid artery of intact geese does induce a transient but inconsistent secretion. In anaesthetized or decerebrate birds 0.53 M potassium chloride infused slowly into a carotid artery was found to induce a profuse secretion even after the secretory nerve had been cut. This implies that the secretory nerve ganglion or the secretory cells can be stimulated by an increase in extracellular potassium concentration. Furthermore, Peaker, Peaker, Phillips & Wright (1971) found that raising plasma potassium levels in ducks enhanced secretion in response to a salt-load. A rapid increase in sensitivity to potassium is therefore a possible explanation for the rapid return of secretion after preganglionic denervation. Another possibility is that while the secretory nerve ganglion is normally insensitive to a raised plasma sodium chloride concentration or hypertonicity, it becomes sensitive after preganglionic denervation and in this sense the receptors are then located in the ganglion. Such a mechanism would be an elegant feat of homeostasis as far as the bird (but not the investigator) is concerned!

It is clear that much work remains to be done on the innervation of the salt gland and while we believe that the secretory nerve is the main efferent pathway and we have an overall picture of the secretory reflex, features remain which are baffling to say the least. So as not to fool the reader into the belief that we can explain everything we are including here some unpublished work by one of us (M.P.) and Dr Alan Wright for which we have no satisfactory explanation.

When Wright *et al.* (1966) studied the effect of adenohypophysectomy on salt-gland secretion in ducks they found that sham operation also had a marked effect in reducing the output of nasal fluid in response to a subsequent salt-load; similar results were later obtained for sham neurohypophysectomy (Wright, Phillips, Peaker & Peaker, 1967). Examination of the data showed a range of values, from a complete lack of secretion to normal amounts. In neither series of experiments did the pituitary appear damaged and the ducks failed to secrete after treatment with adrenocorticotrophic hormone (ACTH), corticosterone and, in later experiments, prolactin (see Chapter 8). We were led to believe that the efferent nervous pathway to the gland may have been damaged since secretion was obtained in response to the administration of cholinomimetics in the ducks in which secretion had been abolished by sham operation. Although the problem with respect to pituitary surgery has now been overcome by using young birds instead of adults (Holmes, Lockwood & Bradley, 1972), we made further studies in adults. The only nerves that were apparently near the transbuccal surgical approach to the pituitary were the palatine branches of the facial nerves on each side. Cutting these however had no effect on secretion. When the base of the skull was drilled to expose only the sella turcica the volume secreted again varied from very little to normal amounts. But when this drilling was combined with sectioning the palatine VIIth branches none of the five birds secreted in response to an intravenous salt-load. However we were unable to discover what structures were being damaged by drilling but if a nervous pathway was involved then it is clear that we seemed to be dealing with two routes to the gland – one in the palatine branch, the other unknown – with damage to both being necessary to inhibit secretion completely. A question raised is, if these experiments constituted preganglionic denervation, why do ducks secrete after the secretory nerve, also believed to be preganglionic, has been cut? Or is there perhaps a second, as yet unknown, preganglionic pathway to the secretory nerve ganglion?

The only electrophysiological study of the secretory nerve was made by Cottle & Pearce (1970). They found three populations of fibres but since a nerve was observed leaving the secretory nerve ganglion to reach an intra-ocular muscle, the faster component (conduction velocity approximately 2 m per second) 'could have been that of fibres passing through the ganglion to innervate a structure other than the salt gland...'. Attempts to record action potentials postganglionically after preganglionic stimulation were unsuccessful; this is not surprising in view of the short length of these nerves to the gland.

The vasodilator and secretomotor roles of cholinergic nerves in the gland are discussed in Chapters 5 and 6 respectively.

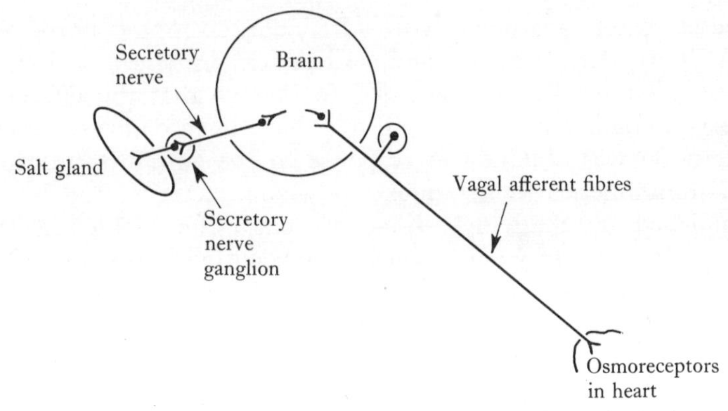

Fig. 4.11. The proposed reflex arc for stimulation of the salt gland in the goose (from Hanwell *et al.*, 1972).

The secretory reflex

With the information on the afferent and efferent pathways described above, Hanwell *et al.* (1972) proposed the secretory reflex shown in Fig. 4.11. The obvious difference between this and the one proposed by Schmidt-Nielsen (Fig. 4.3) is the peripheral osmoreceptors and the afferent fibres running in the vagi. The pathways in the brain have so far not been studied except that, with Dr M. H. Evans, we have found that stimulation of the VIIth motor nucleus initiates secretion. Again it may well prove difficult to work on the afferent side of the presumably multi-synaptic reflex arc because of the inhibition by general anaesthesia.

There is evidence that other nervous centres may affect secretion. McFarland (1965) found in experiments on gulls that an increase in the rate of secretion was occasionally associated with 'alertness' caused by distant moving objects or by sudden noises. To test this further he shone a light into the eyes or subjected them to sound 'made with a quittar (*sic*) using individual notes, chords or a melody'. In both cases secretory rate increased temporarily (Fig. 4.12). He concluded '...the finding that light or sound can enhance salt gland activity suggests that the final common parasympathetic pathway is influenced by neurons besides those from the osmoreceptors'. It is for this reason that in experiments in which secretory rate

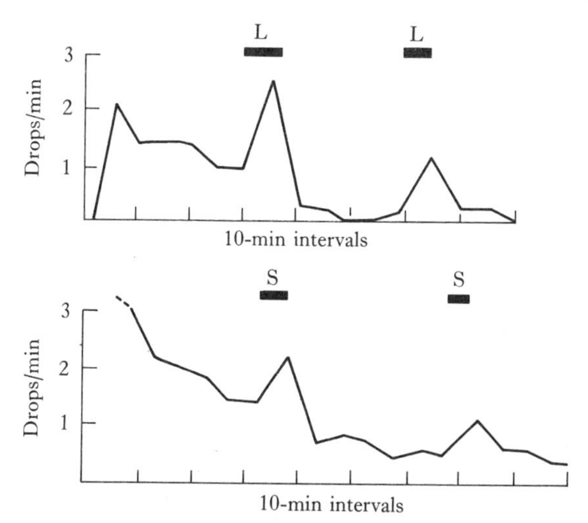

Fig. 4.12. Increase in secretory rate in response to light (L) and sound (S) in salt-loaded gulls (from McFarland, 1965).

is being measured silence should be maintained and the birds should not be subjected to sudden movements. This is also clear evidence of the involvement of the brain in the reflex.

Experiments on the interruption of the afferent pathway clearly show that, as in many mammalian salivary glands (see Burgen & Emmelin, 1961), continuous nervous stimulation is necessary not only for the initiation but also for the maintenance of secretion. In anaesthetized Herring Gulls (*Larus argentatus*) Håkansson & Malcus (1969) found that secretion started in response to electrical stimulation of the secretory nerve after about 15 seconds, and that when stimulation was discontinued secretion rapidly declined to stop completely in about two minutes. The continuation of secretion after stimulation, albeit at a rapidly declining rate, was presumed to be due either to postsynaptic after-discharge or to pressure falling in the gland following physical distension during secretory activity.

There is little information on the detailed workings of the reflex in terms of secretion produced in response to a given osmotic stimulus. Ash (1969) gave hypertonic sodium chloride to ducks to initiate secretion and then after some time administered distilled water orally which, provided that the plasma sodium level and osmolality fell below threshold, abolished secretion. Changes in plasma composition and secretory rate were followed and he found that '...within limits, the magnitude of the response [i.e. rate of secretion] bears some relation to changes in either or both plasma parameters [sodium concentration and osmolality]'. There is of course a maximum rate of secretion. Ash also pointed out that 'it is clear that the mechanism excited does not adapt rapidly to the stimulus'.

Spontaneous secretion

Cases of spontaneous secretion (i.e. not induced by an osmotic stimulus), always at a low rate, have been reported (see for example Ash, 1969; Hughes, 1970*a*; Hanwell *et al.*, 1971*a*; Smith, 1972*b*); this is usually a transient response to handling before an experiment. Since the sodium chloride concentration is usually low we believe that all the secretory cells may not be activated (see p. 120). Since it occurs in ducks and gulls, that require a relatively large increase in plasma tonicity in order to initiate secretion proper, and with a short latency after catching and handling the birds, we suggest that the secretory response is induced within the CNS as a direct result of 'excitement' or 'stress'. This may also be the explanation for secretion in albatrosses in captivity which, if handled and given only fresh water to drink, may die as a result of hyponatraemia (see p. 215).

We should make it clear that the response is not invariably observed when birds are handled. One of us (M.P.) never observed the onset of secretion in such a manner in ducks while working at the University of Hong Kong. This may be because the birds were housed individually in cages well before experimental procedures were begun and they may well have been accustomed to people and to the routine of cage cleaning etc.

5 BLOOD FLOW

Early workers realized that during secretion the rate of blood flow through the salt gland is likely to be very high, in order to supply the glands with the large quantities of sodium, chloride and water excreted. The quantities present in the tissue (see Peaker, 1971*a*) would be exhausted within seconds. Therefore secretion must be accompanied by a rapid rise in blood flow. Indeed Fänge, Schmidt-Nielsen & Robinson (1958), soon after the discovery of the function of the glands, observed the surface of the gland using a low-power microscope and noted a marked vasodilatation during stimulation of the secretory nerve. Interesting findings have since been made on blood flow through the gland and the study of this organ may well throw light on the fundamental but controversial subject of vasodilatation during secretory activity in mammalian salivary glands.

Quantitative measurements

Before quantitative measurements of salt-gland blood flow were made by Hanwell *et al.* (1970*a*, 1971*a*) two attempts were made to calculate what the flow might be. This was because the blood vessels to the salt glands are so inaccessible that direct cannulation techniques, so successfully used for salivary glands, are impossible. Borut & Schmidt-Nielsen (1963) measured oxygen consumption *in vitro* and assumed for their calculation that five equivalents of sodium might be transported per equivalent of oxygen utilized and that the gland might extract as much as 10 ml of oxygen per 100 ml blood. They calculated that this would indicate a blood flow of at least 4.7 ml/g/min but pointed out that even this high flow would result in an arterio–venous difference for sodium of 90 mmoles per litre, and that, in all probability, the rate was even higher. McFarland (1964*a*) and McFarland & Warner (1966) measured the blood volume of salt glands and other organs of gulls and then calculated from the likely 'energy yield' of the gland in terms of ATP production (estimated from the activities of eight enzymes *in vitro* – McFarland, Martin & Freedland, 1965), the likely glucose requirement of the secreting gland and the blood flow that would be needed to supply this, assuming an extraction of 10–25 per cent. They also arrived at a figure of up to 4.7 ml/g/min. However it may be calculated that, at these rates of flow, the extraction of sodium chloride would be so high that even their highest figures would only just be sufficient to account for the rate of secretion quoted (400 μmoles sodium/g/min) and that lower flow rates are clearly impossible. To put these figures into perspective it is known that, in geese, kidney blood flow is approximately 8 ml/g/min and cardiac muscle blood flow 2.2 ml/g/min. In fact, salt-gland blood flow during secretion is much higher than either of these figures (Hanwell *et al.*, 1971*a*, *b*).

Hanwell *et al.* (1970*a*; 1971*a*, *b*) used a method to measure salt-gland blood flow that had previously been employed for other organs with an inaccessible blood supply – the mammary glands of small mammals (Chatwin, Linzell & Setchell, 1969; Hanwell & Linzell, 1973). This involved the combination of two methods. The first, Sapirstein's indicator fractionation technique, measures the proportion of the cardiac output received by each organ while the second, Fegler's thermodilution technique, measures cardiac output simultaneously thus permitting an absolute calculation of flow to be made for every organ of the body. The principle of the indicator fractionation technique is that after the rapid intravenous injection of a freely-diffusible substance there is a period when the indicator, uniformly distributed to the arteries, passes rapidly into the tissues. The volume of tissue into which the indicator diffuses is large so that initially the amount of the indicator in organs is determined by the amount of blood delivered to each. Therefore if the circulation is stopped at this time, the amount of indicator in an organ is proportional to its share of the cardiac output. ^{86}Rubidium is the indicator now most commonly used but it is unsuitable for some organs like the brain which do not permit the free passage of this ion (see Hanwell & Linzell, 1973).

The method was validated for the salt gland by comparison with the fractionation of isotopically-labelled microspheres. These were 25 μm in diameter and nearly all are trapped by capillaries in one passage through an organ. Theoretically they are more satisfactory than a diffusible indicator like ^{86}Rb but their routine use involves a number of difficulties. For example, the injection must be made into the left side of the heart to prevent their removal by the lungs and it can be very difficult to ensure that the spheres are well mixed in the heart and uniformly distributed to all the arteries. Since catheterization of this nature can, and did, cause cardiac distress in the goose, ^{86}Rb was used routinely since the agreement between the two methods, when the microsphere experiments were successful, was good. A disadvantage of the technique is that the animal must be killed one minute after the injection of ^{86}Rb so that only one determination can be made in each bird.

In the experiments of Hanwell *et al.* (1971 *a*) conscious geese were used which had never previously been given salt water. After the catheters and thermistor (for cardiac output determinations) had been inserted into wing veins and arteries under local anaesthesia, different groups of birds were given hypertonic saline intravenously, artificial sea water by stomach tube, or no salt at all. Cardiac output, the rate and composition of nasal secretion, and blood composition were monitored throughout the experiments. When secretion was well established, approximately 30 minutes after loading, ^{86}RbCl was injected intravenously and a final determination of cardiac output carried out. One minute later the bird was killed and the organs and tissues quickly removed. The amount of indicator in each organ was then determined by scintillation spectrometry.

The figures obtained for salt-gland blood flow clearly confirmed that vasodilatation accompanies secretion, and showed a large increase in flow very much greater than that through any other organ (Fig. 5.1). After intravenous loading with salt, cardiac output increases markedly within minutes so that blood flow through all organs increases. In one animal weighing 6 kg, killed five minutes after the injection of 100 ml 0.5 M NaCl, cardiac output rose from 1.06 to 3.68 l/min (in itself an interesting facet of cardiovascular physiology in birds) and the flow through almost all organs was raised by a similar propor-

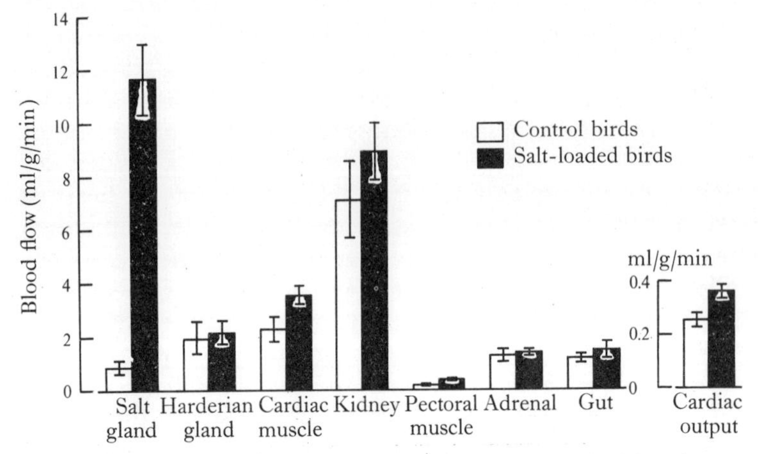

Fig. 5.1. The effect of salt-loading on cardiac output and organ blood flows in geese. Measurements were made in the experimental birds 25–30 min after the intravenous injection of hypertonic saline and in the controls at a similar time after catheterization. Mean±S.E. (drawn from the data of Hanwell *et al.*, 1971*a*, *b*).

tion. In this bird, salt-gland blood flow was the highest recorded at 26.9 ml/g/min; the only other change in the distribution of the cardiac output was to cardiac muscle.

The dilatation of the salt-gland blood vessels was maintained. Cardiac output fell almost to normal again within thirty minutes, at which time salt gland secretion was maximal and blood flow was fourteen times that in non-secreting birds (12.1±1.4 ml/g/min compared with 0.83±0.22 (S.E.) ml/g/min).

The magnitude of these values means that salt-gland blood flow during secretion is one of the highest recorded for any tissue and raises the question of their accuracy. In fact even with these flows the rate of secretion can be so high (i.e. up to 1.9 ml/g tissue/min even in geese unaccustomed to drinking salt water) that the large extractions of ions and water assumed by Borut & Schmidt-Nielsen (1963) in their calculations of likely flow, still apply. This means either that the blood flow is even higher than our figure or, more likely, that the tissue is highly specialized for extracting such quantities. Rates of blood flow of the magnitude measured through the salt glands are not entirely without precedent because Daly, Lambertsen &

Schweitzer (1954) reported that blood flow through the cat carotid body is in the region of 20 ml/g/min.

It should be made clear that the high blood flows obtained by Hanwell *et al.* (1970*a*, 1971*a*) were in geese that had not been acclimated to salt water. By infusing hypertonic saline intravenously, Fletcher, Stainer & Holmes (1967) determined the maximum secretory rate in ducks which had been kept either on fresh water or on salt water for 30 days. Since the rate of sodium and chloride secretion per unit weight of tissue was increased approximately four-fold, it is possible that under these conditions blood flow might be even higher than we measured in geese maintained on fresh water. There is some evidence to suggest that this might be the case. After our work on blood flow in the salt gland was published (Hanwell *et al.*, 1971*a*) we found that Hughes (1962) in an unpublished thesis had examined the fractional distribution of ^{86}Rb in the Glaucous-winged Gull (*Larus glaucescens*), a marine species. She found a 50-fold increase in the proportion of the cardiac output received by the salt gland during secretion but, since cardiac output was not determined, blood flow could not be calculated in absolute terms. However, if it is assumed that the rate of flow in the resting state was similar to that in geese it is obvious that with the 50-fold increase blood flows of approximately 40 ml/g/min are indicated during secretion at a high rate in truly marine birds.

It is a pity that the rate of secretion has not been measured in the ostrich (*Struthio camelus*). This bird can secrete a hypertonic fluid with potassium as the major cation (Schmidt-Nielsen, Borut, Lee & Crawford, 1963; see p. 221) and, since the plasma concentration is low, an extremely high rate of blood flow would be needed to sustain a rate of potassium secretion equal to that of sodium in marine birds. If the secretion is sustained then the potassium must be extracted from the blood; even assuming a total clearance then, with an assumed secretory rate of 200 μmoles of potassium/g/min, a blood flow of approximately 60 ml/g/min would be required. If these assumptions are correct, blood flow will be rate-limiting for potassium secretion by the salt glands of the ostrich and terrestrial reptiles. The matter of the organ with the highest blood flow is under discussion with McWhirter & McWhirter (1972).

Efficiency of extraction from the blood

If salt-gland blood flow, the rate and composition of the secretion, haematocrit, plasma composition and the weight of the glands are known then it is possible to calculate, using the Fick principle, the efficiency of the glands in removing ions and water from the blood and the expected arterio-venous difference across the gland.

In the experiments of Hanwell *et al.* (1971*a*) there was no correlation between the rate of secretion and the rate of blood flow. Therefore extraction, expressed as the quantity secreted as a percentage of that arriving in the arterial plasma in the same period, was variable. Table 5.1 shows values for extraction in some of these experiments and it is clear that, at high rates of secretion, large amounts of ions and water must be extracted from arterial plasma, i.e. up to 80 per cent of chloride, 57 per cent of sodium and 21 per cent of water. Most of the geese used in the experiments did not secrete at such a high rate and with comparable blood flows, the median figures for extraction are lower (chloride 21 per cent, sodium 15 per cent, potassium 35 per cent and water 5.8 per cent.

Table 5.1 also shows the calculated arterio–venous differences across the gland. These are not quite so large for the ions as the extraction figures would infer since water is also removed from the blood perfusing the glands and the calculated venous concentrations are somewhat higher than one imagines at first sight. Nevertheless the magnitude of the extractions in some geese implies a highly efficient mechanism for the removal of ions from the plasma. The infoldings of the basal membrane of the secretory cell may in part account for the high extractions and if this is the case then the marked increase in basal infolding which occurs on adaptation to salt water (see Chapter 9) might indicate that extraction also increases under such conditions. It should also be recalled that individual blood capillaries pass the bases of many cells as they run along the length of the secretory tubule (p. 19). This arrangement would permit the gradual extraction of ions and water, and could lead eventually to a high rate of extraction from the blood. The counter-current arrangement of capillaries and secretory tubules might not be significant in establishing the hypertonicity

TABLE 5.1. *Salt-gland secretion, blood flow and calculated extraction and A–V differences of ions and water in 6 domestic geese following a saline load. Extractions are calculated from the quantity secreted expressed as a percentage of that arriving in the arterial plasma in the same period* (from Hanwell *et al.*, 1971*a*)

	Secretion				Plasma Composition (mM)						Extraction (%)				Calculated concentration in venous effluent (mM)				Calculated A–V difference			
	Rate (g/min)	Composition (mM)						Haematocrit (%)	Blood flow (ml/g/min)	Weight of salt glands (g)												
		Na	K	Cl	Na	K	Cl				Na	K	Cl	Water	Na	K	Cl	Haematocrit	Na	K	Cl	Haematocrit
0.5 M NaCl i.v.																						
1	0.441	480	11.7	495	170	3.9	119	28	15.45	0.70	16	17	24	5.7	152	3.4	96	29	18	0.5	23	+1
2	0.520	465	10.5	462	172	4.1	121	24	8.38	0.39	57	54	80	21	94	2.4	31	29	78	1.7	90	+5
3	0.10	345	13.2	370	173	3.8	121	22.5	10.75	0.53	4.5	6.5	7.0	2.3	169	3.6	115	23	4	0.2	6	+0.5
0.154 M NaCl i.v.																						
4	0.177	385	12.6	400	160	3.2	125	29	8.03	0.53	14	23	19	5.9	146	2.6	108	30	14	0.6	17	+1
5	0.160	426	11.1	428	159	3.1	127	33	4.89	0.37	35	47	45	13	119	1.9	80	36	40	1.2	47	+3
Oral sea water																						
6	0.25	534	30.0	504	154	2.6	124	34	10.45	0.82	15	51	18	4.4	136	1.3	106	35	18	1.3	18	+1

of the secretion (see Chapter 2 and Schmidt-Nielsen, 1960) but it would ensure that the secretory cells most adapted for ion transport and which are situated at the central end of the tubule (p. 23) would receive the arterial blood containing the highest concentration of sodium chloride.

While the measurement of blood flow has permitted calculations of extraction and arterio–venous differences to be made for substances that are secreted by the gland, similar data for oxygen, glucose and other substances of metabolic interest are lacking. This is because actual measurements of arterio–venous differences across the gland as well as blood flow are required to determine the uptake and metabolism of such substances and, so far, nobody has succeeded in obtaining samples of venous effluent from the gland in conscious animals. If this information could be obtained the important relationships between energy metabolism and ion transport in this organ, on which we have had to speculate (p. 106), could be studied directly. Nevertheless, a knowledge of blood flow and arterio–venous differences for sodium and chloride does have some relevance to the mechanism of secretion since it can be calculated that the extraction of sodium and chloride from the extracellular fluid must be greater, at high rates of secretion, than could be achieved by a passive movement of these ions into the secretory cell (see p. 95).

Control of blood flow

The mechanism by which blood flow is increased during glandular activity has interested physiologists ever since Claude Bernard's (1858) experiments on the submaxillary gland of the dog. Mammalian salivary glands have been studied in a bewildering number of experiments and since, in many respects, vasodilatation in salt glands resembles that in salivary glands, where vasodilatation is largely under parasympathetic control, it will be useful to compare their vascular control.

Factors affecting blood flow in the salt gland have been studied in three ways, none of them satisfactory. The simplest is the direct observation of the gland in the anaesthetized bird during nerve stimulation or the administration of drugs (Fänge, Schmidt-Nielsen & Robinson, 1958). The second is the indicator fractionation technique used by Hanwell *et al.* (1971*a*) to measure flow in absolute terms in conscious birds. Its main disadvantage for studying factors that affect flow is that only one determination can be made in each animal. The third method was used by Fänge, Krog & Reite (1963) to assess the effects of nerve stimulation and drugs on salt-gland blood flow in anaesthetized gulls and ducks. These workers measured, polarographically, oxygen tension in the tissue, and assumed that a decrease in P_{O_2} is due to a local vasoconstriction and an increase to a vasodilatation. For these experiments, the oxygen-sensitive electrode (platinum wire) was inserted into the tissue and the reference electrode placed in contact with pectoral muscle. There are three main disadvantages of this technique:

(i) Quantitative information on flow cannot be obtained.

(ii) The administration of drugs can affect systemic blood pressure and arterial P_{O_2} so that it is necessary to distinguish between central and local changes.

(iii) Any change in the rate of oxygen consumption by the tissue is interpreted, perhaps falsely, as a change in the rate of blood flow.

Parasympathetic effects

Fänge, Schmidt-Nielsen & Robinson (1958) observed vasodilatation microscopically in the gland during stimulation of the secretory nerve, and later Thesleff & Schmidt-Nielsen (1962) showed that this occurred about 15 sec before the secretory potential between the duct and blood was established. Since the latter appeared after a latency of 15–30 sec it is obvious that the onset of vasodilatation is rapid. Fänge *et al.* (1963) carried out similar experiments but assessed flow by polarography during a 5 min stimulation of the secretory nerve. The results were by no means clear. Tissue P_{O_2} decreased over the first minute but then slowly increased until the end of the stimulation; after stimulation P_{O_2} increased more rapidly over the next minute to reach a level higher than before stimulation. The complex changes illustrate the difficulties of interpreting P_{O_2} data. Cross & Silver (1962) successfully used this technique in the mammary gland, but this tissue secretes milk at a slow but constant rate and has a commensurate steady oxygen consumption, so that it is unlikely that rapid changes in intercellular P_{O_2} could be due to changes in cellular metabolism. This is by no means true for the salt gland, where there must be a rapid increase in oxygen uptake as soon as secretion starts. Fänge *et al.* (1963) realized that the changes they observed could be due either to an initial vasoconstriction, or to an increase in oxygen uptake by the tissue, followed by a dilatation. They considered that the initial decrease in P_{O_2} was not due to stimulation of sympathetic vasoconstrictor fibres since the rate of change was slower than that observed during stimulation of the cervical sympathetic chain. It is indeed most likely that the initial fall was probably caused by a stimulation of cellular respiration under the influence of acetylcholine (Borut & Schmidt-Nielsen, 1963) so that one cannot draw any firm conclusion from these experiments as to the time of onset of the dilatation.

After the injection of acetylcholine into a carotid artery, Fänge *et al.* (1963) recorded an increase in oxygen tension before peripheral effects on blood pressure and arterial P_{O_2} supervened. This dilatation occurred within 1–5 sec (to judge from their traces) of the start of the injection and confirm the rapidity of the onset observed by Thesleff & Schmidt-Nielsen (1962). The systemic effects of methacholine given intravenously were so marked that it was impossible to be sure whether there was a local rise in P_{O_2} but we should point out that both methacholine and carbachol have been seen to cause a marked reddening of the gland (M. Peaker, unpublished observations).

Additional evidence from denervation experiments supports the theory that both secretion and vasodilatation are elicited by cholinergic nerves. Denervation (see p. 52) almost completely inhibits secretion and vasodilatation in response to salt-loading, and histochemical examination shows an almost total absence of acetylcholinesterase from denervated glands (Fig. 5.2). Similarly, general anaesthesia which almost totally inhibits secretion also prevents vasodilatation (Hanwell *et al.* 1971*a*).

Claude Bernard's experiments on the vasodilatation produced in the dog submaxillary gland by stimulation of the chorda tympani led to the postulation of a special type of nerve fibre – the vasodilator. Heidenhain's (1872) studies appeared

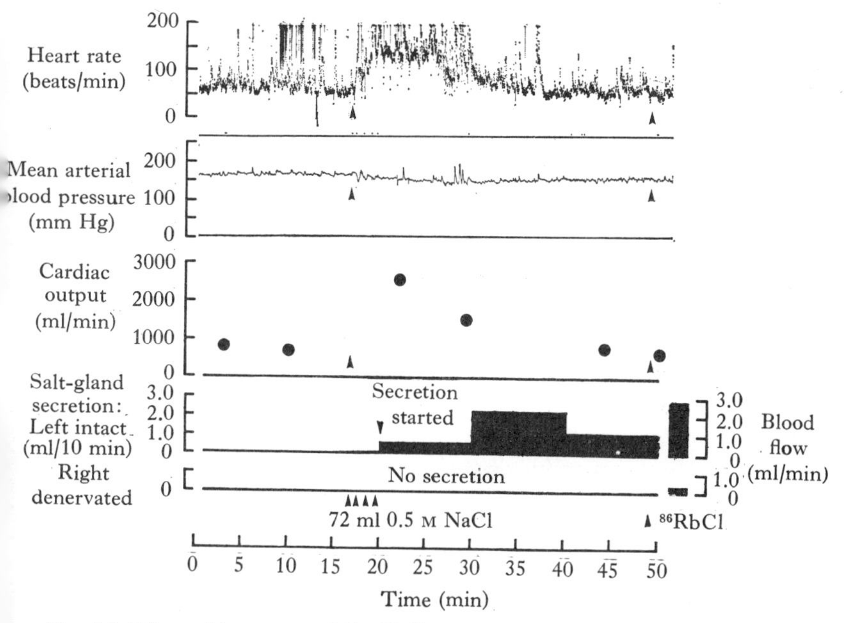

Fig. 5.2. Effects of intravenous 0.5 M NaCl (18 ml/kg) on salt-gland blood flow and secretion, heart rate, blood pressure and cardiac output in a goose with one denervated salt gland. ^{86}RbCl was injected to measure the distribution of the cardiac output (from Hanwell *et al.*, 1971*a*).

to support the theory that such fibres, as well as secretomotor fibres, exist because dilatation was still maintained when secretion was blocked by the injection of atropine but the controversy as to the cause of atropine-resistant vasodilatation has existed ever since. Ninety-nine years later Hanwell *et al.* (1971*a*) published the results of similar experiments on the goose salt gland and showed the same result – atropine-resistant vasodilatation (Fig. 5.3).

The then accepted view that special nerve fibres are responsible for vasodilatation in salivary glands was challenged by Sir Joseph Barcroft and his co-workers (see Barcroft, 1914) who showed that, even when secretion was blocked by atropine, there was an increase in oxygen consumption by the gland during vasodilatation. He therefore proposed that dilatation was secondary to metabolic activity but he did not dismiss the possibility that the initial dilatation may be induced by dilator

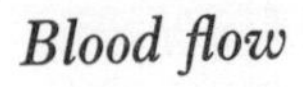

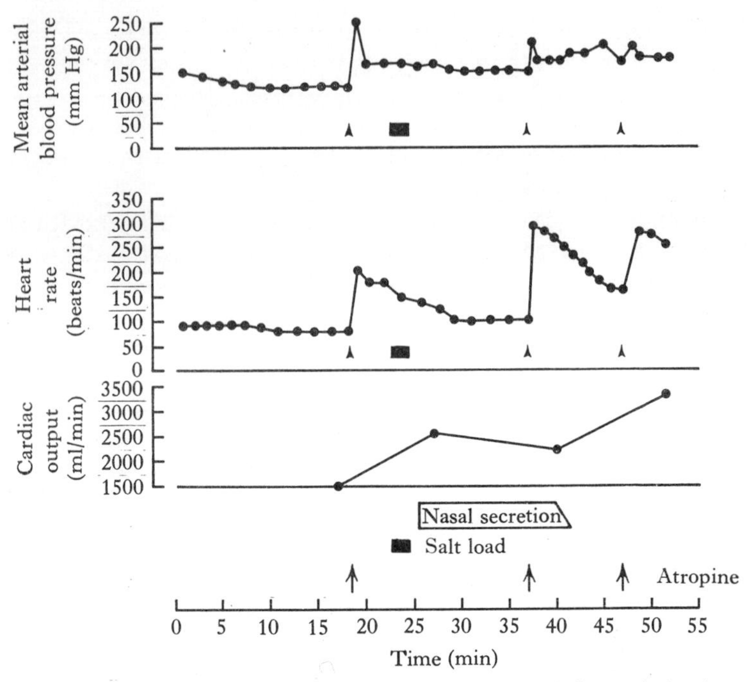

Fig. 5.3. Effects of atropine (0.65 mg) given intravenously before and after intravenous salt-loading, on salt-gland secretion, blood pressure, heart rate and cardiac output in a goose. Salt-gland blood flow (determined by the fractionation of ^{86}Rb) was 18.7 ml/g/min (from Hanwell *et al.*, 1971*a*).

fibres. Indeed the rapid increase in metabolism might demand this. Other workers supported Barcroft's view of secondary vasodilatation and there were therefore two main hypotheses and one compromise solution:

(i) Vasodilator fibres are responsible.

(ii) Metabolites produced by the active gland dilate the blood vessels.

(iii) Vasodilator fibres initiate but metabolites maintain the increase in blood flow.

Dale & Gaddum (1930) accepted that vasodilator fibres are involved and concluded that they are cholinergic in nature. They also made the important finding that whereas secretion, but not vasodilatation, in response to nerve stimulation was blocked by atropine, it also blocked both effects when elicited by exogenous acetylcholine. This led them to conclude that acetylcholine is released in such intimate contact with the receptors on the blood vessels that 'atropine cannot prevent its access thereto'. These results seem incompatible with a suggestion by Henderson & Roepke (1933) that dilator fibres do not exist but that acetylcholine might diffuse to the blood vessels from the terminals of the secretory fibres. As Burgen & Emmelin (1961) point out, if this was the case atropine would be expected to block the effect as in Dale & Gaddum's experiments with injected acetylcholine.

Earlier W. M. Bayliss (1923) in his monograph *The Vaso-Motor System* while admitting that metabolites might assist in the vasodilatation still supported the theory that vasodilator fibres exist. He argued that if metabolites were solely responsible then there should be a correlation between metabolism and the degree of dilatation in the gland. This he failed to find in Barcroft's figures noting that in some experiments there was a larger vasodilatation but a smaller oxygen consumption than in others.

Role of kinins

In 1936 two papers were published which suggested that functional vasodilatation in salivary glands might be mediated by a potent vasodilator agent previously discovered in saliva. Both Werle & Roden and Ungar & Parrot discovered that the activity was due to a substance, kallikrein, described by Frey & Kraut (1928).

Hilton & Lewis (1955*a*, *b*; 1956) in a classic series of experiments on the cat submaxillary gland showed that a dilator substance is released by the active gland even in the presence of atropine and that this substance, after exerting its effect, is eliminated in lymph but not in venous blood. Later work established that two substances are involved in this mechanism. During salivation an enzyme, kallikrein, is produced in the tissue, which itself has no dilator properties, but which acts on plasma proteins (called kininogens) to produce the polypeptide bradykinin, a potent vasodilator. It must be assumed that the process of bradykinin formation occurs in the interstitial fluid which contains some plasma proteins, in view of the absence of dilator substance in the venous blood and its presence in lymph. These experiments therefore were important in that they showed a specific dilator substance was formed in the gland during parasympathetic stimulation and that it was no longer necessary to invoke a non-specific 'metabolite' for the role of dilator.

Hilton & Lewis agreed with Dale & Gaddum (1930) on the presence of a cholinergic secretomotor mechanism since botulinus toxin which selectively inactivates cholinergic fibres abolished both secretion and vasodilatation in response to nerve stimulation but stated '...we do not postulate special cholinergic vasodilator fibres in addition to the known cholinergic secretory fibres'. This polarization towards the theory that vasodilatation is secondary to metabolic activity was vigorously opposed by several groups of workers, some of whom have denied any role for the involvement of kinins.

Schachter and his colleagues have produced most of the evidence against the Hilton & Lewis hypothesis. For example, Bhoola, Morley, Schachter & Smaje (1965) found that the close-arterial injection of saliva (containing kallikrein) into the gland failed to duplicate the pattern of change in blood flow elicited by nerve stimulation, that desensitization of blood vessels to bradykinin by repeated or continuous treatment with that substance did not markedly impair the response to acetylcholine or nerve stimulation and that perfusion with horse serum, which does not form kinins on the addition of cat saliva, did not inhibit vasodilatation in response to nerve stimulation. A great deal of evidence, often contested by the proponents of the bradykinin hypothesis, accumulated along similar lines, convinced many workers that such a scheme was untenable as a sole means of salivary gland vasodilatation. However, few would go so far as the extreme view that the kinin system is in no way involved. The view has been expressed that kallikrein which is found in saliva performs an exocrine rather than a local endocrine role in the submaxillary gland. This was based on histochemical evidence (Heap & Bhoola, 1969) but the fact that a substance is secreted into the glandular lumen does not necessarily indicate that it cannot also be released into interstitial fluid in small amounts.

Many physiologists not directly involved in the controversy felt that neither theory on a mutually exclusive basis explained all the findings and indeed the compromise theory that both mechanisms play a part was favoured and indeed taught by many. It was therefore gratifying to the physiologist not directly involved in this field that Gautvik (1970*a*, *b*, *c*), using the perfused cat submandibular gland, presented convincing evidence that both mechanisms are involved as Barcroft originally suggested in 1914.

Essentially he found that when kininogen was present in the perfusate vasodilatation occurred in two phases, an initial short and a later long one. The first, short phase, which began 1–6 sec after the start of nerve stimulation was abolished by low perfusion pressure while the second, more prolonged phase, was not seen in the absence of kininogen. Gautvik's conclusion is best left in its original form: 'The present experiments strongly indicated that although a direct dilator nerve effect may cause a powerful vasodilatation in the submandibular salivary gland, an additional kinin–dependent mechanism is also part of the complete vascular response to chorda-lingual nerve stimulation. There is no contradiction in the functional hyperaemia of glands involving two vasodilator mechanisms. A nervous mechanism would ensure an immediate increase in flow. Release of kinin-forming enzymes (kallikreins) during activation of glandular elements would make possible a vasodilator mechanism linked to gland activity and suitable for adjustment of local flow according to metabolic needs, as first suggested by Hilton & Lewis (1955*a*).' Karpinski, Barton & Schachter (1971) have also shown that the initial vasodilatation is so rapid (350–900 msec) that there would not be time for the kallikrein–bradykinin mechanism to operate. Gautvik's scheme for vasodilatation is shown in Fig. 5.4.

There is no reason why a similar scheme should not operate in salt glands. However the present evidence suggests that vasodilator fibres may be more important than a control by metabolites. As in salivary glands both secretory cells and blood vessels have a rich cholinergic innervation (see Chapter 2) and the short latency between the injection of acetylcholine and the onset of vasodilatation (Fänge *et al.*, 1963) is certainly indicative of an initial role of vasodilator fibres, as in salivary glands. In the experiments of Hanwell *et al.* (1971*a*) there was no correlation between blood flow and the rate of secretion,

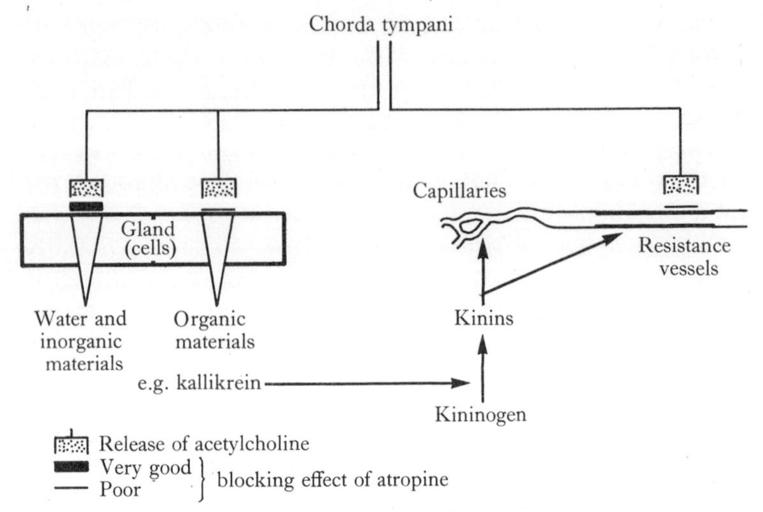

Fig. 5.4. Gautvik's scheme for the regulation of vasodilatation in the cat submandibular salivary gland caused by stimulation of the chorda-lingual nerve. Both cholinergic vasodilator fibres to the blood vessels and secretory fibres to gland cells are suggested but with different sensitivities to atropine (redrawn from Gautvik, 1970*c*).

perhaps suggesting the absence of a direct link between cellular activity and vasodilatation. In contrast, a true functional vasodilatation, presumably related to the production of a metabolite or metabolites, is apparent in the mammary gland which lacks a cholinergic innervation (Linzell, 1959; Hebb & Linzell, 1970) since the rate of blood flow is directly related to the rate of milk secretion (see Linzell, 1971). Mammary tissue contains some kallikrein (Reynaert, Peeters, Verbeke & Houvenaghel, 1968) and bradykinin causes vasodilatation.

What there is in birds is a correlation between the blood flow and the strength of the stimulus. Hanwell *et al.* (1971*a*) found the lowest salt-gland blood flows during secretion in geese given 0.154 M NaCl intravenously even though the rates of secretion per gram of tissue were similar to those obtained in birds given more hypertonic solutions. This would again indicate a direct, nervous control of vasodilatation.

It has also been suggested that kinins assist in increasing capillary permeability during secretion in salivary glands. The marked increase in the inulin space during secretion in salt glands (Bellamy & Phillips, 1966) might indicate that such a change also occurs and the possibility that kinins are involved in mediating this effect (to some workers a more attractive possibility than a role in controlling blood flow) should be investigated.

When we wrote the first draft of this chapter it was not known whether there was a kallikrein–kinin system operating in the salt gland. We therefore stated that this would be an interesting and worthwhile study since, in this case, as opposed to salivary glands, the presence of an activating enzyme (the system is somewhat different in birds – see Werle, Hochstrasser & Trautschold, 1966) could not be related to the exocrine secretion of organic material. However S. Barton, S. Martin and M. Schachter (personal communication) have now examined the goose salt gland using biological and biochemical assay techniques and have failed to find any kallikrein activity. Thus the physiological, anatomical and chemical evidence available at present favours the view that vasodilator fibres are responsible for initiating and maintaining the vasodilatation which accompanies secretion in the salt gland.

Sympathetic effects on blood flow

Although the blood vessels of the salt gland appear to have a dual innervation (p. 38) very little is known about the role of the sympathetic system on blood flow. Stimulation of the cervical sympathetic chain does not induce secretion (Fänge, Schmidt-Nielsen & Robinson, 1958) but does result in vasoconstriction (Fänge *et al.*, 1963). Similarly both *nor*adrenaline and adrenaline cause local vasoconstriction when injected into the carotid artery (Fig. 5.5) but Fänge *et al.* (1963) were of the opinion that adrenaline was somewhat less active in this respect than *nor*adrenaline. Thus it seems fair to conclude that α-receptors are responsible for constriction and that the normal effect of local sympathetic activity would be vasoconstriction (see also p. 211).

There is also the possibility that β-receptors might be present. Isoproterenol which acts almost solely on β-receptors was found by Fänge *et al.* (1963) to cause a local vasodilatation (Fig. 5.5) and this raises the possibility that circulating adrenaline acting through β-receptors could augment vasodilata-

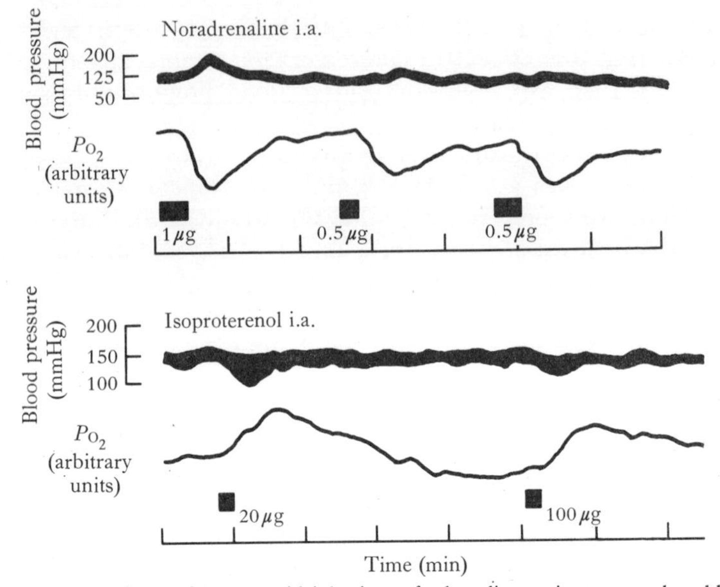

Fig. 5.5. Effects of intracarotid injections of adrenaline or isoproterenol on blood pressure and 'available oxygen' (P_{O_2}) in the salt gland of the duck (redrawn from the data of Fänge *et al.*, 1963).

tion during secretion. However differences between species could be important not only in the possible variation in relative abundance of α- and β-receptors but also in the proportion of adrenaline to *nor*adrenaline released by the adrenal medulla in birds (Sturkie, 1970). While very large doses of adrenaline (0.2 mg/kg) cause a transient inhibition of secretion, presumably because of a vasoconstrictor effect (Fänge, Schmidt-Nielsen & Robinson, 1958), small amounts (2 µg/kg intravenously) can actually enhance secretion in response to a minimal stimulatory salt-load in ducks (M. Peaker, unpublished observations) but the effect on blood flow was not studied. In the goose we found no apparent effect on secretion or blood flow after salt-loading during the infusion of adrenaline (8 µg/kg/min for ten minutes).

It is important that the effects of catecholamines on blood flow and secretion should be fully investigated; their release, locally or systemically, during acclimatization to different environmental conditions (Sturkie, 1970), or during stress, could well influence salt-gland secretion.

Effects of other agents

As in other tissues, 5-hydroxytryptamine acts as a local constrictor in the salt gland, and histamine as a dilator (Fänge *et al.*, 1963).

6 THE SECRETORY MECHANISM

Salt glands in birds were discovered at a time when interest in the mechanism of ion and water movements into and out of cells was increasing rapidly. Since the rate of salt secretion is very high a number of workers became interested in this tissue, particularly those investigating and searching for the biochemical identity of the 'sodium pump'. Apart from these biochemically-orientated investigations very few studies have been made of the salt gland from the point of view of its being a secretory epithelium. Perhaps this is not surprising because the range of techniques that can successfully be applied to this organ is limited, in comparison with those that can be used on the favourite tissues, frog skin and toad bladder. The only studies *in vitro* have been done using salt-gland slices but this technique, when applied to secretory epithelia or organs, suffers from disadvantages additional to those encountered with other tissues. Secretory rate cannot be determined nor can different solutions be applied to the two sides of the epithelium, which are 'short-circuited' by a common incubation medium – a problem appreciated by Ussing (1960). *In vivo* the difficulties are perhaps even greater. The gland will not secrete in response to the normal stimulus in anaesthetized birds and the blood vessels are so small and inaccessible that arterio–venous differences across the gland have not been measured, nor has the gland been perfused in isolation. Although nasal fluid can easily be collected, the intraglandular ducts are also inaccessible except to retrograde injections and we understand that although several workers have attempted micropuncture of the secretory tubules they have not been successful.

In view of these difficulties and the lack of a single technique to provide information on the secretory mechanism we have tried to draw together information from experiments done *in vivo* and *in vitro* in order to present a view of how the secretory cell might work.

Although we shall suggest in Chapter 7 that all cells may not produce a fluid of the same concentration, these differences would be relatively minor and quantitative in nature. For the purpose of this part of the chapter we shall assume that the secretory cells all produce a fluid of similar composition, i.e. hypertonic with respect to plasma and containing mainly sodium and chloride (Table 1.1).

Route of ion movements

It was soon established that the secretory process, although rapid, does not involve classical filtration as in the kidney. Schmidt-Nielsen (1960) found that inulin does not enter the secretion from the blood stream. Later we found that radioactivity could not be detected in the nasal fluid of geese during the intravenous infusion of [^{14}C]sucrose even though very high concentrations were achieved in plasma (Peaker, 1971*c*).

Komnick & Komnick (1963; see also Komnick, 1965) suggested, on the basis of the histochemical localization of ions as seen in the electron microscope, that while sodium passes through the secretory cell, chloride may enter the lumen of the secretory tubules through the junctional complex which joins neighbouring cells. They observed more chloride than sodium in the lateral intercellular spaces. However their interpretation is open to question on at least two grounds. The techniques of ion localization by histochemistry have been criticized by a number of workers on the grounds that they are not specific and that ion movements may occur during the processing. The basic premise, that concentrations of ions in compartments between plasma and the lumen of the tubules provide information on the route of ion permeation, can also be questioned.

There now seems little doubt that in some epithelia the 'tight junctions' or zonulae occludentes which join neighbouring cells are not in fact physiologically 'tight'. Substances can pass between cells and this route constitutes a paracellular or 'shunt' pathway. Frömter & Diamond (1972) have brought together the evidence on 'leaky' and 'tight' tight junctions and although such electrical information as the sum of the basal and apical membrane resistances compared with the epithelial resistance, which enable conclusions on 'tightness' to be reached, have not been obtained for the salt gland, it would appear that on other criteria the salt gland tight junctions are effectively 'tight'. There is of course a marked difference in tonicity between nasal fluid and plasma (i.e. the two sides of the epithelium) and this appears not to be the case in epithelia with 'leaky' junctions. Similarly there is a large potential difference across the salt gland which is not evident in leaky epithelia. Frömter & Diamond (1972) also believe there may be a correlation between 'leakiness' and the permeability to small non-electrolytes like sucrose and we have already seen that the salt gland does not allow the passage of sucrose into the secretion. Thus we are led to the conclusion that there are true 'tight junctions' in the salt gland although we lack definitive electrical evidence. It is clear of course, as Frömter & Diamond point out, that 'leakiness' of junctions is incompatible with the maintenance of steep solute gradients across the epithelium because a passive 'shunt' pathway in the salt gland would permit sodium chloride to pass back into interstitial fluid, and water to enter the fluid in the lumen.

We therefore believe that sodium, chloride, water and potassium must cross the two membranes of the secretory cell in order to enter the lumen of the tubule.

Ion transport

We now consider the mechanisms by which ions enter and leave the secretory cell. Since the junctional complexes between cells are situated on the luminal side of the epithelium, sodium and chloride must enter the cell across the basal and lateral membranes and leave across the apical or luminal membrane. For the sake of convenience we shall use the term basal membrane to include the lateral membranes also.

Concentration gradients and potential differences

Electrical potentials in the salt gland of the Herring Gull, *Larus argentatus,* were measured by Thesleff & Schmidt-Nielsen (1962). Even though the birds were anaesthetized they found that during stimulation of the secretory nerve the interior of the duct becomes positive with respect to the blood by 40–60 mV. This maximal response was obtained by stimulation at 20 impulses per second for 30 seconds or more. The potential developed after a latency of 15–30 seconds, which coincided with the onset of secretion, remained high during stimulation and afterwards declined, together with secretion, over several minutes (see also p. 75) (Fig. 6.1). Similar potentials were recorded after the administration of methacholine intravenously. This evidence indicates that active sodium transport is involved and this view was strengthened when the same workers found that the blood–duct potential and secretion were abolished by the injection of ouabain into the duct. This evidence would also imply that a sodium pump is located on the luminal membrane of the secretory cell. Most workers therefore believed that the cell operates in the manner originally suggested for transport in frog skin by Ussing and his colleagues with the passive entry of sodium into the cell and active extrusion from the opposite side.

A quite different scheme was later proposed by M. R. Hokin (1967; 1969). She analysed goose salt-gland slices after incubation in the presence of isotopically-labelled inulin to measure the extracellular space and calculated the intracellular concentrations of sodium, potassium and chloride which were very high indeed. She suggested that intracellular sodium is 354 mmoles/l intracellular water, potassium 236 and chloride 236. Sodium and chloride concentrations were even higher in slices treated with acetylcholine plus eserine and similar to the concentrations in nasal fluid. Since these levels were unaffected by ouabain she proposed that the concentration gradient is established across the basal cell membrane by a process insensitive to ouabain, and possibly involving exchanges with hydrogen and bicarbonate ions. However in view of the findings of Thesleff & Schmidt-Nielsen (1962), M. R. Hokin (1969) also included in her scheme a sodium pump on the apical membrane as well as one on the basal membrane acting to maintain the intracellular potassium concentration (see later).

However the figures for tissue and intracellular sodium that Hokin obtained seemed not to be in agreement with measurements of tissue composition made by Borut & Schmidt-

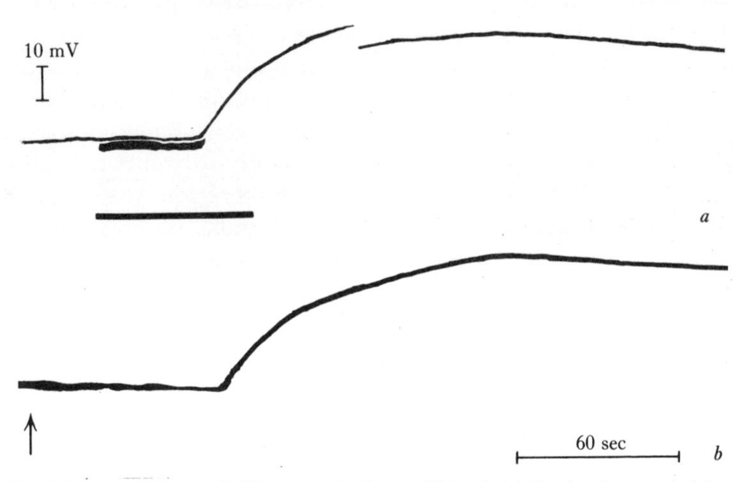

Fig. 6.1. Potentials recorded between the duct and blood. (*a*) the development of duct positivity in response to stimulation of the secretory nerve at 20 Hz (the duration of stimulation is shown by the horizontal white line, and the cathode ray beam was shifted by 10 mV to prevent it from leaving the screen). (*b*) A similar potential after an intravenous injection (at arrow) of 0.25 mg methacholine chloride (from Thesleff & Schmidt-Nielsen, 1962).

Nielsen (1963), Van Rossum (1966) and Fourman (1969). Peaker (1971*a*) therefore repeated the experiments and was unable to confirm Hokin's findings. The disagreement was not in the size of the extracellular compartment but simply in the amounts of sodium and chloride in the tissue. The figures obtained by Peaker for intracellular composition were: sodium 31 mmoles/l intracellular water, potassium 105 and chloride 38. Acetylcholine plus eserine (or methacholine in later experiments) did not raise the cell concentrations but ouabain had its usual effect of markedly decreasing intracellular potassium and increasing sodium concentrations. Later experiments in which intracellular composition was estimated *in vivo* showed that sodium and chloride concentrations were somewhat lower than these figures (Peaker & Stockley, 1973). The reason why Hokin obtained such high figures is unknown and neither *in vitro* nor *in vivo* have we been able to obtain figures even approaching hers.

We conclude therefore that the cells of the salt gland resemble those of other tissues in having a high potassium and

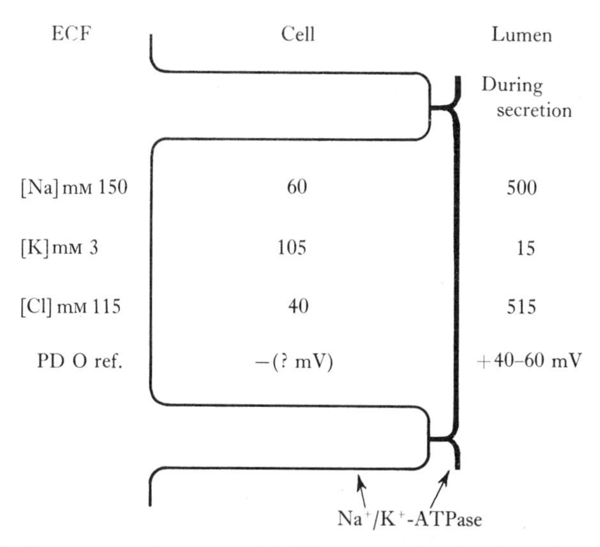

Fig. 6.2. Ion concentrations, potential differences and sodium/potassium activated ATPase distribution in the secretory cell. Subsequent work *in vivo* suggests somewhat lower intracellular sodium and chloride concentrations (Peaker & Stockley, 1973). ECF, extracellular fluid; PD, potential difference; O ref, reference electrode (from Peaker, 1971*c*).

low sodium and chloride content, and that the electrochemical gradient is established across the luminal membrane as other workers had suggested (Fig. 6.2).

Stimulation by acetylcholine

Borut & Schmidt-Nielsen (1963) and M. R. Hokin (1963; 1966) clearly showed that cholinomimetics increase oxygen consumption in salt gland slices; this implies a direct effect on the secretory cells. Nevertheless Bonting, Caravaggio, Canady & Hawkins (1964) suggested an entirely different mechanism to account for the stimulation of secretion by acetylcholine. They proposed that sodium diffuses passively from plasma into cells and that the low blood flow in inactive glands limits the supply of sodium and metabolic substrates to the luminal sodium pump. In their scheme the cholinergic nerves act only as vasodilators permitting more sodium and substrates to reach the pump. However, their hypothesis must be rejected for the following reasons. The clear effects of cholinomimetics on slices *in vitro* in terms of stimulating oxygen consumption, sodium efflux (Van Rossum, 1966) and other biochemical processes (see later) are clearly not compatible with this process. Similarly the demonstration of cholinergic nerve endings in the secretory epithelium as well as on blood vessels (Chapter 2) implies a direct innervation of these cells. Direct physiological evidence against the hypothesis was obtained by Hanwell *et al.* (1971*a*) who showed that while secretion is blocked by atropine, the greatly increased rate of blood flow through the gland which occurs during secretion is not. All the available evidence therefore indicates that acetylcholine *in vivo* has a direct stimulatory effect on the secretory cells, as it does in the salivary glands and pancreas of mammals.

Luminal and basal sodium pumps

As well as a sodium-extruding pump on the luminal membrane, Van Rossum (1966) pointed out that as in all cells a sodium pump would be required to extrude sodium and accumulate potassium across the basal membrane in order to maintain the intracellular potassium concentration during both activity and inactivity, and the sodium concentration during inactivity; the same conclusion was reached by Peaker (1971*a*). Efforts to determine the factors which stimulate the luminal pump *in vitro* are therefore complicated because both pumps are inhibited by ouabain and activity of one might tend to mask the activity of the other. However Van Rossum (1966) found that ^{24}Na efflux from salt-gland slices was temporarily stimulated by methacholine. This occurred even in the absence of external potassium but not in the presence of ouabain. However, evidence of potassium stimulation was obtained but this could represent stimulation of the basal sodium–potassium exchanging pump as well as stimulation of the luminal pump. Evidence for some coupling between sodium and potassium movements was obtained but this was not absolute and is compatible with the hypothesis that two types of sodium pump are present on the two poles of the cell although not necessarily with the same coupling characteristics. (i) The first, on the luminal membrane, is stimulated by methacholine and acts electrogenically to pump sodium into the lumen of the tubule with chloride following passively; this type is probably not coupled to inward movements of potassium (see Keynes, 1969). In addition it may operate even in the absence of potassium in the external medium. Lithium appears to be transported by this mechanism across the luminal membrane but not to the same extent as sodium (Peaker & Stockley, 1973). For the formation of the hypertonic secretion it must of course be assumed that the passive permeability of this membrane to ions and water is low – indeed this is the basis of its function as a salt-secreting gland. (ii) The second, on the basal membrane, is, as Keynes (1969) points out, apparently present in all animal cells for the 'regulation of their internal ionic composition and volume' and presumably operates in a non-electrogenic or only slightly electrogenic manner. As we shall discuss later, the presence of these pumps, which are probably similar in nature (both of the type I pumps of Keynes, 1969), on both poles of the secretory cell leads to difficulties when trying to detect by histochemical methods enzymes associated with transport mechanisms.

Ion movements across the basal membrane and the local action of acetylcholine

Those workers who assumed, or later showed, that the intracellular sodium concentration is low, suggested that sodium may move passively across the basal membrane down the electrochemical gradient (see Bonting *et al.*, 1964; Van Rossum, 1966; Peaker, 1971*a*). This is not an unattractive hypothesis because all that acetylcholine would have to do is increase the permeability of the basal membrane to sodium to an extent beyond that with which the basal sodium pump could cope (Van Rossum, 1966; Peaker, 1971*a*). Intracellular sodium concentration would then increase and together with an increased supply of ATP (respiration being stimulated by an increase in intracellular sodium or by some other mechanism) would stimulate the luminal pump; this scheme is shown in Fig. 6.3. It might be expected that cholinomimetics would increase the intracellular sodium concentration, which they do not (see above). However the rise

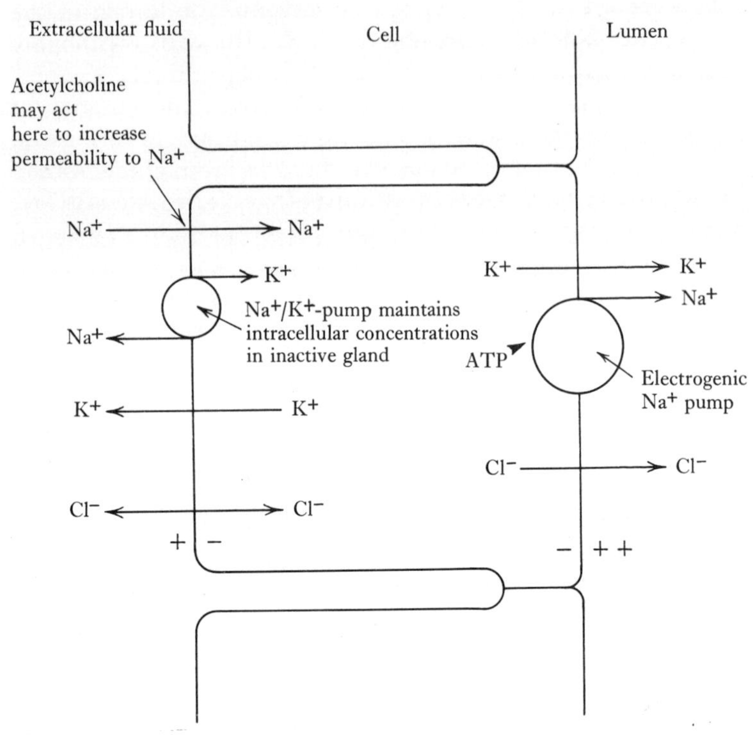

Fig. 6.3. Scheme for ion transport in the salt gland involving passive sodium and chloride movements across the baso-lateral membrane (from Peaker, 1971*a*).

might only be small when the luminal pump is simultaneously stimulated.

Van Rossum (1966) produced very good evidence that cholinomimetics facilitate the entry of sodium into the cells, even though they do not alter the intracellular concentration. Slices were loaded with ^{24}Na by keeping them in cold medium. They were then incubated at 25 °C and, towards the end of incubation, methacholine was added to some of the slices. The specific radioactivity was found to be lower in the tissue treated with methacholine which implies that unlabelled sodium had entered the cells to a greater extent than in the control slices. These results could be accounted for by either passive or active mechanisms of sodium transport into the cells. In fact there are a number of reasons to suggest that the mechanism we have outlined (Fig. 6.3) may not in fact be operative and that sodium and chloride may enter by an exchange process.

The available electrophysiological data are incompatible with an effect of acetylcholine on the permeability of the basal membrane to sodium. Although Thesleff & Schmidt-Nielsen (1962) experienced difficulty in obtaining reliable measurements of intracellular potentials in the salt gland (an initial value of 40–80 mV inside negative with respect to blood falling to 20–30 mV within a few seconds of impalement), they could find no consistent difference in this potential during stimulation of the secretory nerve and when, as we have discussed earlier, the blood–duct potential difference increased to +40 to 60 mV. One might expect if basal sodium permeability is increased during secretion that there would be some charge separation and that the intracellular potential would be affected accordingly i.e. by depolarization of the basal membrane. Furthermore calculations made from secretory rate and blood flow through the gland measured simultaneously in conscious geese (Chapter 5) strongly suggest that the gland can extract, at high rates of secretion, more sodium and chloride from the plasma than a simple process of passive entry down an electrochemical gradient would allow. Therefore Peaker (1971*c*) suggested that, as in a number of other secretory epithelia (see Keynes, 1969), an additional non-electrogenic mechanism transporting sodium and chloride into the cell might be operating during secretory activity.

Like M. R. Hokin (1969) but on different grounds, Peaker (1971*c*) suggested that sodium may enter in exchange for hydrogen ions, and chloride for bicarbonate; this is the type II pump of Keynes (1969) (Fig. 6.4). The effects of the carbonic anhydrase inhibitors acetazolamide and methazolamide in blocking secretion *in vivo* when injected intravenously (Fänge, Schmidt-Nielsen & Robinson, 1958; Nechay, Larimer & Maren, 1960; M. Peaker & S. J. Stockley unpublished) support this hypothesis. Since large amounts of carbonic anhydrase were found in the gland by Nechay *et al.* (1960) and by Bonting *et al.* (1964) it would seem that the scheme shown in Fig. 6.4 could be operating. Carbon dioxide from the cell's metabolism would rapidly form carbonic acid in the presence of

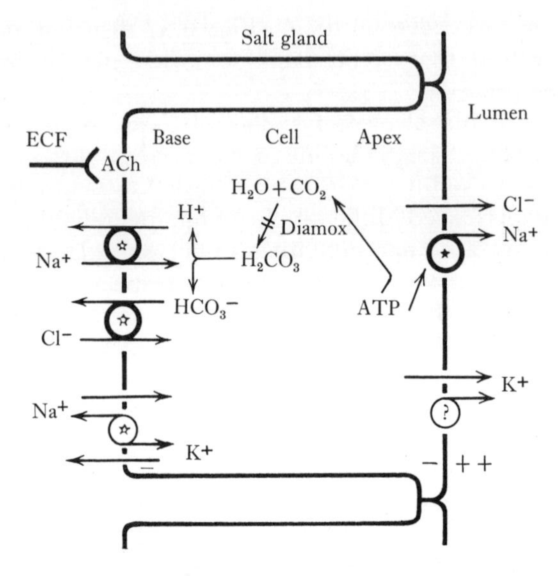

Fig. 6.4. Scheme for ion transport in the salt gland involving coupled exchange mechanisms on the baso-lateral membrane (from Peaker, 1971*c*).

carbonic anhydrase. This would ionize to form hydrogen and bicarbonate ions, which could then be available for exchange with sodium and chloride across the basal membrane. An inhibition of the rapid formation of carbonic acid by carbonic anhydrase inhibitors would therefore be expected to block secretion. Nechay *et al.* (1960) found that the inhibition of secretion could be overcome by the intravenous administration of large amounts of sodium bicarbonate i.e. metabolic alkalosis. Under these conditions when blood carbonic acid concentrations rise it is possible that this acid, in the unionized form, may enter the cells and thus by-pass the block in the pathway brought about by acetazolamide (Peaker, 1971*c*).

Cholinomimetic drugs can induce some secretion from a gland inhibited by acetazolamide (Fänge, Schmidt-Nielsen & Robinson, 1958) and this led Schmidt-Nielsen (1960) to suggest that carbonic anhydrase inhibitors may not affect the gland directly. However cholinomimetics in large quantities may stimulate the luminal pump at the expense of intracellular sodium and chloride levels (see below). Nechay *et al.* (1960) and Bonting *et al.* (1964) suggested that carbonic anhydrase may serve only to regulate intracellular pH and the effect of inhibitors may be to affect cellular metabolism in this way. However the pH of nasal fluid did not change after methazolamide was given intravenously (Nechay *et al.*, 1960) even though it increased after sodium bicarbonate and decreased during acidosis. This might suggest that intracellular pH is unaffected. Similarly oxygen uptake of salt gland slices is unaffected by methazolamide (Borut & Schmidt-Nielsen, 1963) which may also suggest that the effect of carbonic anhydrase inhibitors is on ion transport acting via an inhibition of carbonic acid formation rather than on cell pH and metabolism since an alteration in pH might be expected to affect respiration.

Acidosis, induced by the administration of hydrochloric acid or by giving high carbon dioxide concentrations in oxygen to breathe, was found to reduce the output of nasal fluid in gulls by Nechay *et al.* (1960). Under these circumstances nasal fluid pH fell and if, as seems likely, this reflects a similar change in intracellular pH, the equilibrium between carbon dioxide and bicarbonate ions would be in favour of carbon dioxide and away from the ionized form. In other words, the overall effect would be less sodium and chloride entering by exchange across the basal membrane.

Efforts to test the carbonic anhydrase inhibitors *in vitro* have been largely unsuccessful because the pH of the medium must be altered markedly in order to dissolve them. These changes in pH can affect the ionic composition and therefore it is difficult to interpret any effect of acetazolamide itself (M. Peaker & S. J. Stockley, unpublished).

In the scheme proposed (Fig. 6.4) it is clear that a close relationship between carbon dioxide production and ion transport would be expected. It seems likely that stimulation of the luminal pump is responsible for the increase in respiration of slices when incubated with cholinomimetics (Borut & Schmidt-Nielsen, 1963; Hokin, M. R., 1963; 1966) but it is also possible that acetylcholine may have more direct effects on respiration (Van Rossum, 1964*a*, *b*; 1965*a*, *b*; 1968). The mechanism by which ion transport stimulates respiration is by no means clear but there is evidence that control may not involve changes in ADP concentration or the ratio of ATP to ADP, as seems to be the case in some tissues (Van Rossum, 1964*b*, *c*; Hokin, M. R., 1966). With the luminal pump operating in step with respiration and sodium and chloride influx through the basal membrane, the intracellular ionic composition would be unchanged during stimulation by acetylcholine *in vivo* and by cholinomimetics *in vitro* (see above).

Peaker & Stockley (1974) argued that if the increase in respiration could be dissociated from stimulation of the luminal pump, then an effect on intracellular composition might be seen as the basal exchange pumps bringing sodium and chloride into the cell operated in the absence of luminal pump activity; such evidence would strongly favour the presence of these type II pumps (Keynes, 1969) on the basal membrane. It seemed that lithium could be an agent which might cause this dissociation since Borut & Schmidt-Nielsen (1963) had found that low concentrations of lithium in the external medium stimulated oxygen consumption in salt-gland slices; the addition of methacholine had no further effect. Moreover it was found that while hypertonic lithium chloride will stimulate secretion in the conscious goose, presumably by activating the osmoreceptors, it is ineffective in the anaesthetized bird in which the secretory reflex is blocked. Thus it can be inferred that lithium has no direct stimulatory effect on the luminal sodium pump (Peaker & Stockley, 1974). When 10 mM lithium chloride was added for 10 or 20 minutes to the Krebs-bicarbonate medium incubating goose salt-gland slices it was found that intracellular sodium and chloride concentrations were increased whereas potassium was not significantly affected. The mean differences for both intracellular sodium and chloride levels between slices incubated in lithium chloride and lithium chloride plus methacholine were identical (16 mmoles/l), which implies that under these conditions sodium and chloride movements were affected equally while the potassium concentration remained unchanged. These results are clearly compatible with the scheme for secretion shown in Fig. 6.4. The increase in respiration with lithium would lead to more carbon dioxide being available for the basal exchanges to occur.

The equimolar decrease in intracellular sodium and chloride concentrations which occurred when methacholine was also added implies that cholinomimetics activate the luminal sodium pump and that in these experiments this then extruded the additional sodium (and chloride) and thus restored the balance between influx and efflux across the two poles of the cell.

The effects of methacholine on slices incubated with lithium in the medium also imply that a rise in the intracellular sodium concentration is unlikely to be the stimulus for activation of the luminal sodium pump because, although the level of sodium in the cell was raised when lithium was present, methacholine still had an effect compatible with activation of the luminal pump. A similar conclusion, that a raised intracellular sodium concentration does not initiate transport across the apical membrane of the cat submandibular gland secretory cell, has recently been reached by Poulsen (1973). If this interpretation is correct, the question arises of how acetylcholine, arriving at the basal side of the cell *in vivo*, stimulates the pump on the luminal membrane. This is important because this may well be the primary site of activation of the whole secretory mechanism. There seem to be three possibilities; firstly acetylcholine may enter and cross the cell, which is perhaps unlikely; secondly that an organic 'second messenger', released from the basal membrane into the intracellular fluid, is involved; thirdly that the permeability of the basal membrane could be altered to allow other substances to enter, divalent cations for example, which could then activate the luminal pump. In this connexion it is interesting to note that aminophylline decreased the rate of secretion (Nechay *et al.*, 1960) but that we have been unable to elicit salt-gland secretion by infusing that ubiquitous 'second messenger' cyclic-AMP (in the dibutyryl form) into the carotid artery of geese.

Keynes (1969) tentatively suggested that 'all tissues on which vasopressin and related peptide hormones exert a stimulatory effect have pump II [exchange of sodium for hydrogen or ammonium ions, and chloride for bicarbonate or hydroxyl ions] at one surface', and it is interesting to note that arginine-vasotocin, a naturally-occurring neurohypophysial hormone in birds, affects secretion (see p. 155).

The secretory mechanism

The fact that the salt gland can concentrate other anions – iodide (Carey & Schmidt-Nielsen, 1962) and thiocyanate (Douglas, 1966*b*) – does not necessarily imply that chloride moves passively across the secretory epithelium because anionic transport mechanisms are notorious for their relative lack of specificity. However, the salt gland does not seem to discriminate between thiocyanate and chloride but certainly does between chloride and iodide since higher secretion: plasma ratios were obtained for chloride. Other anions have not been studied but this would clearly be of interest.

Potassium secretion

The potassium concentration of nasal fluid is usually low (Table 1.1). However there are small differences between species and the maximum ratio of sodium: potassium in marine birds is about 40:1. The passage of potassium from the cell could occur by diffusion or, more likely, by the acceptance of some potassium instead of sodium by the luminal pump. Hughes (1970*a*) has shown that when Glaucous-winged Gulls (*Larus glaucescens*) are given sea-water to drink, the potassium concentration collected after an intravenous salt-load, may be as high as 68 mM. In other words the sodium: potassium ratio falls from 24:1 in birds given only fresh water to drink to 12:1 in birds acclimated to salt water. It may therefore be necessary to postulate an active potassium-extruding process, on the luminal cell membrane subject to influence by the salinity of the environment (Fig. 6.4) (Peaker, 1971*c*).

Biochemical and histochemical aspects of ion transport

Apart from early suggestions on the mechanism of secretion, for example membrane flow from the base to the apex of the cell (Doyle, 1960), an elution-type mechanism (McFarland & Sanui, 1963) and even a suggestion, in abstracts, of active water movements in salt glands (Bernard & Wynn, 1963*a*; 1964), only two mechanisms have been seriously considered for sodium transport. One of these, based on work in the salt gland soon after its discovery, was proposed as the mechanism of the 'sodium pump' in all cells. Since this scheme – the phosphatidic acid cycle of Hokin & Hokin – for the universal sodium pump has been abandoned by workers in this field we shall consider later the clear effect of acetylcholine on phosphatide turnover in the salt gland. Most of the other work has been based on the $Na^+/K^+/Mg^{++}$-activated ATPase, which following the work of Skou in the 1950s, is now recognized as being a key part of the sodium pump of all cells. In addition because the salt gland is clearly a highly active ion-transporting organ it has also been included in comparative studies of membrane constituents and enzymes that might be associated with ion movements.

Na^+/K^+-activated ATPase

M. R. Hokin (1963) was the first to demonstrate the presence of this enzyme, which is inhibited by ouabain, in high activity in the salt gland, and this was soon confirmed by Bonting *et al.* (1964) (see also Bonting, 1964; 1970). As we shall consider in detail in Chapter 9 all authors except Hokin (1963) have found that the amount of enzyme activity in the gland increases when birds are given salt water to drink.

There are serious complications in interpreting data on 'transport' ATPase levels in salt gland tissue, particularly when one tries to calculate the energy requirements of the gland. One of the problems is that once this enzyme has been shown to be present in a tissue, many authors assume it is the only type of active transport mechanism present. Keynes (1969) in his review of secretory epithelia made this point in the following way: 'It should be pointed out that since all animal cells are apparently dependent on pump I [i.e. sodium–potassium exchange] for regulation of their internal ionic composition and volume, *all* secretory epithelia must inevitably contain some ouabain-sensitive Na^+/K^+-activated transport ATPase. The demonstration that extracts from a tissue contain this enzyme system is therefore not adequate proof, as it is sometimes thought to be (Kuijpers & Bonting, 1969), that pump I is the chief or only active transport mechanism present.' This leads us to the second point, which was made earlier, that we should expect this enzyme to be present on both poles of the secretory cell. The levels measured in homogenates therefore might reflect the activity of both luminal and baso-lateral membranes. Since during adaptation to salt-water conditions an enormous development of the baso-lateral membrane occurs (Chapter 9), it is possible that much of the increase in transport ATPase is associated with this membrane. At present therefore it is impossible to determine the relative distribution of the enzyme on the two sides of the cell.

The same considerations apply to the localization of the enzyme by histochemical methods, but in this case the situation is greatly complicated by arguments concerning techniques, for example, the specificity of the methods, the site of activity, and the inhibition of activity by the reagents used in the processes. Indeed Tormey (1966) when considering this field in a wide variety of tissues concluded 'There is at the present time no warrant for drawing physiological conclusions from the histochemical localization of ATPase in active transport systems.' Bearing these uncertainties in mind we can do no more than to consider the evidence. Some authors have failed to detect any ATPase activity in the secretory epithelium by histochemical methods (e.g. Ellis *et al.*, 1963). However Ballantyne & Wood (1968; 1970*b*), who have been successful in applying a number of histochemical techniques to the salt gland, employed the Wachstein–Meisel technique to duck salt-gland sections incubated in the presence and absence of ouabain. When the preparations were examined in the light microscope they found ATPase associated with the secretory cells and, as one might expect, the reaction was most intense on the luminal side; activity was clearly reduced in the presence of ouabain, so it would seem that at least some, if not most, of the activity was due to Na^+/K^+-ATPase and located in sites compatible with the physiological evidence. (An interesting observation Ballantyne & Wood (1970*b*) also made was that activity could be detected chemically in the nasal fluid. Eighty-one per cent of this activity was ouabain-sensitive, which might imply a loss of enzyme from the luminal membrane during secretion.)

The situation becomes more complicated when localization of ATPase at the ultrastructural level is considered. The fixatives used undoubtedly inhibit the enzyme's activity but

Abel (1969) was able to show ATPase on the baso-lateral membrane and in some cases on the luminal membrane. By contrast Ernst & Philpott (1970) who made no attempt to distinguish transport ATPase from other ATPase activity found deposits only on the basal membrane. In attempts to make the reaction more specific, Ernst (1972*a*) devised a method to show the potassium-dependent, ouabain-sensitive phosphatase which is associated with 'transport' ATPase in a number of tissues. Since the substrate normally used to determine its activity is *p*-nitrophenyl phosphate (NPP), the enzyme is often called K^+-dependent NPPase. When this technique was applied to the salt gland (Ernst, 1971, 1972*b*), the reaction product was again confined to the baso-lateral membranes. This would of course imply that a sodium pump is not present on the luminal membrane. However this is difficult to accept when the physiological evidence is considered. What may have been demonstrated was the Na^+/K^+ exchanging type of sodium pump rather than the strongly electrogenic type which is not linked to potassium movements. After all, Van Rossum's (1966) evidence we have considered above does suggest that the luminal pump may be stimulated by methacholine in the absence of external potassium. In this confusing field it seems to us that Tormey's (1966) opinion, based on similar disagreements between physiological and biochemical evidence on the one hand and histochemical findings on the other, is still true.

Other ATPases

Those workers who studied the Na^+/K^+-ATPase of the salt gland also reported the presence of a ouabain-insensitive enzyme, and showed that its activity also increases when birds are given salt water to drink. This might be expected in view of the increased number of mitochondria and might not be directly concerned with ion transport. Over recent years anion-activated ATPases have been discovered in a number of tissues where there is strong evidence for the type of exchange pump we have suggested might well be present on the basal membrane of the salt gland (e.g. chloride for bicarbonate) (Kasbekar & Durbin, 1965; Simon, Kinne & Knauf, 1972;

Simon, Kinne & Sachs, 1972). So far no one has looked for such an enzyme in the salt gland, but this should clearly be done.

Phosphatidic acid

L. E. Hokin & M. R. Hokin (1959) found that when acetylcholine is added to salt-gland tissue *in vitro*, the incorporation of ^{32}P into phosphatides, and phosphatidic acid in particular, is markedly increased. Since these substances had previously been shown to form lipid-soluble salts with cations, these workers proposed that phosphatidic acid is a sodium carrier and they put forward a biochemical pathway to suggest how such a sodium pump would work; this is shown in Fig. 6.5. Because they obtained similar effects in a wide variety of tissues when stimulating with natural agents (for example, pancreas by acetylcholine or pancreozymin, submaxillary gland by acetylcholine or adrenaline, cerebral cortex by acetylcholine) they also suggested their proposed phosphatidic acid cycle may well be the mechanism of the ubiquitous sodium pump. Later work showed that the effect is confined to a membrane-bound pool of phosphatidic acid in the microsomal fraction of salt gland homogenates. Their work is described in the following papers: L. E. Hokin & M. R. Hokin (1959); L. E. Hokin & M. R. Hokin (1960); L. E. Hokin & M. R. Hokin (1963*a*, *b*, *c*, *d*); M. R. Hokin & L. E. Hokin (1964*a*, *b*); Santiago-Calvo *et al.* (1964); M. R. Hokin & L. E. Hokin (1966).

Later workers however showed in a number of tissues, that sodium pump activity and transport ATPase activity could be dissociated from the labelling of phosphatidic acid. Moreover the coupling with inward potassium movements cannot be explained by this mechanism for the sodium pump and it has also been questioned whether the cycle would work for sodium extrusion since the return of orthophosphate the scheme demands, would be against the electrochemical gradient. On grounds such as this the scheme has been abandoned as being a mechanism of, or showing close involvement with, the sodium pump (see for example, Caldwell, 1970).

The question remains, if phosphatidic acid is not involved with the typical 'sodium pump' then what is its role and why

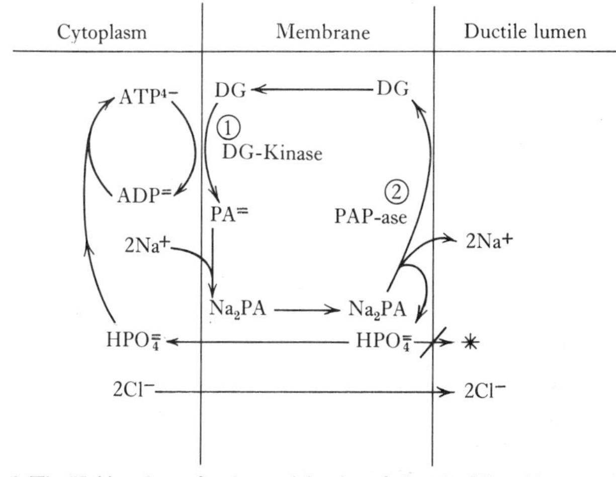

Fig. 6.5. The Hokin scheme for the participation of phosphatidic acid as a carrier in the active transport of sodium ions across the apical membrane of the secretory cell. ECF, extracellular fluid; DG, diglyceride; PA, phosphatidic acid; PAP-ase, phosphatidic acid phosphatase. * HPO_4^{-} does not leave the external surface of the membrane (from Hokin, L. E. & Hokin, M. R., 1960).

does acetylcholine stimulate phosphorylation of phosphatides? In the salt glands it is likely that the microsomal fraction in which the phosphatides are located is derived from the basal and lateral cell membranes. Is it perhaps more likely that it is involved in the response to chemical transmitters in a wide variety of tissues? Or might it even be involved in inward sodium movements across the baso-lateral membrane during secretory activity?

Effects of inhibitors

Apart from ouabain and inhibitors of carbonic anhydrase the effects of other drugs on secretion have been tested. The mercurial diuretic meralluride was found by Nechay *et al.* (1960) to decrease the rate of secretion. Aminophylline had a similar effect but it is not known whether the change was mediated by an involvement in a cyclic-AMP system. In both cases the concentration of the nasal fluid (measured as chloride) decreased but since the collection was made by holding a vial over the gull's beak, the possibility of dilution by fluids from oral glands must not be overlooked.

The secretory mechanism

Ethacrynic acid, which affects ion transport in many tissues but whose site of action is still open to question, was found to lower the rate of secretion in ducks (Smith, 1972*a*). As might be expected of an active mechanism, cooling the salt glands *in vivo* using an ice-pack resulted in a lowering of the rate of secretion without affecting concentration (Staaland, 1967*a*).

Ion transport and energy metabolism

Direct calculations on the amount of sodium transported by the gland in relation to the uptake of oxygen and metabolic substrates have never been made because no one has succeeded in obtaining pure samples of venous blood from the gland. This is because the venous drainage is complex and seemingly inaccessible to surgery under local anaesthesia. Although secretion can be induced under general anaesthesia by stimulation of the secretory nerve or by the administration of cholinomimetics it is doubtful if sufficiently stable conditions of secretion could be maintained in order to fulfil the criteria laid down by Zierler (1961) on the measurements of arterio–venous differences, even if a reliable sample of venous blood could be obtained.

Therefore a number of authors have tried to calculate the energy utilization of the gland during secretion by measuring such variables as oxygen consumption of slices *in vitro* (Borut & Schmidt-Nielsen, 1963), enzyme activities and the levels of metabolic intermediates (Hokin, M. R., 1963; Bonting *et al.* 1964; Chance, Lee, Oshino & Van Rossum, 1964; McFarland *et al.*, 1965); estimates vary from 4 to 8 Na^+ transported per mole of ATP. However all of these studies suffer some disadvantages. The main one is that measurements made *in vitro* do not necessarily reflect performance *in vivo*, and of course in none of these studies could sodium transport be measured. Some authors merely took figures on secretory rate from the literature which is clearly unwise because it cannot be assured that the maximum rate was always achieved. Secondly of course is the difficulty of extrapolating from enzyme activities of homogenates to the working of a whole organ. Not only may the enzymes be exposed to different substrate concentrations than in the intact cell but the assay conditions can also influence the apparent activity.

However it is important to try to obtain some estimate of what the energy requirement might be in order to assess the feasibility of the mechanism involving exchanges of sodium and chloride for hydrogen and bicarbonate ions across the basolateral membrane in terms of energy metabolism and the stoichiometry of these processes. This part of the chapter is therefore an attempt to predict what the uptake of oxygen might be if the basal exchanges operate with a stoichiometry similar to that measured in other epithelia, and if the calculated oxygen uptake could actually be achieved *in vivo*.

The question of how much work salt-gland cells must perform in order to transport sodium from the intracellular fluid to the lumen of the secretory tubule against an electrochemical gradient can be approached in two ways. Firstly, by calculating, from a knowledge of the concentration and potential gradients, the amount of work required and, secondly, by calculating the oxygen consumption from the ratio of the number of moles of Na^+ transported per mole of ATP broken down to ADP.

An estimate of the work required to pump Na^+ against the concentration and potential gradients can be obtained using Zerahn's (1956) formulae. This was found to be 16.2 J/mmole Na^+; the data and assumptions made for this calculation are as follows:

(1) [Na] in secretion 460 mM (in ducks, Fletcher *et al.*, 1967).

(2) [Na] in secretory cell intracellular fluid 25 mM (the concentration *in vivo*, Peaker & Stockley, 1973, being lower than that obtained *in vitro*, Peaker, 1971*a*).

(3) Na^+ secretory rate (maximum) 5.5 mmoles/g tissue/hr (in ducks Fletcher *et al.*, 1967; similar in geese, A. Hanwell, J. L. Linzell & M. Peaker, unpublished).

(4) Body temperature 41 °C.

(5) P.D. across apical membrane 90 mV (lumen positive) (+50 mV blood-lumen, −40 mV blood-ICF; approximate figures from Thesleff & Schmidt-Nielsen, 1962).

(6) Blood flow 13 ml/g tissue/min (Hanwell *et al.*, 1971*a*).

(7) O_2 content of arterial blood 20 ml/100 ml (see Sturkie, 1965).

(8) Blood [glucose] 100 mg/100 ml (Peaker, Peaker, Phillips & Wright, 1971 in ducks, similar in geese, M. Peaker, unpublished).

In all cases birds were given fresh water to drink and secretion was elicited by salt-loading or secretory nerve stimulation.

However the total requirement cannot be calculated because the amount of work needed to overcome the internal resistance of the apical membrane is not known. For this to be done a knowledge of the 'flux ratio' (one-way tracer flux; back-flux of Na^+) (Ussing, 1949) is required, but it is unlikely that with the techniques presently available this could be determined across the apical membrane of the salt-gland secretory cell.

Estimates of the number of moles of Na^+ transported per mole of ATP utilized by the salt gland vary. The most reliable figure is that obtained by Fletcher *et al.* (1967) because, unlike other workers, they measured maximum secretory rate *in vivo* and salt-gland weight in the same ducks in which they determined Na^+/K^+-activated ATPase activity *in vitro.* They obtained a ratio of approximately 4 Na^+ : 1 ATP but since there is always some difficulty in relating enzyme activities measured *in vitro* to physiological performance *in vivo*, and since in other tissues a ratio of 2–3 Na^+/ATP is generally accepted (see Glynn, Hoffman & Lew, 1971), calculations have been made for ratios other than 4:1. We are also assuming, with some trepidation (see p. 102), that all the Na^+/K^+-ATPase is associated with the luminal pump.

Assuming that glucose is the metabolic substrate, that each mole of oxygen yields approximately 6 moles of ATP (i.e. a P:O ratio of 3) and that 1 mmole O_2 yields approximately 420 J, then the amount of Na^+ transported per mmole O_2 utilized and the work in joules required to transport one mmole of Na^+ can be calculated for different Na^+:ATP ratios. If the figure obtained from Zerahn's formulae on the amount of work required to overcome the electrochemical gradient is subtracted from those obtained from Na^+:ATP ratios an estimate of the work required to overcome the internal resistance of the membrane can be calculated. It is clear that if the data used to calculate the work requirement are reliable, then the Na^+:ATP ratio cannot be higher than about 4.5:1 because above this value the work needed to overcome internal membrane resistance would be less than zero. On the other hand, the term for membrane resistance increases exponentially as the Na^+:ATP ratio is decreased and calculations, again using Zerahn's formulae, indicate that even at 4 Na^+/ATP the Na^+ flux into the lumen would exceed the back-flux by approximately 10^{20}:1.

Thus the ratio of 4.48 Na^+/ATP obtained by Fletcher *et al.* (1967) may well be a very close estimate of the efficiency of the apical Na^+ pump.

If the secretory cell consumes oxygen only at the rate determined by the efficiency of the luminal Na^+ pump, the stoichiometry of the postulated basal exchange mechanisms would, with an RQ of 1, be equal to the ratio of Na^+ transported per mole of oxygen utilized for the apical pump i.e. basal $Na^+:H^+$ or $Cl^-:HCO_3^-$ equals apical $Na^+:O_2$; these vary with Na^+:ATP ratios from 12 to 24:1. These are similar to the figures obtained in other tissues, e.g. frog skin, rabbit gall bladder, during experiments in which ion transport and oxygen consumption were determined *in vitro* (see Vieria, Caplan & Essig, 1972; Martin & Diamond, 1966). Since during secretion the rate of blood flow through the salt gland is very high (Chapter 5), the extraction of oxygen (for ion transport) from the blood perfusing the gland would only amount to approximately three per cent at a secretory rate of 5.5 mmoles/g tissue/hr.

Three factors must now be considered which together might imply that the overall process of secretion, i.e. passage of ions across the basal and apical membranes, operates at a lower $Na^+:O_2$ efficiency than a consideration of only the luminal pump would indicate, in order to extract sodium and chloride from the blood at a high rate. Firstly, the rate of extraction which Hanwell *et al.* (1971*a*) found could be as high as 57 per cent for Na^+ and 80 per cent for Cl^-; secondly, the availability of oxygen is far greater than that required for the operation of the luminal Na^+ pump, and thirdly, the coupling ratio between Na^+ and Cl^- influx and H^+ and HCO_3^- efflux through the basal membrane, calculated from the efficiency of the apical pump only, could be very high (24:1 at 4 Na^+/ATP). In fish gills a coupling ratio of 3–4 Na^+ or Cl^- to 1 H^+ or HCO_3^- has been found (see Maetz, 1971). If this coupling ratio applies to the salt gland then the $Na^+:O_2$ ratio of the apical pump must be divided by a factor which depends on the Na^+:ATP ratio; at 4 Na^+/ATP the factor to bring $Na^+:O_2$ to 3:1 is 8, and this would therefore mean that oxygen consumption is eight times higher than was calculated for the operation of the luminal pump alone. However, this would still only bring the extraction of oxygen from arterial blood to 25–30 per cent which is a reasonable figure in comparison with other active organs, for example the mammary gland which has an oxygen extraction of approximately 40 per cent (see Linzell, 1968). Thus it is possible that the secretory mechanism of the salt gland could operate in the manner outlined with an overall efficiency of about 3–4 Na^+/O_2 and this would agree not only with physiological evidence but also with the stoichiometry for Na^+–H^+ and Cl^-–HCO_3^- exchanges observed in fish gills.

The question remains, if the secretory process operates in this way, what use does the salt gland make of the energy produced in excess of that required by the apical pump? There are ATPases other than the Na^+/K^+-activated (ouabain-sensitive type) in the gland and some authors have found these to be a much greater proportion of the 'total' ATPase activity. For example, Ballantyne & Wood (1970*b*) found that the ouabain-sensitive component accounted for only 17 per cent of ATPase activity, and Fletcher *et al.* (1967) obtained a figure of 29 per cent. It is possible that the basal exchange pumps need a supply of ATP for them to operate, and also that glycogen is synthesized in the gland during secretory activity (Benson & Phillips, 1964). Another energy-demanding process is the hypertrophy of the secretory cells which occurs when birds are first given salt water (see Chapter 9); this mainly involves the synthesis of large quantities of cell membrane materials, protein and mitochondria.

If, as argued, a great deal of cellular respiration is concerned with the uptake of Na^+ and Cl^- across the basal membrane of the cell by producing carbon dioxide, then the presence of a greater number of mitochondria and carbonic anhydrase in this region of the cell (Chapter 2) would be compatible with this scheme, and might also infer that the basal exchange processes require energy.

Considerations similar to those applied to oxygen utilization can also be made for glucose uptake. At the efficiency of the apical pump the extraction of glucose from the blood at 4 Na^+/ATP would amount to less than one per cent of that arriving at the gland. At the lower overall efficiency compatible with the stoichiometry of the basal exchanges being similar to that in fish gills, the glucose extraction would be seven to eight per cent, which again indicates that the supply of metabolic substrates (oxygen and glucose) would be sufficient.

Using the figure of 3 Na^+/O_2, it is of interest to make further calculations on the energetic cost of salt-gland secretion to the whole animal. Taking figures from Prange & Schmidt-Nielsen (1970) for the oxygen consumption of ducks at rest, salt-gland weight and maximum secretory ability from Fletcher *et al.* (1967) and the calculated oxygen consumption (see above), salt-gland secretion at maximum rate would require one per cent of the available energy. When ducks are given salt water to drink, as opposed to fresh water, maximum secretory ability increases more than seven-fold (Chapter 9). Although information on blood flow is lacking under such circumstances, it is clear that if the secretory mechanism operates in the same manner then the energy requirement would be seven per cent of the metabolic rate when the bird is at rest. In this connexion it is interesting to note that Krista, Carlson & Olson (1961) found that the growth of ducklings is impaired when hypertonic sodium chloride solutions are given for drinking. Although a decreased food intake may partly explain this effect (Ellis *et al.*, 1963) the additional burden of osmoregulation could demand the use of energy that might otherwise be used for growth.

7 FACTORS AFFECTING THE CONCENTRATION OF NASAL FLUID

The sodium chloride concentration in the nasal fluid of marine birds and the factors which affect it are important aspects of the physiology of the salt glands because together with the volume of secretion, the concentration is one of the main determinants of the extent to which a bird can drink sea water and survive under marine conditions. We must therefore consider the basic mechanisms of water movement across the secretory epithelium, the possible way in which environmental factors are involved in controlling concentration and the relations between the composition of the nasal fluid and the rate of secretion.

Water movements across the secretory epithelium

Prior to the late 1960s we had a simple notion of water movements across the secretory tubules of the salt gland. One could envisage either that water movement is coupled in a direct manner to the transport of sodium across the luminal membrane (i.e. solute drag), or that water passes osmotically into the lumen. In the latter case, which is the mechanism favoured for isotonic fluid transport by Diamond (1965), the osmotic permeability of the membrane to water must of course be low and the process would not reach equilibrium because the fluid would flow from the site of secretion to the exterior. In either case, osmotic equilibrium between the interstitial and intracellular fluids would be maintained across the baso-lateral membranes.

However no discussion on water movements across epithelia is complete without reference to the hypotheses of Diamond & Bossert (1967; 1968) on the mechanism of such processes and the establishment of 'standing osmotic gradients' in extracellular channels. It is clear that the salt gland with its extensive basal infolding, at least in the fully-adapted state, constitutes a 'backwards-facing epithelium', i.e. the extracellular channels face towards the side of the epithelium from which transport of solute and water occurs. In such a system, a standing gradient could be established as Diamond & Bossert (1968) explained in the following way: 'If solute is actively taken up *out of* the channel lumen and transported across the channel wall [as may well be the case, Chapter 6], the channel lumen will become hypotonic rather than hypertonic (Fig. 7.1*b*). A standing osmotic gradient will be maintained in which channel osmolarity progressively *decreases* from isotonicity at the open end to a minimum value at the closed end. Water will leave the channel across its walls by osmosis, at increasing rates along its length, because the lumen is increasingly hypotonic. In the steady state a fluid of fixed osmolarity (isotonic or hypertonic, depending upon the parameters of the system) will continually pass into the mouth of the channel and be taken up across the channel walls.' Computer simulation of this system showed that the transported fluid would be hypertonic if pumps are situated along most of the length of the channel. In other words sodium chloride would diffuse down a concentration gradient from plasma to the membranes of the cell and then be transported into the cell in the same concentration as the fluid entering the channel mouth. There are however so many unknown variables, for example the osmotic water permeability of the membrane, that direct proof that such a mechanism operates is lacking. If the scheme does operate in the manner outlined we feel that it may act firstly to preserve the balance between the osmolarity of the fluid entering the cell and that leaving as secretion so that the volume of the cell is unaffected and, secondly, to ensure a steep concentration gradient between plasma and the cell membranes in the interstitial fluid so that the efficiency of the uptake process is high, rather than to actually determine the concentration of the nasal fluid, which, we believe is governed by the properties of the luminal as opposed to the baso-lateral membranes.

There is virtually no folding of the luminal membrane (see Chapter 2) and if water does accompany sodium movements by osmosis such folding 'would seem not to be desirable in glands forming a hypertonic secretion since if the pump was situated in

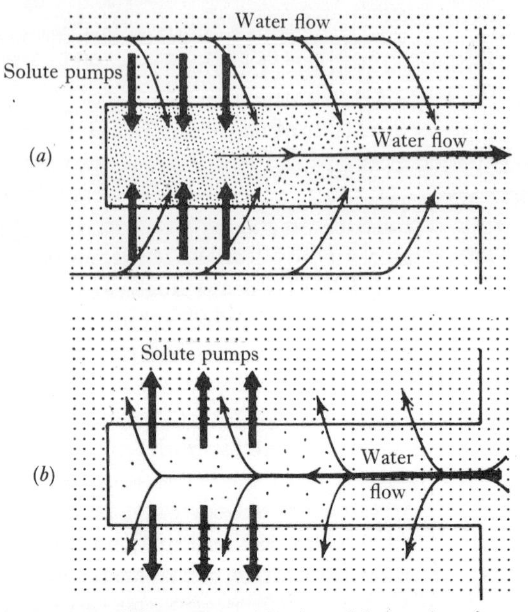

Fig. 7.1. Comparison of 'forwards' and 'backwards' operation of a standing gradient flow system, which consists of a long narrow channel closed at one end (for example, a basal infolding, lateral intercellular space). The density of dots indicates the solute concentration.

Forwards operation (*a*): solute is actively transported into the channel across its walls, making the channel fluid hypertonic. As solute diffuses down its concentration gradient towards the open mouth, more and more water enters the channel across its walls due to the osmotic gradient. In the steady state a standing osmotic gradient will be maintained in the channel by active solute transport, with the osmolarity decreasing progressively from the closed end to the open end; and a fluid of fixed osmolarity (isotonic or hypertonic, depending upon the values of such parameters as radius, length, and water permeability) will constantly emerge from the mouth. Backwards operation (*b*): solute is actively transported out of the channel across its walls, making the channel fluid hypertonic. As solute diffuses down its concentration gradient towards the closed end, more and more water leaves the channel across its walls owing to the osmotic gradient. In the steady state a standing osmotic gradient will be maintained in the channel by active solute transport, with the osmolarity decreasing progressively from the open end to the closed end; and a fluid of fixed osmolarity (isotonic or hypertonic, depending upon the parameters of the system) will constantly enter the channel mouth and be secreted across its walls. Solute pumps are depicted only at the bottom of the channels for illustrative purposes but may have different distributions along the channel (from Diamond & Bossert, 1968).

these folds water from the cells would have more chance of passing into and thereby diluting the solution on the lines suggested by Dr Diamond for the formation of isotonic fluid in forwards-facing channels' (Fig. 7.1*a*) (Peaker, 1971*c*).

Factors determining the concentration of sodium chloride

In general, the concentration of sodium chloride in the secretion of a particular species is related to the salinity of the drinking water and to the type of food consumed. The type of food is important because the body fluids of many marine invertebrates are in osmotic equilibrium with sea water; their salt content is therefore high, and birds that eat marine invertebrates are subject to a greater intake of salt than those that eat fish or other vertebrates. Schmidt-Nielsen (1960) gives several examples to illustrate these relationships. Cormorants, which are found mainly in estuaries, have relatively low concentrations of sodium in the secretion (approximately 500 mM). The Herring Gull (*Larus argentatus*) occurs in similar habitats but eats more invertebrates and can achieve a sodium concentration of 600–800 mM while the Greater Black-backed Gull (*Larus marinus*), which is more marine, can reach 700–900 mM. Leach's Petrel (*Oceanodroma leucorrhoa*) appears to hold the record for the highest concentration in a marine bird with a sodium concentration of 900–1100 mM; this species, which is pelagic, eats only invertebrates and returns to land only to breed.

Zaks & Sokolova (1961) have also made an excellent ecological study of birds in the Kandalaksha Reserve on the White Sea and Sevastopol Bay on the Black Sea. They classified the food of different species into hypertonic (marine invertebrates) or hypotonic/isotonic (sea fish, terrestrial vertebrates, birds and their eggs, insects, berries) and then related the percentage of hypertonic food to the sodium concentration of the secretion. Their results showed clearly that the highest concentrations are achieved by birds which eat marine invertebrates almost exclusively (Table 7.1).

It seems likely that concentration is genetically determined but a number of workers have provided clear evidence that the concentration of the nasal fluid within a species can be altered by changing the salinity of the drinking water. Schmidt-Nielsen & Kim (1964) slowly raised the concentration of sodium chloride in the drinking water of ducks to 0.17–0.51 M and then gave them a standard intravenous salt-load. The birds

TABLE 7.1 *Extra-renal excretion and food composition* (from Zaks & Sokolova, 1961)

Species	Quantity of secretion in (ml/min/100 g body-weight)	Concentration (mM) Na	K	Sodium excretion (mmoles/min)	Proportion of food eaten (%) Hypotonic	Hypertonic
Oyster-catcher	0.0270	830	90	0.022	—	Almost 100
Herring Gull	0.0174	775	34	0.013	60	40
Tern	0.0133	765	33	0.010	?	?
Guillemot	0.0039	675	22	0.0026	78	22
Common Gull	0.0018	743	56	0.0013	85	15

acclimated to salt water were found to secrete far more fluid of a higher sodium concentration (525 mM) than the ones on fresh water (435 mM). Fletcher *et al.* (1967) and Stainer, Ensor, Phillips & Holmes (1970) carried out similar experiments, in the course of their excellent work on the changes induced by salt adaptation in this species (Chapter 9), and found a marked rise in the concentrations of sodium, potassium and chloride as well as in secretory rate. These changes were reversed by returning the birds to fresh water (Fig. 7.2). Most ducks cannot tolerate sea water but Hughes (1970*a*) used the marine Glaucous-winged Gull, *Larus glaucescens*, for her experiments and kept them either on fresh water, sea water or mixtures containing one-third or two-thirds sea water. The sodium, chloride and potassium concentrations in the secretion obtained in response to an intravenous salt-load, increased in birds on one-third sea water but no further change occurred in those given two-thirds sea water. On full-strength sea water however, potassium and chloride (but not sodium) concentrations again increased and the rise in potassium resulted in a marked lowering of the sodium:potassium ratio in the nasal fluid (Fig. 7.3) (see also p. 161).

In ducks and geese kept on fresh water there is often considerable variation in the concentration of the nasal fluid secreted in response to salt-loading, both between individuals or groups of individuals within the same domestic strain of bird. W. N. Holmes (personal communication) found that the domestic ducks he obtained in British Columbia secreted nasal

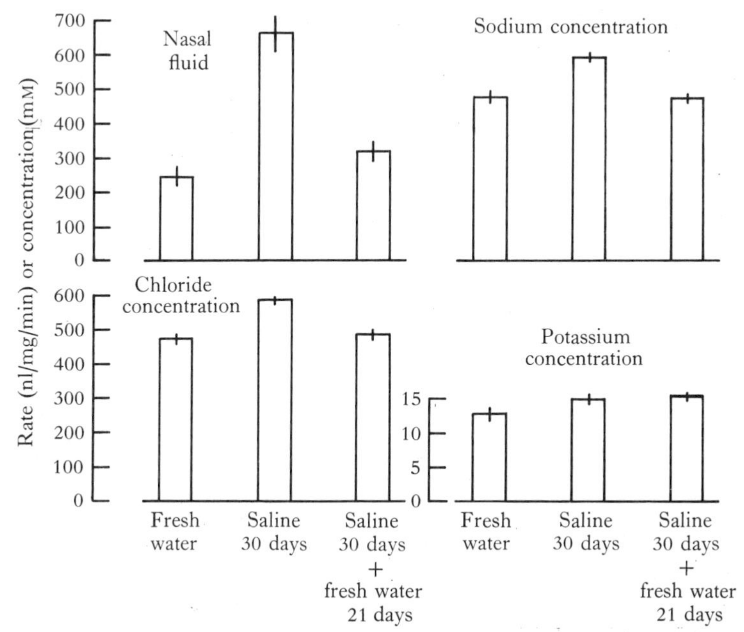

Fig. 7.2. Concentration and rate of salt-gland secretion in ducks given fresh water and hypertonic saline (284 mM sodium, 6 mM potassium) to drink. Secretory rates were determined by the intravenous infusion of ten per cent sodium chloride, and expressed per mg salt gland tissue (mean ±S.E.) (drawn from the data of Fletcher *et al.*, 1967).

fluid of a higher concentration than those of the same White Pekin variety from California. Even lower concentrations were recorded in ducks of this breed obtained in Hong Kong (M. Peaker, unpublished observations). Perhaps these variations are not surprising. Wild ducks and geese are found in different habitats all over the world and can survive on fresh or brackish water. Domesticated varieties would not be selected for their ability to survive on salt water so that different breeds might well have ancestors differing in their concentrating capacity.

Relationships between secretory rate and concentration

In many exocrine glands there is a relation between the rate of secretion and the concentration of a number of solutes in the secretion. Recently this has shown to apply to the salt gland also. When birds are adapted to salt water both the total output

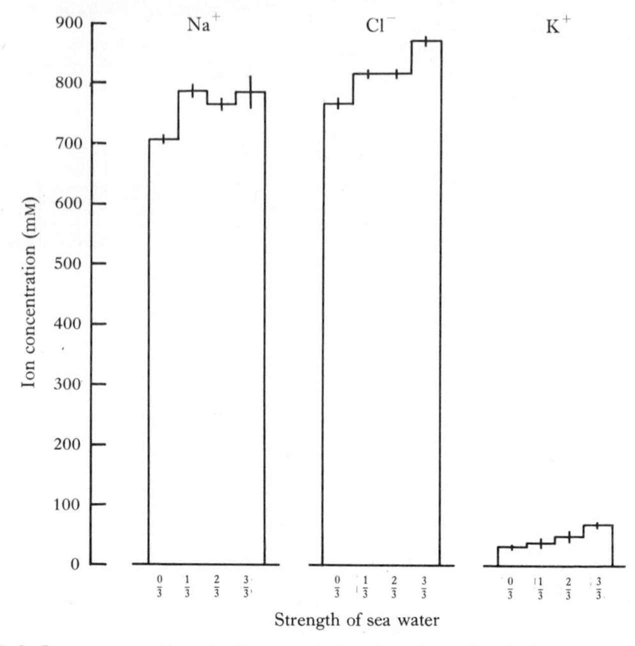

Fig. 7.3. Ion concentrations in the nasal fluid of salt-loaded gulls (*Larus glaucescens*) previously given increasing concentrations of sea water to drink. Full-strength sea-water (3/3) concentration: sodium 402, chloride 447, potassium 10 mM; 0/3 = fresh water (drawn from the data of Hughes, 1970*a*).

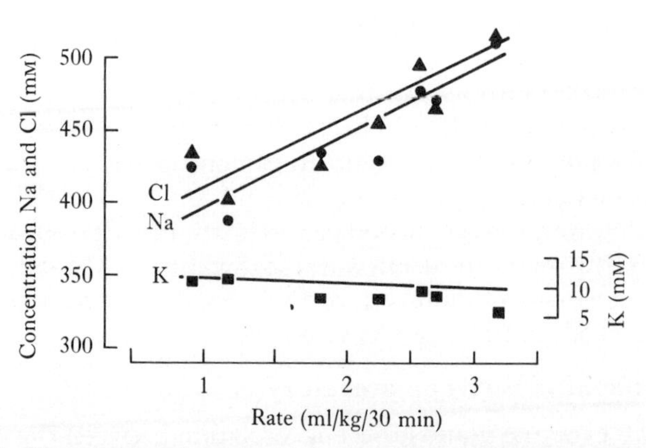

Fig. 7.4. Concentration of nasal fluid and rate of secretion in geese given 0.5 or 0.154 M sodium chloride intravenously. Details of the regressions: Na $r = 0.762$, $P < 0.05$; Cl $r = 0.837$, $P < 0.02$; K $r = -0.503$ (from Hanwell *et al.*, 1971*a*).

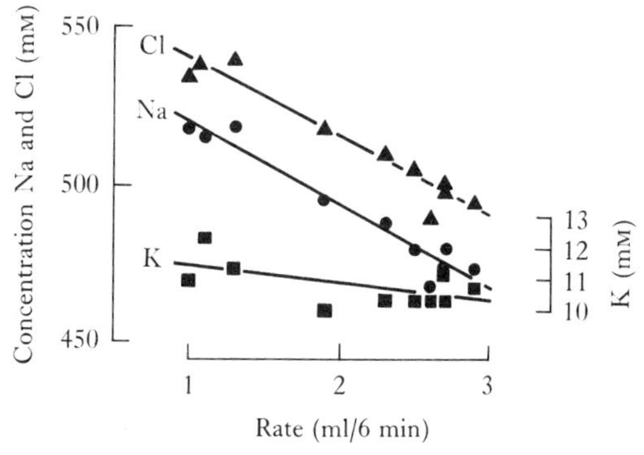

Fig. 7.5. Concentration of nasal fluid and rate of secretion in a goose given 100 ml sea water orally. Nasal fluid was collected over 6 min periods. Details of the regressions: Na $r=-0.97$, $P<0.001$; Cl $r=-0.96$, $P<0.001$; K $r=-0.53$ (from Hanwell *et al.*, 1971*a*).

of secretion and the concentration of sodium chloride in the nasal fluid increase; in other words, between birds, concentration is positively related to the rate of secretion. Hanwell *et al.* (1971*a*) showed that this also applies in geese kept on fresh water and given a single salt-load. In individual birds there was considerable variation in the volume of fluid secreted during the first 30 minutes after loading and when this volume was plotted against concentration, a positive, significant correlation was found for both sodium and chloride. In contrast the slope for potassium was negative but the correlation coefficient was not statistically significant (Fig. 7.4). Thus birds that secrete more fluid are not only more efficient at extra-renal excretion but, since the concentration of the secretion is also higher, can obtain more free water. Most workers have assumed that the relatively small variations in the concentration of nasal fluid are independent of the rate of secretion in individual birds. However in most experiments relatively long collection periods have been employed and such integrated samples may have masked any changes that occur. For example, Hanwell *et al.* (1971*a*), using a more efficient method, where samples were collected over six minute periods, found an *inverse* relationship between rate and the concentrations of sodium, potassium and chloride in geese kept on fresh water when given an intravenous or oral salt-load (Fig. 7.5). These different relationships i.e. *between* different birds or *within* an individual

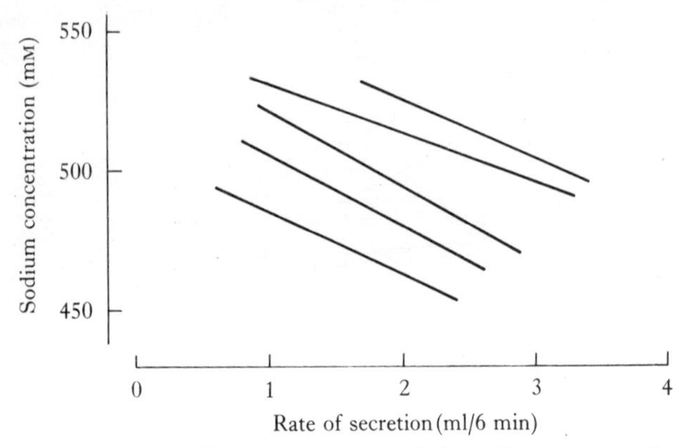

Fig. 7.6. Regression lines (not extrapolated) for sodium concentration and rate of secretion in five geese given either 0.5 M sodium chloride intravenously (18 ml/kg) or 100 ml sea water orally. Note that at a given rate of secretion the sodium concentration is higher in a bird that can secrete at a higher rate.

imply that at a given rate of secretion, the concentration of the fluid is higher in the bird that can secrete at a higher rate. This is in fact the case as Fig. 7.6 shows.

Hughes (1970*b*) has also examined possible relationships between the rate of secretion and concentration in the Glaucous-winged Gull, *Larus glaucescens.* She plotted concentration against secretory rate and obtained a relationship similar to that in Fig. 7.4 in birds kept either on salt water or, to a lesser extent, in birds kept on fresh water. Unfortunately, data were included for different secretory rates in a number of different birds so that comparison with our results in geese is impossible.

An entirely different relationship was obtained by Smith (1972*b*) in individual ducks. He measured fluid but followed secretion for long periods and obtained a positive correlation between rate and concentration. However his data included spontaneous secretion, which is produced at a very low rate in response to handling (p. 66), and rates measured while secretion was starting and declining as the salt-load was eliminated. Similar relationships are apparent in data from the experiments of Ash (1969) who gave distilled water to ducks secreting in response to a salt-load, when the rate of secretion rapidly declined. Examination of Smith's (1972*b*) data sug-

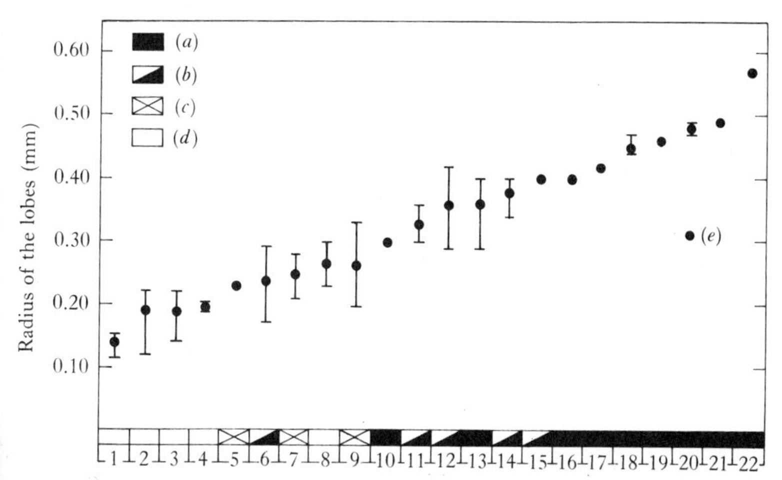

Fig. 7.7. The length of the secretory tubules in relation to the habitat of wading-birds (Charadriiformes). The radius of the lobe is a measure of the length of the tubules. (*a*) marine birds, (*b*) birds preferring marine environments, at least during migration, (*c*) birds found both in fresh water and marine environments, (*d*) fresh-water birds, (*e*) a captive black guillemot presumably given fresh water. Vertical lines are the ranges of variation and the dots the mean values. (1) Common Sandpiper chicks, (2) Wood Sandpiper, (3) Green Sandpiper, (4) Snipe, (5) Little Stint, (6) Common Sandpiper adults, (7) Dunlin, (8) Golden Plover, (9) Bar-tailed Godwit, (10) Brünnich's Guillemot, (11) Ringed Plover, (12) Black-headed Gull, (13) Little Auk, (14) Knot, (15) Sanderling, (16) Kittiwake, (17) Herring Gull, (18) Guillemot, (19) Puffin, (20) Black guillemot, (21) Razorbill, (22) Glaucous Gull (from Staaland, 1967*b*).

gests that when the rate was near the maximum a similar relationship to that obtained by Hanwell *et al.* (1971*a*) was evident i.e. a negative correlation. Thus any hypothesis to account for changes in concentration with secretory rate must take all these findings into account.

Morphological considerations

Staaland (1967*b*) carried out an excellent study on European charadriiform birds (waders, gulls and the like) which gives important clues to the mechanism by which the concentration of the secretion is determined. He found that the radius of the lobes of the salt gland, which is approximately equal to the length of the secretory tubules, is directly related both to the habitat of the bird and nasal fluid concentration of a particular

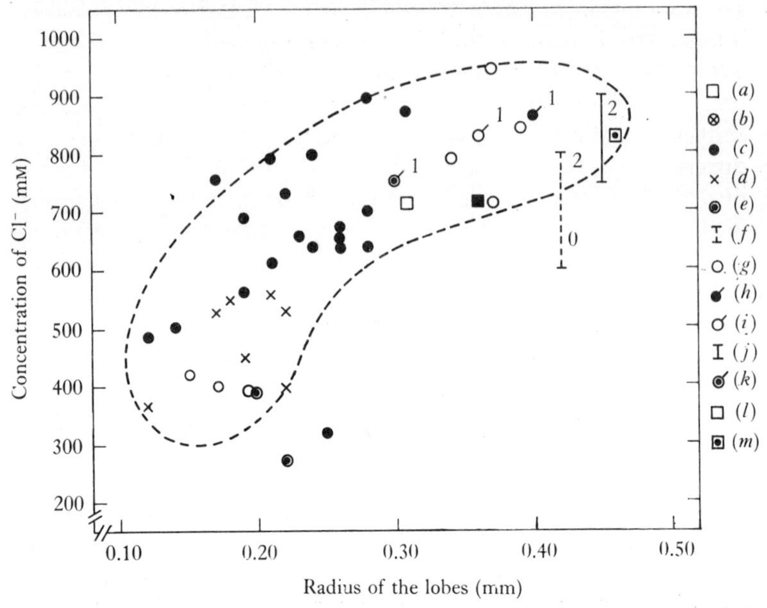

Fig. 7.8. Correlation between the length of the secretory tubules and the nasal-fluid chloride concentration. (1) anatomical and physiological data from different specimens, (2) concentrations in mM sodium from Schmidt-Nielsen (1960). (*a*) Ringed Plover, (*b*) Snipe, (*c*) Common Sandpiper, (*d*) Wood Sandpiper, (*e*) Green Sandpiper, (*f*) Herring Gull, (*g*) Black-headed Gull, (*h*) Kittiwake, (*i*) Little Auk, (*j*) Guillemot, (*k*) Brünnich's Guillemot, (*l*) Black Guillemot, (*m*) Puffin (from Staaland, 1967*b*).

species. Thus more marine birds had larger secretory lobes and secreted fluid with a higher sodium chloride concentration (Figs. 7.7, 7.8). Staaland therefore suggested that concentration may be determined by the length of the secretory tubule and that '...each cell along the secretory tubules may possess the capability of transporting a small number of sodium ions into a primary low-concentrated secretion and thus gradually increase the secretion concentrations towards the central canal'.

Komnick & Kniprath (1970) found that the secretory tubules are shorter in Herring Gulls (*Larus argentatus*) given fresh water than in those drinking salt water, a situation in which the sodium chloride concentration in the secretion is lower. A most definitive ultrastructural study has also been carried out on salt-adaptation in domestic ducklings by Ernst & Ellis (1969). They not only showed that the tubules increase in length during adaptation to salt water but also that the type of cell in the secretory tubule altered (Fig. 7.9). Apart from the generative cells at the blind end of the tubule (p. 15), Ernst & Ellis found differences in the secretory or principal cells and classified them into two main types – partially-specialized and fully-specialized cells. Partially-specialized cells, characterized by their almost flat basal cell membranes, short interdigitations of the lateral cell membranes between neighbouring cells and few mitochondria, were found to be predominant in birds kept on fresh water. However, in birds on salt water such cells were only found adjacent to the generative cells and, instead, fully-specialized cells in two stages of development predominated in the secretory tubules. These fully-specialized cells have many mitochondria, marked folding of the basal membrane and lateral interdigitations, and correspond to the type of cell present in truly marine birds (Chapters 2 and 9).

Thus Ernst & Ellis showed that there is a gradient of cell types along the secretory tubule, the more highly developed types being adjacent to the central canal of the lobes. They believe that the secretory cells are formed by division of the peripheral generative cells and develop into the more specialized types in birds given salt water, there being an increase in the rate of cell division which leads to an increase in the length of the tubule.

In adult geese kept on fresh water, a gradient of cell types can also be seen along the tubule. In such birds the most specialized type of cell corresponds to the first-stage, fully-specialized cell described by Ernst & Ellis in that there is some basal infolding, lateral interdigitation and numerous mitochondria (N. Ackland & M. Peaker, unpublished observations).

As noted in Chapter 2 there is a counter-current arrangement of blood capillaries and secretory tubules and the more specialized cells receive blood of the highest salt concentration. However, as Schmidt-Nielsen (1960) has pointed out, there is no evidence to suggest the presence of a 'hairpin loop' to constitute a counter-current multiplier system, as in the mammalian kidney, and it seems highly unlikely that the

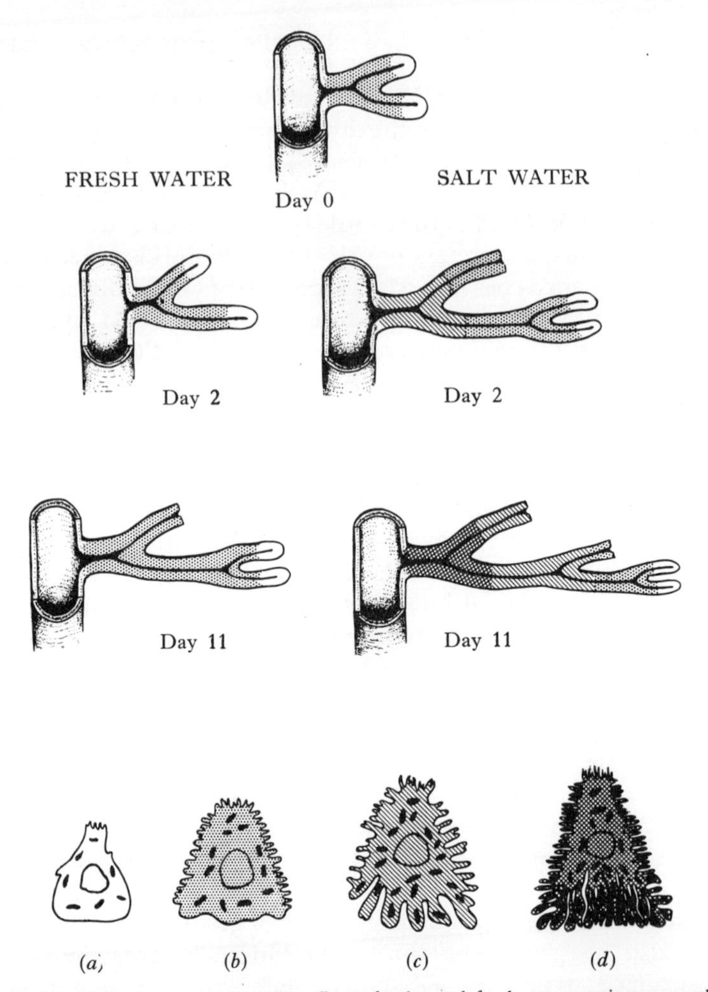

Fig. 7.9. This diagram shows the effect of salt- and fresh-water regimens on the development of cellular specialization in the secretory epithelium of the salt gland. Peripheral cells (*a*) exhibit little specialization of their cell surfaces and contain few mitochondria. Partially-specialized secretory cells (*b*) are characterized by short folds along their lateral surfaces and flat basal membranes; the distribution of mitochondria in this cell type is similar to that of the peripheral cells. Two stages in the development of the fully-specialized secretory cell type (*c* & *d*) are shown. In the first stage (*c*) the cells exhibit some folding of the basal surface in addition to the lateral plications; mitochondria, more numerous in this cell type than in the partially-specialized secretory cell, are fairly evenly distributed in the cytoplasm. In the second stage (*d*) the specialized secretory cell is fully developed: both the lateral and basal membranes are extensively infolded, forming complex intracellular compartments and extracellular spaces: the mitochondria are increased dramatically in number and pack the basal labyrinths. The distribution of these cell types in fresh-water- and salt-water-adapted glands on various days of the regimen is shown for a single secretory tubule as it grows out from the central canal. In addition to the development of cellular specialization, the tubules of birds given salt water elongate more rapidly and show more branching than birds kept on fresh water (from Ernst & Ellis, 1969).

determination of concentration can be explained by such a mechanism in the salt gland.

Hypothesis to account for the determination of concentration

In view of the relationships between the rate of secretion and the sodium chloride concentration, and alterations in the type of secretory cell in birds kept on salt water, Hanwell *et al.* (1971*a*) extended Staaland's (1967*b*) suggestion, that the cells nearer the central canal of the lobes add sodium to a primary secretion from the periphery, into a hypothesis that could account for these relationships.

The hypothesis is that all the secretory cells secrete in response to salt-loading but the more specialized cells secrete fluid of a higher sodium chloride concentration than the less specialized peripheral cells. Therefore to the less concentrated primary secretion would be added a more concentrated fluid as it flows along the tubule towards the central canal. At low secretory rates the primary secretion would be in contact with the more central cells for a longer period than at high rates. Thus a more concentrated final secretion would be formed at low rates (Fig. 7.10), and this, of course, fits the inverse relationship between rate and concentration shown in Fig. 7.5. The different relationship seen between individuals (Fig. 7.4) and the changes that occur upon adaptation to salt water may then be explained by the existence of longer secretory tubules with a greater preponderance of the more specialized type of cell in birds that can secrete at higher rates. Thus more fluid of a higher concentration would be secreted in birds with longer secretory tubules (Fig. 7.4).

An alternative possibility is that each secretory cell produces fluid of a variable concentration as the rate of secretion varies. This is unlikely since Diamond & Bossert (1968) in their model of a 'backwards-facing' epithelium such as the salt gland (see p. 112) could predict no relationship between the rate of uptake into the cell and the concentration of the fluid entering. However one might infer from their work that all other things being unchanged, the presence of the large basal infoldings in the fully-specialized cell could indicate that they might secrete

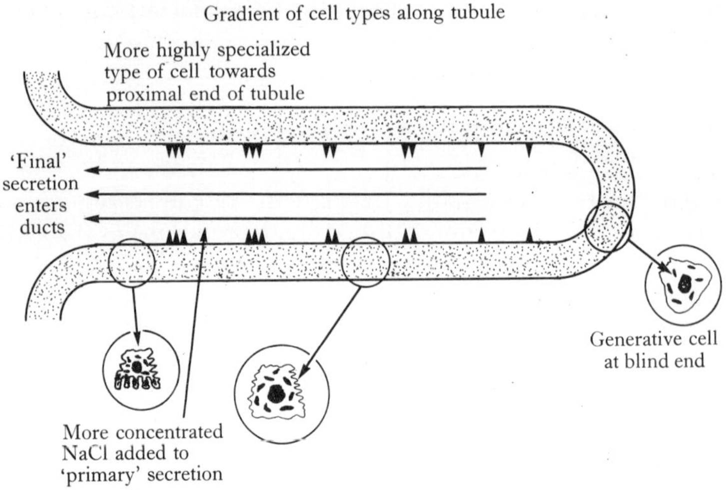

Fig. 7.10. A scheme to account for the relationship between secretory rate and nasal-fluid sodium chloride concentration in geese kept on fresh water. The cell types (ringed) are from N. Ackland & M. Peaker (unpublished).

fluid of a higher concentration, provided there was a concomitant change in the properties of the luminal membrane. However, this certainly does not mean to say that the more peripheral, less specialized cells with almost flat basal membranes cannot secrete a hypertonic solution. Hally, Buxton & Scothorne (1966) showed in ducks that the salt glands have the ability to secrete even three days before hatching. On the day after hatching the ability to secrete, assessed as the output of chloride on a gland weight basis, was 50 per cent of that of the adult and yet at this stage Ernst & Ellis (1969) found only the partially-specialized type of cell to be present. Therefore it can be argued that the primary secretion is still hypertonic and that the fully-specialized cells, when present downstream, can increase the concentration still further.

When tubules consist mainly of fully-specialized cells, as in glands of marine birds, one might predict that the sodium chloride concentration in the secretion of individuals will be high and vary less than in birds kept on fresh water and given a salt-load. However, experiments to explore the relationship between rate and concentration in truly marine birds have not been carried out and this is clearly an aspect that should be investigated.

A further extension of this hypothesis is necessary to account for the low sodium chloride concentration of 'spontaneous' secretion and the positive relationship observed between rate and concentration at the beginning and end of secretion induced by salt-loading (Smith, 1970; 1972*b*). It seems possible that, at these times, all the cells in a tubule are not secreting at a similar rate, and that the more peripheral cells begin to secrete sooner and continue for a little longer than the more central cells of the tubule. This might even depend on the osmotic stimulus with the nervous pathways to the peripheral cells being slightly more sensitive to an increase in plasma tonicity than those to the central cells. If the stimulus is maximal for the response however and then the rate of secretion is altered by other means, stress for example, the inverse relationship between concentration and rate is still apparent (Fig. 7.11).

It should be pointed out that in a few geese we have observed the positive relationship, as described by Smith (1972*b*) in ducks, but only at the end of the secretory response. This may be a reflection of the relative sensitivity of the receptors in the two species (p. 51) because it would seem that only a very small increase in plasma tonicity is sufficient to evoke secretion at nearly maximal rates in geese. In our experience secretion ceases abruptly in these birds when the salt-load has been excreted, but only slowly in ducks.

Direct evidence on the hypothesis we have outlined is difficult to obtain. Micro-puncture would be an ideal way to approach this problem but unfortunately we understand that the salt gland is not one of the organs that is 'susceptible' to this technique as it now stands.

Ducts

The function of salt-gland ducts has not been studied, chiefly because the main ducts are small and buried in the beak, and the intraglandular ducts are inaccessible. However predictions can be made from their structure and the function of ducts in other glands.

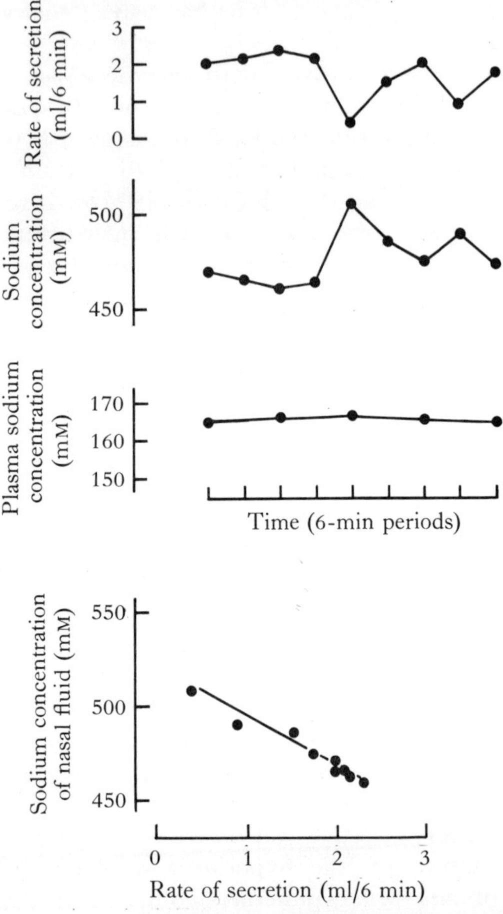

Fig. 7.11. Relationship between the concentration and rate of secretion in a goose that was restless during the experiment. Note the decrease in secretory rate but that sodium concentration was still related to rate rather than to plasma sodium concentration. The first collection period shown was 78 min after the start of secretion.

In some glands the ducts not only carry the secretion, but play an active part in its elaboration (e.g. salivary glands, pancreas). In some the ducts may be simple conduits (e.g. ureter) but in other organs they also store the secretion without any change in composition (e.g. mammary gland) or store it and alter it (e.g. gall bladder). In all cases the structure is commensurate with function. Where active alteration of secretion occurs, the lining cells appear highly specialized with much cytoplasm, a spherical or ovoid nucleus, many mitochondria and marked foldings of the cell membrane. Simple conduits, by contrast, are lined by one or two layers of unspecialized cuboidal or flattened cells with few organelles, little cytoplasm, and an indented nucleus. Some of these unspecialized epithelia are almost completely impermeable to small ions (e.g. mammary gland) and others to ions and water (e.g. mammalian ureter and bladder).

The structure of salt-gland ducts (see Chapter 2) resembles that of the ureter. There is a transitional epithelium of at least two layers of unspecialized cells. It seems likely that salt-gland ducts are impermeable to ions and water. The ureter, which carries a hypertonic fluid very different to plasma, is impermeable to both ions and water, thus preventing both ionic and osmotic equilibration. The case is similar in the salt glands, whose ducts carry markedly hypertonic sodium chloride.

Effects of drugs and hormones on concentration

Since there is a relationship between secretory rate and concentration of nasal fluid, it is necessary to investigate both in order to assess whether concentration is affected *per se.* In most experiments this has not been done because it was not realised that such a relationship existed. However it is now possible to explain the results of some earlier experiments in which concentration was affected by drugs, because there were also changes in rate. For example, large doses of adrenaline result in an increase in nasal fluid concentration but the secretory rate decreases (Fänge, Schmidt-Nielsen & Robinson, 1958) – an effect completely in agreement with the type of relation shown in Fig. 7.5.

It is relatively simple to detect direct effects on concentration from data which show decreases both in secretory rate and concentration in any one bird. Thus atropine (Fänge, Schmidt-Nielsen & Robinson, 1958), pentobarbitone (Cooch, 1964) and the diuretics meralluride and aminophylline (Nechay *et al.*, 1960) lower concentration and flow rate, presumably acting directly on water permeability of the cells. Nevertheless, even

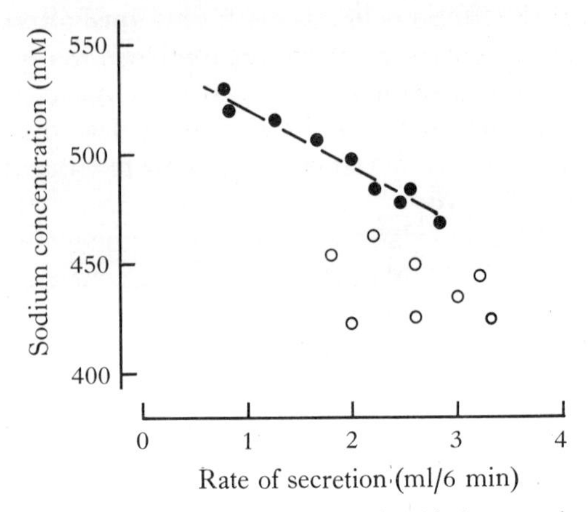

Fig. 7.12. The disruption of the normal relationship between the rate and concentration of secretion in a goose by arginine-vasotocin (AVT). The regression line and closed circles are for points obtained before, and the open circles after, the intravenous injection of 40 μg AVT.

when concentration falls and rate rises it is still possible to detect effects on concentration by plotting rate versus concentration for samples of secretion taken before and after the administration of a substance. The lowering of concentration by arginine-vasotocin can be detected in this way (Fig. 7.12). (M. Peaker, unpublished observations).

While it is possible to lower concentration by the administration of drugs, no agent has been found so far that will increase the concentration of the secretion by a direct effect without reducing the rate of flow. This is in agreement with the view that cellular changes are necessary for this to occur i.e. development of fully-specialized cells, and that salt-adaptation is the only means by which this can be brought about.

Inoue (1963) made interesting observations in the duck which, without more information, are not easy to interpret in terms of a single effect. He injected mercuric chloride into the ducts of the salt gland and then stimulated secretion by injecting methacholine. He found that while the rate of secretion was unchanged, the concentration fell markedly, and claimed that the mechanisms which determine secretory rate and concentration are independent. However another explanation could be that the mercuric chloride, previously known as corrosive sublimate, poisoned the duct epithelium. We have already suggested that the ducts are normally impermeable to sodium chloride and water and the effects of destroying the epithelium would be to allow water in and sodium chloride out, which is what he observed.

8 HORMONES AND SALT-GLAND SECRETION

It is easy to see why several workers actively searched for evidence of endocrine control of the salt glands. In 1957, the same year that Schmidt-Nielsen and his associates discovered the function of the nasal gland, the eminent zoologist I. Chester Jones published his comprehensive monograph on the adrenal cortex. Chester Jones, in his experiments and reviews, clearly showed that the hormones of the adrenal cortex play a vital role in homeostasis in all vertebrates, from fish to mammals. It is not surprising that many of his pupils, and other zoologists, were influenced by his work to an extent such that they investigated the effect of adrenocorticosteroids and proposed a hormonal mechanism for the initiation and maintenance of salt-gland secretion, in apparent opposition to Schmidt-Nielsen and his colleagues who had carried out classical physiological experiments and reached the conclusion that secretory nerves are responsible.

As late as 1970 Bern was still upbraiding physiologists for ignoring hormones in the following manner:

> The endocrinologist is often commendably more cognizant of the influence of non-endocrine control systems upon his target organs, or target processes, than the physiologist, who is not endocrinologically disposed, seems aware of endocrine control systems. The history of our knowledge of salt-gland physiology in birds, to refer to these organs again, is especially revealing in this regard. The students of nervous control of the salt glands have acted as if the endocrine system has not yet been invented. The students of endocrine control are somewhat less recalcitrant *vis-à-vis* the nervous system.

While there is some truth in these remarks the boot was, in fact, on the other foot and the comparative endocrinologists were guilty of ignoring until quite recently most of the evidence Fänge, Schmidt-Nielsen & Robinson (1958) obtained, which clearly showed that the salt gland is activated by secretory nerves (Chapter 4).

Perhaps it would be fair to admit that the differences in approach to the functioning of the gland are the result of the background of the investigators. Hormones, and in particular adrenocorticosteroids, were in vogue at the time the hormonal theory for the activation of the gland was proposed and a very wide range of direct actions was claimed for these compounds by both mammalian and comparative endocrinologists. It is probably true to say that these were largely ignored by physiologists but in endocrinological circles at that time, claims for a controlling role of adrenocorticosteroids in salt-gland secretion would not have been unexpected. On the other hand, to the mammalian physiologist, the activation of the salt glands is so like that of the salivary glands – organs on which many undergraduate physiologists cut their experimental teeth – that they would immediately think of a nervous secretomotor mechanism.

Adrenocortical hormones and salt-gland secretion

Holmes, Phillips & Butler (1961) were the first to study the effects of adrenocortical hormones on salt-gland function. They did this because evidence was accumulating which suggested that these hormones were intimately involved in the adaptation of many vertebrates to different environmental salinities. They gave ducks either cortisol, deoxycorticosterone ('cortexone'), aldosterone or mammalian ACTH, prior to oral loading with 20 ml of 20 per cent sodium chloride. They noted two phases of excretion; the first which lasted for about an hour they called the 'renal' phase and the second, which began during the first, they called the 'extra-renal' phase (i.e. nasal secretion). Treatment with all these steroids and ACTH enhanced both the rate and duration of salt-gland secretion, aldosterone being the most effective; in addition, the latency between the administration of salt and the onset of nasal secretion was shorter in the hormone-treated birds and it was noted that, although plasma osmolality was not measured, 'serum' (in fact plasma because the birds were heparinized) sodium was significantly lower in birds given aldosterone or ACTH when secretion started.

As a result of these experiments Holmes, Phillips & Butler (1961) considered that salt-gland secretion is controlled by adrenocorticosteroids, and continued '...the initiation of nasal secretion in each experimental group preceded the time at which this event occurred in the controls. This would suggest that the increased time-lag in the controls marks a period during which the endogenous mobilization of adrenal steroids takes place to the threshold level necessary for the precipitation of nasal secretion. In other words the nasal gland had been previously "primed" and was galvanized into action at a lower level of osmotic stimulation.' In further experiments using the same technique, Phillips, Holmes & Butler (1961) found that complete adrenalectomy abolished and unilateral adrenalectomy reduced, secretion in response to a similar salt-load. The administration of cortisol restored secretion in the totally-adrenalectomized ducks.

Phillips *et al.* (1961) also suggested that the renal phase of excretion was terminated by antidiuretic hormone, released, as in mammals, in response to an increased plasma osmolarity. They further postulated that adrenal corticoid hormones stimulate the salt glands directly and are secreted by the adrenals under the influence of either ACTH or antidiuretic hormone. They envisaged the following sequence of events to occur in marine birds: 'ingestion of sea water, elevation of the sodium chloride content of the blood, rise in the level of adrenocorticosteroids, nasal gland secretion, reduction of the sodium chloride content of the blood, fall in the level of circulating adrenocorticosteroids and cessation of nasal gland secretion', and concluded, '...we suggest that adrenocorticosteroids of the glucocorticoid type are essential for the functioning of the extra-renal salt glands in the duck and together with antidiuretic hormone determines the characteristic diphasic response to hypertonic saline loads. These hormonal mechanisms together effectively restore homeostasis by causing a dynamic shift from the renal route to the physiologically more economical extra-renal route of excretion. The survival value of these mechanisms is self-evident and is especially advantageous to marine birds, who by virtue of their ecological niche are subjected to unusually high intakes of sea water or marine invertebrates.'

Curiously, Phillips, Holmes & Butler (1961) did not attempt to fit their scheme into what was already then known about the nervous initiation of secretion (Chapter 4). They dismissed the cholinergic nerves as merely causing vasodilatation although admitting that it was accompanied by secretion. Thus there was an apparent conflict between the two schemes, or in other words, nerves versus hormones.

The experiments of Holmes, Phillips & Butler were more complicated than at first appears. For example it is not certain that their 'renal' phase was in fact solely, or even mainly, related to renal excretion. They collected fluid from the cloaca and, as all those who have been the target of a passing bird will remember, birds can forcibly project a mixture of urine and faeces from this orifice. Douglas (1970) studied the transit of polyethylene glycol through the gastro-intestinal tract of gulls and pointed out that the high sodium concentrations in 'urine' during the 'renal' phase in their experiments could have been 'largely due to intestinal transit of a portion of the 20% NaCl loading solution'. In our work we obtained graphic evidence which dramatically confirmed Douglas's (1970) view that solutions can pass through the alimentary tract of birds very rapidly. Whilst searching for the location of the receptors that initiate secretion (see Chapter 4), we at one time thought that they might lie in the gut wall. We therefore mixed barium sulphate with sea water and radiographed the bird at the moment secretion started. In one bird, the whole of the intestine was filled two minutes after oral loading (Plate 8.1) and, in fact, fluid containing barium sulphate was projected from the cloaca some distance onto the feet of the radiographer! We now thank him for his part in what was to the onlookers an hilarious experiment.

Another important consideration is the use of such strong salt solutions to load the birds. Twenty per cent sodium chloride is an emetic in mammals and this strength is not only unnecessary in birds but probably causes stress. In fact sea water (or a sodium chloride solution of similar strength, three per cent) given orally is an effective stimulus in cormorants (Schmidt-Nielsen *et al.*, 1958), ducks (Ash, 1969) and geese (Hanwell *et al.*, 1971*a, b*) and in all these species secretion starts within a few minutes. By contrast, in the experiments of Holmes, Phillips & Butler (1961) the mean time was 52.8 min in ducks given 20 per cent sodium chloride. When we gave three geese such a solution, only one secreted within an hour and the other two did not respond even to an intravenous load next day. In other birds given oral sea water the median latent period was six minutes (Hanwell *et al.*, 1971*b*). We suggest that the strong salt solution inhibits secretion.

Phillips *et al.* (1961) considered it odd that aldosterone, which is secreted in mammals during sodium deficiency, should actually increase sodium loss by the salt glands. They suggested that in the earlier experiments, aldosterone may have reduced sodium excretion by the kidneys and that this extra sodium resulted in an increased secretion by the salt glands. Therefore Phillips & Bellamy (1962) gave ducks an intravenous salt-load. Under these conditions no fluid was voided from the cloaca during the experimental period so that this was a better way of investigating the effects of hormones on salt-gland secretion. In this preparation aldosterone had no effect. However both cortisol and deoxycorticosterone were still effective in enhancing secretion and later experiments showed that corticosterone, which is the major circulating adrenal steroid in birds, was also effective (Holmes, Phillips & Chester Jones, 1963). Thus it was concluded that, in this respect, glucocorticoids rather than mineralocorticoids were the active component of the adrenocortical secretion.

While it was admitted by most workers at that time that these results suggested a 'permissive action' or a possible involvement in modifying salt-gland secretion, few appeared to agree that they actually initiate secretion. We feel that this was because it was realized that secretion can be elicited so quickly after salt is given intravenously that there would be insufficient time for activation of the pituitary–adrenal axis. Moreover, the experiments of Fänge, Schmidt-Nielsen & Robinson (1958) had clearly provided a satisfactory explanation for the onset of secretion (Chapter 4).

The pituitary–adrenal axis after salt-loading

Perhaps at this stage it is convenient to explain in endocrinological terms what the hypothesis of Holmes, Phillips & Butler demanded. It was necessary to postulate that the increase in plasma tonicity induced the release of ACTH from the adeno-

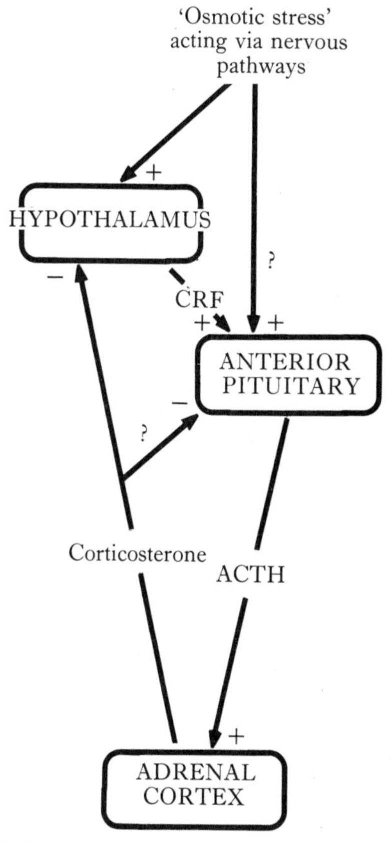

Fig. 8.1. Simplified diagram showing two mechanisms that control ACTH and corticosterone secretion (+ = stimulation, – = inhibition): CRF, corticotrophin releasing factor.

hypophysis which, in turn, increased the circulating concentration of corticosterone. This implies, as in stress, a direct hypothalamic stimulation of ACTH output as opposed to the regulation of ACTH and corticosterone secretion by a negative feed-back mechanism which operates to maintain a fixed concentration of circulating corticosterone (Fig. 8.1). Some evidence is certainly indicative of a direct stimulation of ACTH release. Marine birds were found to have larger adrenals than terrestrial species or those which inhabit fresh water (Fig. 8.2). Similarly the weight of the adrenal was found to increase when gulls and ducks were given only salt water to drink (Fig. 8.3)

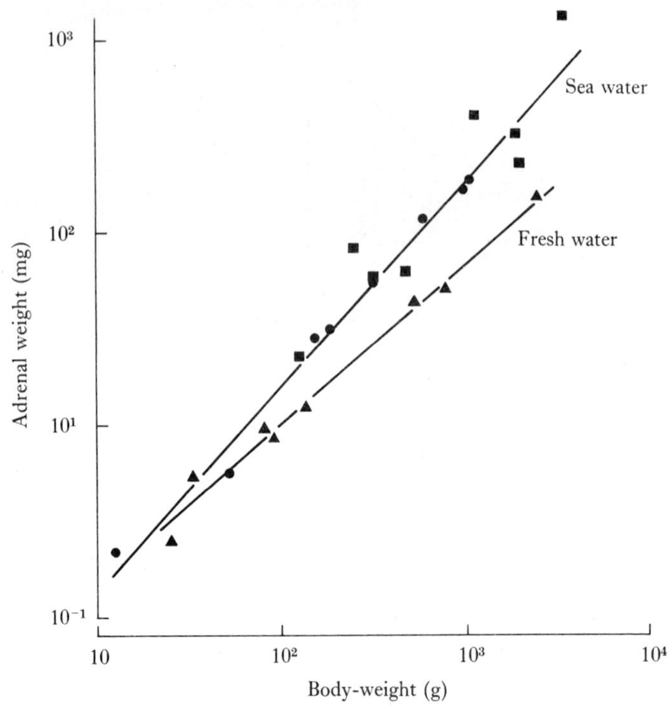

Fig. 8.2. Relation between adrenal weight and body-weight in: marine ■, brackish ●, fresh-water and terrestrial ▲ birds. x = body weight (g), y = adrenal weight (mg).

Fresh-water/terrestrial species:

$y = 0.20 + x^{0.85}$ $r = 0.994$ $P < 0.001$

Brackish-water species:

$y = 0.12 + x^{1.04}$ $r = 0.992$ $P < 0.001$

Marine species:

$y = 0.16 + x^{1.05}$ $r = 0.936$ $P < 0.001$

Marine and brackish-water species (combined data):

$y = 0.11 + x^{1.09}$ $r = 0.973$ $P < 0.001$

Analysis of variance shows a significant difference ($P < 0.05$) between the exponent of body weight (i.e. the slope) for the fresh-water/terrestrial species and that for the combined data of brackish and marine species (calculated from the data collected by Holmes, Butler & Phillips, 1961).

(Holmes, Butler & Phillips, 1961; Benson & Phillips, 1964). Furthermore, liver glycogen concentrations were raised under such conditions implying that an increase in the rate of corticosterone secretion was manifested in hepatic gluconeogenesis (Phillips & Bellamy, 1967). However, Macchi, Phillips, Brown & Yasuna (1965) and Macchi, Phillips & Brown

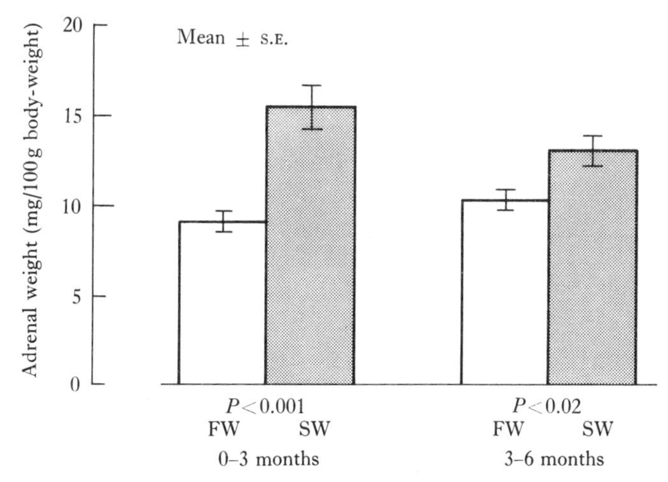

Fig. 8.3. The effect on adrenal weight of drinking sea water for up to six months in Glaucous-winged Gulls (*Larus glaucescens*): FW, fresh water; SW, sea water (drawn from the data of Holmes, Butler & Phillips, 1961).

(1967) found no significant change in peripheral corticosterone concentrations after salt-loading, but since plasma volume was found to increase they pointed out that there must be an increase in the rate of corticosterone production in order to maintain blood concentration; this increase could of course be achieved by a normal self-regulating feed-back mechanism. The data of Donaldson & Holmes (1965) appeared to support these findings to some extent since they could find no significant difference in plasma corticosterone concentrations (measured by a double-isotope derivative assay) between ducks maintained on salt water and those on fresh water even though the adrenals were larger in the former. Donaldson & Holmes (1965) also studied the kinetics of corticosterone secretion *in vivo*, by injecting [^{14}C]corticosterone intravenously, and its apparent volume of distribution. Their kinetic calculations were made on the assumptions that there was a single compartment of distribution for the hormone and that endogenous peripheral corticosterone concentrations were at steady state. They concluded, on the basis of such calculations, that secretory rate is increased after salt-loading in birds maintained either on fresh water or on salt water but only, it would appear, as a consequence of an increased apparent volume of distribution in the body. This would imply that the increase in corticosterone production is mediated by a normal feed-back mechanism, as Macchi *et al.* (1967) found.

However there are some reasons for suggesting that these experiments should be repeated. Macchi *et al.* (1967) used a brief fluorometric method to determine corticosterone in venous plasma. Such methods are now known to be subject to gross errors (see Frankel, 1970) and it would seem necessary to repeat these measurements using a more reliable method. The experiments of Donaldson & Holmes (1965) are difficult to interpret because of the method they used to determine secretory rate *in vivo* and the lack of experimental proof for the assumptions they had to make in their calculations. There is very good evidence in sheep that the rate of cortisol production cannot be measured reliably following the single intravenous injection of labelled cortisol (Harrison & Paterson, 1965). It has also been shown by Paterson & Harrison (1968) that kinetics based on a single compartment of distribution are inadequate and that their data fitted a two-compartment system. If these considerations can be extrapolated from sheep to birds, and there is no reason to believe that they cannot, then it is clear that the experiments of Donaldson & Holmes (1965) should be repeated using kinetic methods similar to those developed by Paterson & Harrison (1967; 1968). However Thomas & Phillips (1973), in an abstract, have been unable to show any differences in the turnover of corticosterone or aldosterone between birds kept on fresh water or salt water. Another important consideration is the binding of adrenocorticosteroids to plasma proteins which can of course influence the concentration of free hormone in the circulation. The concentration of free steroid rather than the total concentration is the main determinant of biological activity for most tissues and it would seem worthwhile to investigate the concentrations of free and bound corticosterone in the plasma of ducks maintained either on salt water or on fresh water.

There is certainly evidence that, following ingestion of salt water, there is an increase in the rate of corticosterone production mediated by ACTH released as a direct consequence of this 'osmotic stress'. The increases in adrenal weight and liver glycogen content as well as adaptive changes in the gut, which can be attributed to adrenocortical hormones (see p. 193), would indicate that this is the case. Furthermore, Abel, Rhees & Haack (1971) have found that there are marked cytological changes in the anterior pituitaries of ducks given salt water and that these changes are indicative of a stimulation of ACTH production. Similarly there is a marked increase in the concentration of ACTH in the peripheral plasma of ducks at a time after salt-loading when plasma volume has returned almost to normal (S. J. Peaker, unpublished observations, using the rat adrenal ascorbate depletion assay) and this could also suggest a specific release of ACTH. Perhaps Holmes, Phillips & Wright (1969) were thinking along similar lines when they included in their scheme a direct link between the afferent nervous pathway for salt-gland secretion and a stimulation of ACTH release via a releasing factor, although they did not detect a rise in plasma corticoids in their earlier experiments.

It is possible that plasma corticosterone concentrations increase not immediately after salt-loading as Phillips *et al.* (1961) expected, but after hours or days of exposure to saline conditions as part of the process of adaptation. Dr Marthe Vogt has pointed out to us that there is some evidence that in the adaptation to many stressful situations the involvement of the pituitary and adrenal cortex is not immediate and may be quite brief, triggering cellular changes which themselves persist after the activation of ACTH and corticosteroid production. Thus the determinations of Macchi *et al.* (1967) made up to 45 minutes after salt-loading may have been too soon and those of Donaldson & Holmes (1965), after one week of exposure to salt water, may have been too late to detect any changes. It is clear that to study this problem regular blood samples are required during the initial period of adaptation. In support of this suggestion is the fact that when eels (*Anguilla anguilla*) are transferred from fresh water to sea water there is a transient rise in plasma cortisol concentrations, seen from one to five days in the 'yellow phase' of this species in winter, and from one to three days in the 'silver phase' in spring (Ball *et al.*, 1971). In this teleost fish there is good evidence that cortisol has effects on salt and water balance during this change in environmental salinity, and it seems clear that a 'pulse' of cortisol rather than continuously high levels in plasma is responsible.

An alternative hypothesis

Faced with the findings of Macchi *et al.* (1967), that there was no change in plasma corticosterone concentrations after salt-loading, Phillips & Bellamy (1967) proposed a mechanism by which the salt glands could receive more corticosterone even in the absence of raised concentrations. They argued that if the salt gland is highly efficient in extracting corticosterone from the blood then with a vasodilatation induced by nerves, more corticosterone would be available for uptake by the cells of the gland. The results they presented suggested that such a mechanism could operate.

Bellamy & Phillips (1966) measured tissue radioactivity and inulin space in a number of tissues after the injection of [^{3}H]corticosterone into ducks loaded with either 0.154 M sodium chloride (which did not induce salt-gland secretion) or 0.5 M sodium chloride. Tissue radioactivity was higher in the salt gland than in any other organ (Fig. 8.4) which certainly indicates a highly efficient trapping mechanism. They also found an increased amount of tritium in the salt gland as well as an increase in the inulin space during secretion, and continued,

> The change in inulin space gives an estimate of the maximum possible increase in tissue blood volume. As the concentration of blood radioactivity was known, it was possible to calculate (assuming that the increased inulin space was entirely due to extra blood) the maximum possible amount of additional steroid in blood in each tissue. The quantities of radioactivity calculated in this way were less than the amount of extra steroid found in the liver and nasal glands of birds which received 0.5 M NaCl. There was a particularly large difference between the calculated and observed values in the nasal gland. As the concentration of extracellular steroid in the interstitial space is likely to be much lower than that in blood (Bellamy, Phillips, Chester Jones & Leonard, 1962) it may be concluded that most of the additional steroid was associated with the cells. Thus, there appears to be a mechanism – of particular importance in the nasal gland – by which the amount of cellular steroids may be increased without a change in the level of circulating steroids.

Phillips & Bellamy (1967) also showed that [^{14}C]cortisol was taken up in a similar manner and they again stressed that a major role of the cholinergic nerves to the gland was to increase the rate of blood flow.

Wood & Ballantyne (1968) (see also Ballantyne & Wood, 1967) have found that β-glucuronidase is present in the cyto-

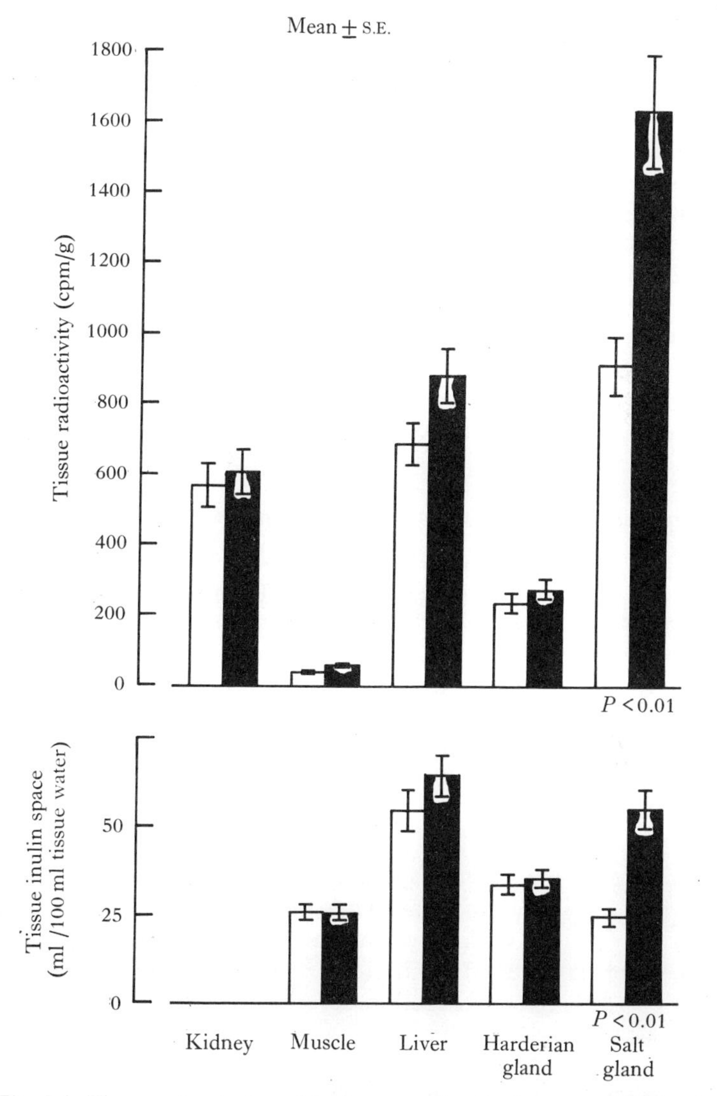

Fig. 8.4. Tissue radioactivity and inulin space after the injection of [^{3}H]corticosterone and either 0.154 M (open columns) or 0.5 M sodium chloride (closed columns) intravenously, in the duck (drawn from the data of Bellamy & Phillips, 1966).

plasm of the secretory cells and that the activity per unit weight of gland increases in ducks kept on salt water. Ballantyne (1967) has found this enzyme to be present in a number of tissues where active sodium transport occurs and these workers have suggested that β-glucuronidase may liberate free steroid from glucuronide conjugates. If this were the case then one might not expect changes in free corticosterone in plasma. However, more recently, Helton & Holmes (1973) have found that corticosterone-glucuronide conjugates are not present in the duck so that this suggestion seems untenable.

Having reviewed the main lines of evidence which suggest some involvement of adrenocortical hormones in the functioning of the salt glands we must now consider the actual role of these hormones in the light of current evidence.

Adrenocorticosteroids in the initiation and maintenance of secretion

There is no evidence to suggest that adrenocorticosteroids are in any way involved in the normal initiation or maintenance of secretion. As we have explained in Chapter 4 the secretomotor activity of the cholinergic nerve fibres to the gland is essential for both the initiation and maintenance of secretion. Furthermore since secretion appears so rapidly after intravenous salt-loading it is clear that the hormonal mechanism for activation proposed by Phillips *et al.* (1961) on the basis of experiments involving the oral administration of a very strong sodium chloride solution cannot operate. Peaker, Peaker, Phillips & Wright (1971) found that while secretion appears in the nostrils of ducks within two minutes of an intravenous salt-load, approximately 15 minutes are required for exogenous ACTH to enhance secretion after a minimal stimulatory salt-load.

Holmes, Phillips & Butler (1961) found that a very potent synthetic corticoid, 9α-fluorocortisol, evoked secretion at a low rate in the absence of a salt-load. However, the latent period was later found to be 25 minutes and there are marked systemic effects (increases in body temperature, respiratory rate, and plasma potassium) so that a specific action on the salt glands seems unlikely (M. Peaker, unpublished observations).

The necessity for adrenocortical hormones – a permissive action?

There is no doubt that a reduction in the circulating concentration of corticosterone results in an inhibition of salt gland secretion. The results of adrenalectomy clearly show this (Phillips *et al.*, 1961). Similarly, adenohypophysectomy results in a marked diminution of salt-gland secretion in response to salt-loading. Although replacement experiments were not done, Wright *et al.* (1966) suggested that the lack of ACTH (and, therefore, of corticosterone) was responsible.

Holmes, Lockwood & Bradley (1972) have also adenohypophysectomized ducks and studied the effects of salt-loading. These experiments were more satisfactory than those on the adrenalectomized animals, which die within hours of the operation or of the cessation of replacement therapy. The effects of loss of cortical hormones are so widespread and severe that one must be cautious in ascribing specificity of action, so that in adrenalectomized ducks it is difficult to be certain that the general condition of the animal was not responsible for the failure of secretion. By contrast, Holmes *et al.* (1972) found that the adenohypophysectomized ducks remained in good condition even though peripheral plasma corticosterone concentrations fell from 3.6 to 0.29 μg/100 ml. Salt-gland secretion in response to intravenous salt-loading was greatly reduced but both the secretory response and the plasma corticosterone concentrations were restored by the administration of mammalian ACTH (Fig. 8.5). Adenohypophysectomized birds are somewhat hypoglycaemic (Bradley & Holmes, 1971) and since it has been suggested that exogenous ACTH may exert its effect on secretion in the intact bird partly by raising the concentration of glucose in the blood (see below), glucose was also infused to see if it would restore salt-gland secretion; this it failed to do.

Although adrenal insufficiency in mammals results in systemic changes, other than hypoglycaemia, which could secondarily affect salt-gland secretion, for example acidosis (Nechay *et al.*, 1960), possible failure of nerve conduction (Secker, 1938; 1948), peripheral vasoconstriction and renal fluid loss, there is no reason to believe (except for some experiments with

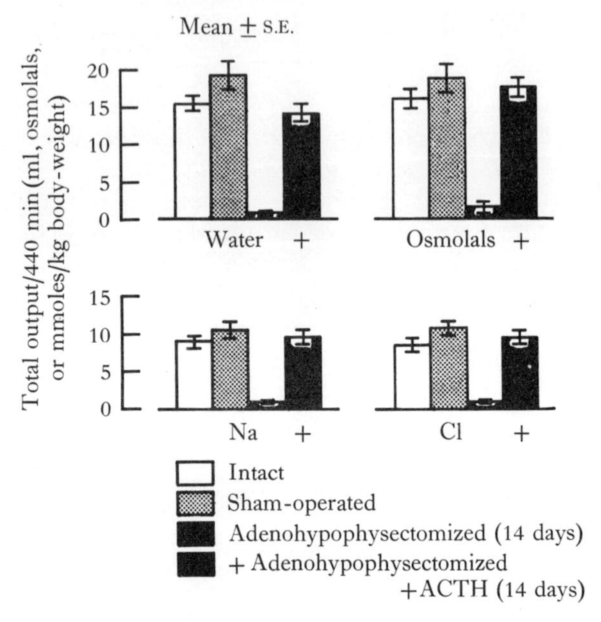

Fig. 8.5. The effect of adenohypophysectomy and ACTH replacement on salt-gland secretion in response to an intravenous salt-load in ducks (drawn from the data of Holmes *et al.*, 1972).

prolactin, see later p. 154), that the cells of the salt gland, like perhaps all vertebrate tissues, do not require adrenocortico-steroids for the maintenance of normal function. Certainly organ culture studies with a wide variety of tissues suggests this. If one accepts that adrenocortical hormones are essential for the normal functioning of the cells of the salt gland this does not mean that these hormones normally initiate, or maintain secretion. In other words we feel that while a 'permissive' action of corticosteroids has probably been demonstrated, this alone in no way indicates a 'control' by these hormones.

Exogenous hormones – an indirect action?

Peaker, Peaker, Phillips & Wright (1971) have questioned whether exogenous glucocorticoids affect the cells of the salt gland directly to enhance secretion. They found that as in the experiments of Holmes, Phillips & Butler (1961) pre-treat-

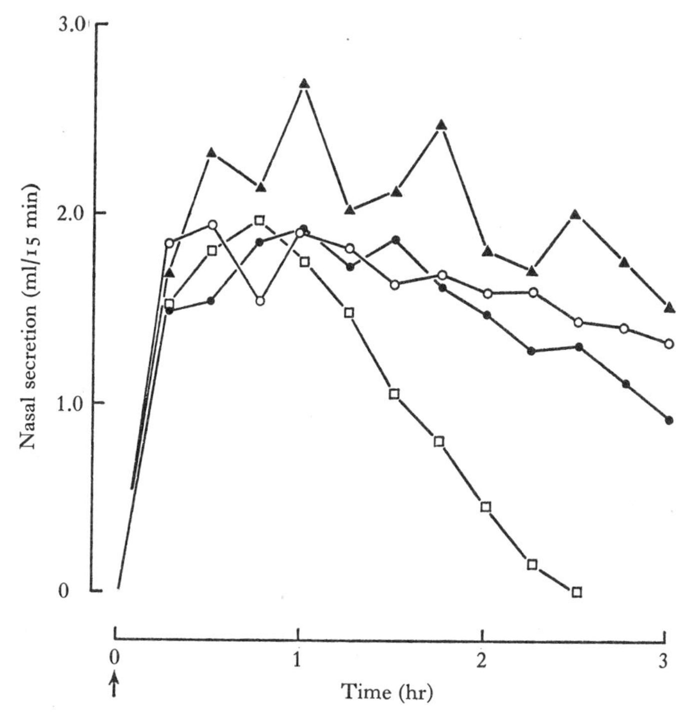

Fig. 8.6. Mean effects of corticotrophin (ACTH) (○), glucose (●) and potassium chloride (▲) on nasal secretion in ducks given 0.5 M sodium chloride intravenously at time 0 (arrow). ACTH was given for five days (20 i.u./day). Glucose (0.12 M) or potassium chloride (0.06 M) was dissolved in the 0.5 M sodium chloride, which was given as a single injection (18 ml/kg): □, untreated (from Peaker *et al.*, 1971).

ment of ducks with ACTH resulted in a prolongation of secretion in response to an intravenous salt-load. It was also found that blood glucose and plasma potassium concentrations were raised in the birds given ACTH (Table 8.1). When these changes in blood composition were mimicked by injecting glucose or a little potassium chloride at the same time as the salt-load, secretion was prolonged in an exactly similar manner to that observed with ACTH (Fig. 8.6). Similarly ACTH was found to prolong secretion in response to a minimal stimulatory salt-load (Fig. 8.7) and this prolongation coincided with a rise in blood glucose concentration. Thus it was not necessary to postulate a direct action of glucocorticoids in enhancing secretion. However a direct effect could not be entirely ex-

TABLE 8.1. *Effects of corticotrophin (ACTH) (20 i.u./day for 5 days) on nasal salt-gland function in response to 0.5 M NaCl, blood composition before salt loading and adrenal weight (means ±S.E.; 5 birds in each group)* (from Peaker, Peaker, Phillips & Wright, 1971)

	Sham-treated	ACTH-treated	*P*
Nasal secretion in 3 hr			
Weight (g)	11.0±1.1	19.7±0.9	<0.001
Na^+ (mmoles)	4.9±0.6	7.5±0.8	<0.05
K^+ (mmoles)	0.15±0.03	0.22±0.05	NS
Nasal fluid concentration			
Na^+ (mM)	445.0±10	380±10	<0.01
K^+ (mM)	13.9±0.6	11.4±0.5	<0.01
Haematocrit (%)	43.8±1.0	39.6±1.2	<0.05
Blood glucose (mg/100 ml)	105.0±3.8	132.0±6.0	<0.01
Plasma K^+ (mM)	3.4±0.06	3.75±0.10	<0.05
Plasma Na^+ (mM)	145.3±2.1	147.2±2.2	NS
Adrenal weight (mg)	126.1±8.2	184.5±3.7	<0.001
(mg/kg body-weight)	60.2±2.6	73.9±2.5	<0.01

NS = not significant.

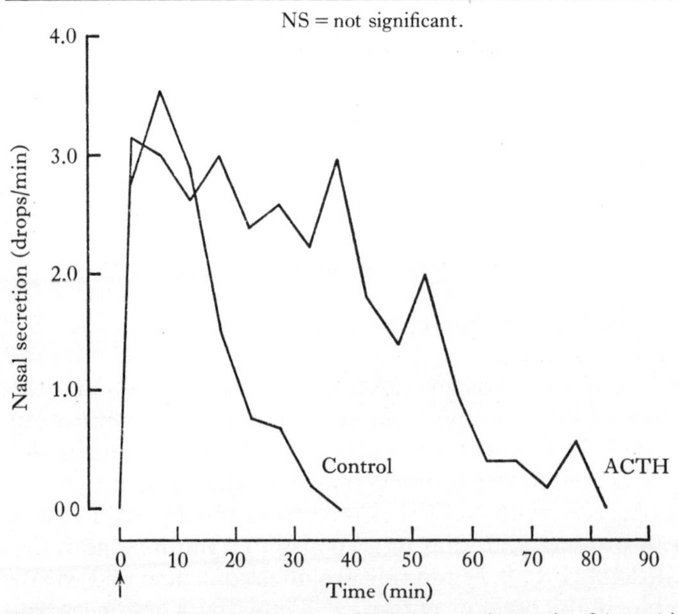

Fig. 8.7. Prolongation of secretion by ACTH (25 i.u.) given at time 0 (arrow) in a duck given a minimal stimulatory salt-load (ten per cent sodium chloride was infused intravenously at 1 ml/min until nasal secretion started at time 0) (from Peaker *et al.*, 1971).

cluded since, although the concentrations of sodium and potassium in the secretion were reduced in the ACTH-treated birds, this effect was not mimicked by the injection of glucose or potassium chloride.

A possible role of corticosterone

If it is accepted that adrenocorticosteroids only affect salt-gland secretion by an indirect action, what is the purpose of the uptake of corticosterone by the gland found by Bellamy & Phillips (1966)? As well as *in vivo* this also occurs *in vitro* (M. Peaker and S. J. Stockley, unpublished) and preliminary experiments have indicated that a large proportion of labelled corticosterone taken up is bound (S. J. Peaker, unpublished). More recently Sandor & Fazekas (1973) have found that corticosterone is bound to macromolecules in the high-speed supernatant fraction of salt gland homogenates; this binding was much greater for corticosterone than for deoxycorticosterone or aldosterone, and the salt gland bound more corticosterone than cytosol from kidney, lung, brain, intestine and liver. These findings would suggest *prima facie*, at least, that the salt gland is a target organ.

It seems to us that corticosterone is not important in controlling nasal secretion in the short-term. It would seem far more likely that corticosterone might be involved in the maintenance of normal function and in the induction of the long-term adaptive changes which occur in the tubules of the gland when birds drink salt water, as suggested by Holmes *et al.* (1969). It is likely that corticosterone would act in concert with other humoral and local factors (p. 177) associated with secretion rather than alone because the full range of adaptive changes is not seen when birds are treated with ACTH. For example, the concentration of the nasal secretion is not increased, as during adaptation (p. 161) even after treatment with ACTH for five days (Peaker, Peaker, Phillips & Wright, 1971). Adaptive changes also occur in the intestine and it seems likely that adrenocorticosteroids are involved at this site (Crocker & Holmes, 1971*a*, *b*). The possibility therefore exists that corticosterone could affect a number of physiological systems which are concerned in the response of the whole organism to a saline environment.

Hypophysial hormones

Before we go on to discuss the hormones individually we should mention that there is a complicating factor in studying pituitary function in ducks, which should not be overlooked. This is that the surgery necessary to ablate the whole or parts of the pituitary gland can itself result in a marked reduction in salt-gland secretion in response to a subsequent salt-load. Thus sham operation for adenohypophysectomy or neurohypophysectomy in ducks can inhibit secretion to a variable extent (Wright *et al.*, 1966; Wright, Phillips, Peaker & Peaker, 1967; Phillips & Ensor, 1972). Subsequent work has shown that variable damage to the pituitary or brain is not involved and that, in all probability, damage to a nerve or nerves in the track of the transoral (transbuccal) surgical approach to the gland is responsible (p. 62). Thus it is important to ensure that either sham operation has no effect in such experiments or that replacement therapy with pituitary hormones will completely restore any effect of removal of the pituitary or any of its parts. Holmes *et al.* (1972) have also noted that such damage can occur but have overcome the problem by using younger ducks, seven to eight weeks old, in which there is less ankylosis between the basisphenoid and basitemporal bones than in more mature birds; sham operation was then without effect.

The effects of ACTH on salt-gland secretion have already been dealt with and the only other adenohypophysial hormone that has been studied is prolactin.

Prolactin

Comparative endocrinologists have found that this ill-named hormone and its counterparts in other vertebrate groups can affect a wide variety of physiological processes in an equally wide variety of animals (see reviews by Bern & Nicoll, 1968 and Nicoll & Bern, 1972). Amongst these processes are ones in which prolactin is involved in osmoregulation or salt and water movements, for example, in teleost fish. Therefore Peaker *et al.* (1970) studied the effects of ovine prolactin on salt-gland secretion in the duck (see also Peaker & Phillips, 1969). They found that secretion in response to a minimal stimulatory

TABLE 8.2. *Effect of prolactin on nasal secretion by the minimally stimulated salt gland of the duck*

(Four ducks were infused intravenously with 10% NaCl at 1 ml/min until the first drop of nasal fluid appeared when 20 i.u. ovine prolactin were administered i.v. Number of observations in parentheses: NS = not significant. (From Peaker *et al.*, 1970.)

Min after appearance of first drop of nasal fluid	Mean nasal fluid output (drops/min) Control treatment (4)	Prolactin treatment (4)	S.E. (diff.)*	*P*
0–5	1.90	2.75	±0.015	< 0.001
5–10	2.25	3.00	±0.015	< 0.001
10–15	1.67	2.25	±0.534	NS
15–20	1.17	2.10	±0.653	NS
20–25	0.80	1.35	±0.820	NS

* Because of marked biological variation between individuals, differences between means were compared by the paired *t*-test, each animal acting as its own control.

salt-load was enhanced during the first ten minutes after a small amount (20 i.u.) was given intravenously (Table 8.2). In those birds that were still secreting after about 25 min, a second peak of secretion was observed in response to prolactin and this coincided with a rise in the concentration of glucose in blood. As in similar experiments where ACTH was injected (p. 148), this second peak could have been the result of an indirect action mediated by a rise in blood glucose. However the initial increase, not observed after ACTH, could not be attributed to any observed effect on blood composition and in view of its rapidity it was suggested that prolactin might exert a direct effect on the salt glands. Peaker *et al.* (1970) concluded that 'in view of the small dose of prolactin administered (approximately 1 mu/g body weight), a physiological role for prolactin in the normal control of extra-renal excretion in the duck is a possibility. The effect of prolactin on extra-renal excretion may be related to the change in habitat of the mallard (from which the domestic duck is descended) at the end of the breeding season. These birds breed and rear their young on fresh water and then tend to migrate to estuaries for the winter. A mechanism which would assist the duck to adapt immediately to salt water would be advantageous. If prolactin is associated with migration in the duck as it seems to be in *Zonotrichia* [the White-crowned Sparrow] (Meier, Farner & King, 1965), this hormone may predispose the duck to the sudden change from a fresh-water to an estuarine environment by enabling the previously inactive salt gland to start secreting at a high rate immediately the duck drinks sea water and thereby maintain the bird in temporary ionic equilibrium. Subsequently, continual ingestion of salt water may stimulate growth and proliferation of the nasal gland (Chapter 9) which may then become independent of prolactin unless prolactin itself be a factor in stimulating hypertrophy and hyperplasia of the nasal salt gland in the duck'.

Ensor & Phillips (1970*a*) have also suggested that prolactin may be released from the pituitary during adaptation to salt water. They have shown that in a number of vertebrates prolactin concentrations in the pituitary change in a manner that would suggest a release of this hormone during dehydration (i.e. lack of osmotically-free water). They showed a marked rise in pituitary prolactin content two to three days after ducks were first given 0.3 M sodium chloride to drink. This they interpreted as an indication of an increase in the rate of synthesis and release of the hormone by the anterior lobe of the pituitary. After this time the prolactin content declined to subnormal values but by then the ducks were not maintaining normal plasma sodium concentrations and they interpreted this to mean that release from the pituitary was in excess of the rate of synthesis (Fig. 8.8). Ensor & Phillips (1970*b*) have also found that there is a circadian rhythm in pituitary prolactin content which parallels changes in salt-gland secretion. Therefore they have suggested (Phillips & Ensor, 1972) that in marine birds 'prolactin might be secreted in excess amounts to offset periods of isolation from free water such as occurs on nesting sites and it has been suggested that it additionally might serve to "anticipate" periods of hypersalinity when adults and fledglings move from the dehydratory conditions of the nesting sites to the more saline conditions of the marine/estuarine feeding locations; in this sense the role of prolactin might be considered as a by-product of reproductive activities'.

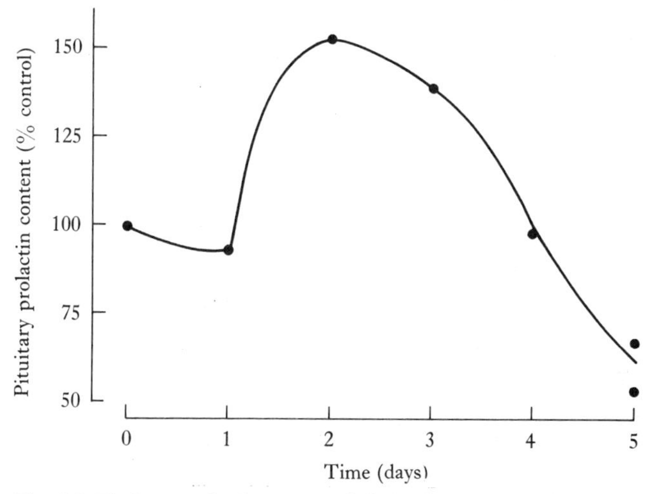

Fig. 8.8. Pituitary prolactin content of ducks given 0.3 M sodium chloride to drink beginning on day 0 (drawn from the data of Ensor & Phillips, 1970*a*).

In view of the fact that prolactin is essential for maintaining sodium balance in a number of euryhaline teleost fish in fresh water, Peaker (1971*c*) suggested that, since replacement experiments were not carried out by Wright *et al.* (1966) in their adenohypophysectomized ducks (see p. 145), the absence of prolactin and not ACTH might have been responsible for the inhibition of salt-gland secretion. Later, Phillips & Ensor (1972) and Ensor, Simons & Phillips (1973) reported that in fact this was the case since both mammalian and avian prolactin restored secretion in adenohypophysectomized birds. However there is a puzzling feature. Holmes *et al.* (1972) have found that mammalian ACTH is also effective in such birds and we have to consider whether mammalian ACTH is acting like bird prolactin or alternatively whether prolactin is acting like ACTH; in other words did prolactin restore the corticosterone concentration in plasma to normal? Recent evidence suggests that the administration of mammalian prolactin to intact ducks does not increase the plasma corticosterone concentration (Bradley & Holmes, 1972*b*) so we may have to conclude that secretion can occur normally in hypophysectomized ducks even though plasma corticosterone concentrations remain low. If this is the case then it must be inferred that the salt gland is affected secondarily and not primarily by the lack of adrenocorticosteroids. Ensor *et al.* (1973) found that following hypophysectomy food intake fell practically to zero, water intake also decreased and, not surprisingly, body-weight decreased. It was also noted that such birds tended to produce a very dilute urine. Since prolactin in intact ducks acts as an antidiuretic and, as well as restoring salt-gland secretory ability in hypophysectomized birds, also resulted in a return of food and water intake to normal, it would seem as these authors suggest, that the hypophysectomized bird is dehydrated. Ensor & Phillips (1972*b*) had previously shown that dehydration by water deprivation inhibits secretion (see p. 211) so it would appear that the secondary effect on secretion in hypophysectomized (and possibly adrenalectomized) birds is a general effect on fluid balance. However this does not mean that prolactin does not have actions of its own as we have indicated above. It remains to be seen how prolactin exerts its effect on secretion and to what extent pituitary prolactin levels reflect plasma concentrations of the hormone. Our own view is that prolactin might be a candidate for a role in the induction of the cellular changes which occur in the salt gland when birds are transferred from fresh water to salt water. Such a role would be compatible with that played by prolactin in some other tissues, for example the mammary gland, where a specific tissue has to develop allometrically.

Neurohypophysial hormones

Phillips, Holmes & Butler (1961) considered it likely that, as in mammals, antidiuretic hormone (ADH) is released after salt-loading. ADH would then act to reduce the renal loss of water thus ensuring that osmotically-free water gained by the action of the salt glands is not lost by the kidneys. If such is the case then it is not unreasonable to suppose that ADH, which affects ion and water movements in many other tissues, might also influence salt-gland secretion.

Attempts to study possible actions of ADH have been made in two ways: (1) the administration of ADH; (2) the effect of neurohypophysectomy on secretion. The first approach has been hampered by the lack of a readily-available supply of

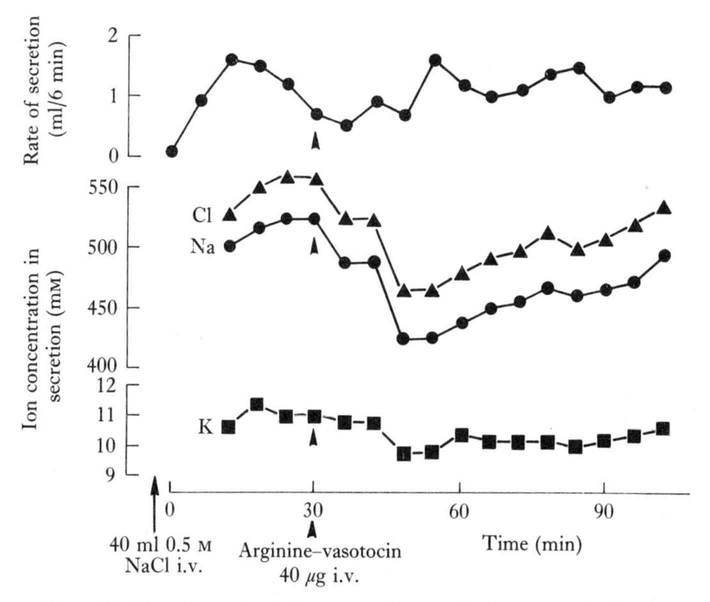

Fig. 8.9. The effect of arginine-vasotocin on salt-gland secretion in a goose.

arginine-vasotocin (AVT) – the antidiuretic hormone of birds, and most workers have had to use one of the mammalian vasopressins instead. However it is now known that the action of vasopressin does not, in all respects, resemble that of AVT. Thus Hughes (1962) and Gill & Burford (1969) found that vasopressin decreases the rate of salt-gland secretion; in addition, the latter authors noted a decrease in the sodium concentration of the nasal fluid. In this latter respect AVT has a similar effect but in addition actually enhances secretion in response to a salt-load in ducks and geese (Fig. 8.9). At high doses AVT is one of the few substances that can induce secretion in the absence of a salt-load, albeit at a very low rate (Peaker, 1971*b*). In three out of four birds given AVT but no salt there were no apparent changes in blood composition which would have indicated an osmotic stimulation of secretion and it was suggested that in view of these effects AVT might have a direct effect upon the secretory cells, or upon the nervous pathway responsible for secretion. This view was strengthened when it was found that a smaller dose of AVT was required to induce secretion when infusions were made into an indwelling catheter in a carotid artery.

While these experiments suggest the possibility that AVT could modify secretion by the stimulated salt glands, it should be made clear that the decrease in salt concentration obtained after the administration of AVT is hardly of adaptive value to a bird already short of free water. Clear evidence has also been obtained that the presence of the posterior lobe of the pituitary is not essential for secretion in ducks. Neither the rate nor the concentration of the secretion were significantly affected in neurohypophysectomized ducks when compared with sham-operated controls even though the birds were suffering from diabetes insipidus (Wright *et al.*, 1967).

Exogenous oxytocin has been found to have similar effects to those obtained with AVT. Thus the 'tocins', as opposed to the 'pressins', are active, and this implies that the presence of isoleucine rather than phenylalanine in the ring of amino acids of the octapeptide is the requirement for biological activity in this respect.

Phillips & Ensor (1972) have made the interesting suggestion that AVT might release prolactin from the anterior pituitary since pituitary prolactin levels are altered during dehydration and the administration of AVT results in a reduction in the amount of prolactin in the pituitary. There is therefore the possibility that some of the effects of AVT on salt-gland secretion might be mediated by prolactin.

Thyroid hormones

Although removal of the thyroid gland altered the pattern of salt-gland secretion in ducks given an oral salt-load, no effect on secretion was apparent after an intravenous load. It was suggested that no direct action on the salt glands could be claimed but that the intestinal absorption of salt was modified by thyroidectomy (Ensor *et al.* 1970).

Future work

It is clear that the methods used to study the effects of hormones on salt-gland secretion have not been capable of yielding unequivocal results. In addition it is unfortunate that, in some cases, critical simple experiments have not been carried out. For example, if a direct effect of adrenocorticosteroids is claimed, is it not vital to investigate whether or not parasympathomimetics will elicit secretion in adrenalectomized or adrenal-insufficient birds? In terms of techniques an in-vitro system to study both the effects and uptake of hormones would seem desirable, and it might be worthwhile to try tissue slices and perfusion of the isolated decerebrate head for short-term experiments and organ culture in the presence of hormones for long-term studies of morphogenesis.

Although it remains to be convincingly demonstrated that hormones affect the salt glands directly, this does not mean to say that this does not occur nor that endocrine adjustments are not associated with the changes in a number of systems; for example intestinal absorption, kidney function and energy metabolism that occur during adaptation to hypersalinity of the environment. Reviews, apart from those mentioned above, dealing with the effects of hormones on the salt gland are: Phillips & Bellamy (1963); Holmes & Phillips (1965); Phillips (1968); Holmes, Chan, Bradley & Stainer (1970); Holmes (1972); Chester Jones *et al.* (1972).

9 ADAPTATION OF THE GLAND

Although the salt glands begin to secrete promptly when a bird ingests salt water for the first time, other processes are initiated within the glands which, in time, lead to an overall increase in the efficiency of their secretory ability. This process of adaptation occurs in a saline environment but the reverse change or 'de-adaptation' is also seen if a marine bird is given fresh water to drink. It is these changes in the state of the salt gland and factors controlling them that concern us in this chapter. In view of the interest in the mechanisms by which functional hypertrophy and hyperplasia in many organs are brought about, these considerations may be of wider interest.

The physiological, biochemical and structural changes that occur during adaptation have been studied in two ways. Firstly, by comparing the glands of birds living on either fresh water or salt water, a number of workers have made useful observations on the final results of the adaptive process. Secondly, some groups have studied the time-course of such changes, and this approach has provided important information on the rate and sequence of the various phases of adaptation and de-adaptation. In the latter type of study two rather different experimental regimes have been employed. In the first type, ducks reared on fresh water have been transferred to hypertonic salt water, and then in some cases, back again to fresh water. In the second, the salinity of the drinking water has been changed a few days after hatching so that the adaptive process is superimposed on the normal growth and development of the gland in the young bird. Essentially similar results have emerged from both approaches but we feel that the former, which involves the study of the transition from one steady-state condition to another, is easier to interpret. In the wild, of course, some birds live at sea whilst others, having spent some time on fresh water, fly to estuaries or the coast at certain times of their lives.

Under laboratory conditions adaptation to salt water has been achieved in several ways. As we point out in Chapter 11, the salt tolerance of a species is limited and in order to follow the process of adaptation in ducklings, Ellis *et al.* (1963) employed the following regime starting four days after hatching. Sodium chloride solutions of 0.17 or 0.21 M were offered for twelve hours each day starting at 09.00 hr, and fresh water for the remaining twelve hours. This procedure, followed in the group's subsequent work, was found to result in adaptation without exceeding the salt tolerance of young ducklings. In contrast, Holmes and his co-workers have given seven to nine week old birds only hypertonic saline (284 mM sodium, 6 mM potassium) to drink, without giving periods of relief. Other methods have included regular administration of hypertonic sodium chloride by stomach tube while fresh drinking water was also available (Ballantyne & Wood, 1969), and slow acclimation to increasing concentrations of sodium chloride (Schmidt-Nielsen & Kim, 1964).

Changes in size and secretory ability of the glands

Changes in the size of the nasal glands related to the salinity of the drinking water were observed before the function of the glands was discovered. Heinroth & Heinroth (1928) noticed that the size of the glands decreased when marine Eider Ducks (*Somateria mollissima*) in the Berlin Zoological Gardens were given only fresh water. Later Schildmacher (1932) found that in the domestic duck the glands increased in size when salt water was supplied for drinking and decreased again when fresh water was available. Since that time a number of workers have confirmed this effect in gulls as well as in ducks (Holmes, Butler & Phillips, 1961; McFarland, 1963*a*; Ellis *et al.*, 1963; Benson & Phillips, 1964; Schmidt-Nielsen & Kim, 1964; Fletcher *et al.*, 1967; Holmes & Stewart, 1968; Ballantyne & Wood, 1969; Komnick & Kniprath, 1970). We have reported in Chapter 11, that salt-gland weight is related to body-weight in marine birds. In ducks drinking salt water the points fall on or close to the regression line describing all species. However when drinking

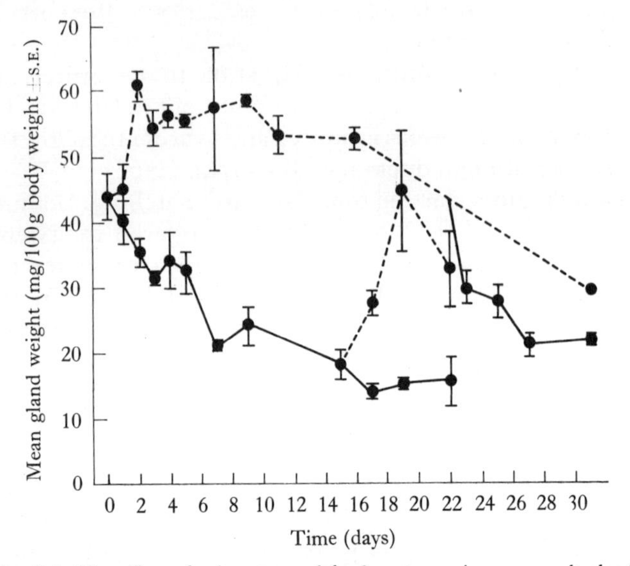

Fig. 9.1. The effect of salt-water and fresh-water regimens on salt-gland weight in ducklings. Mean±S.E. Day 0 = 4 days after hatching. ●—●, birds on fresh water; ●- - -●, birds on salt water (0.12 M sodium chloride to drink for 12 hr each day). Note the effect of changing the salinity (from Ernst *et al.*, 1967).

fresh water the points lie substantially below, i.e. the glands are smaller than would be expected. Similarly in young ducklings growth of the salt glands is accelerated by giving only salt water to drink. (Fig. 9.1) (Ellis *et al.*, 1963; Ernst, Goertemiller & Ellis, 1967; Levine *et al.*, 1972).

Following the transfer of ducks to hypertonic saline Fletcher *et al.* (1967) and Holmes & Stewart (1968) found an 80 per cent increase in the weight of the glands during the first day of adaptation and by five days the weight was approximately doubled but then continued to rise only slowly (Fig. 9.2). Since these workers also found that the water content of the glands remained almost constant during this period they concluded that the increase in weight 'must have reflected an accumulation of synthesized materials and did not simply result from fluid imbibition'. Thus it would appear that synthesis of these materials is at its maximum during the first day of adaptation. After 28 days on salt water, the ducks were returned to fresh water and salt-gland weight then decreased within three days but to a level which was higher than that obtained before adaptation to salt. Twenty-two days later when the birds were returned to salt water, salt-gland weight again increased (Fig. 9.2). A similar conclusion, that the main increase in the formation of dry matter occurs between one and 24 hours, was reached by Ballantyne & Wood (1969; 1970*a*) and clearly this remarkable phenomenon is worthy of study.

In addition to the increase in the weight of the glands adaptation also affects both the rate of secretion and its composition. This evidence has been discussed in Chapter 7 and in this section we must consider the time-course and nature of these changes. Fletcher *et al.* (1967) studied them by infusing hypertonic sodium chloride intravenously at intervals after the ducks had been transferred to salt water, and then recording the maximum rate of secretion and the ionic composition of the nasal fluid in response to the additional load. The effect of acclimation was to increase the maximum rate of secretion in these birds by a factor of approximately six. In addition the concentrations of sodium, chloride, and to a lesser extent, potassium, increased so that the overall effect was a great increase in the efficiency of extra-renal excretion (see Fig. 7.2). The increases in the rate of secretion of fluid, sodium and chloride were greater than the increase in the weight of the glands so that the amount secreted per unit weight of gland was also markedly increased by more than three-fold (Fig. 7.2). This increase in maximum secretory ability per unit weight of glandular tissue occurred mainly in the first five days of adaptation to salt water but continued to rise at a lower rate for the following five days (Fig. 9.2). When the birds were returned to fresh water these changes were reversed (Fig. 7.2).

It should be noted that Goertemiller & Ellis (1962) and Ellis *et al.* (1963) failed to find any tendency for the concentrations of sodium or potassium to change when one-day old ducklings were given salt water for 12 hours every day for 13 days. However they did not measure either how much salt water was drunk or the rate of secretion at the time of sampling, which it is now known affects concentration (Chapter 7).

An interesting series of experiments, so far only published as an abstract, have shown that the development of the gland can be accelerated in the embryo. Hally *et al.* (1966) have shown

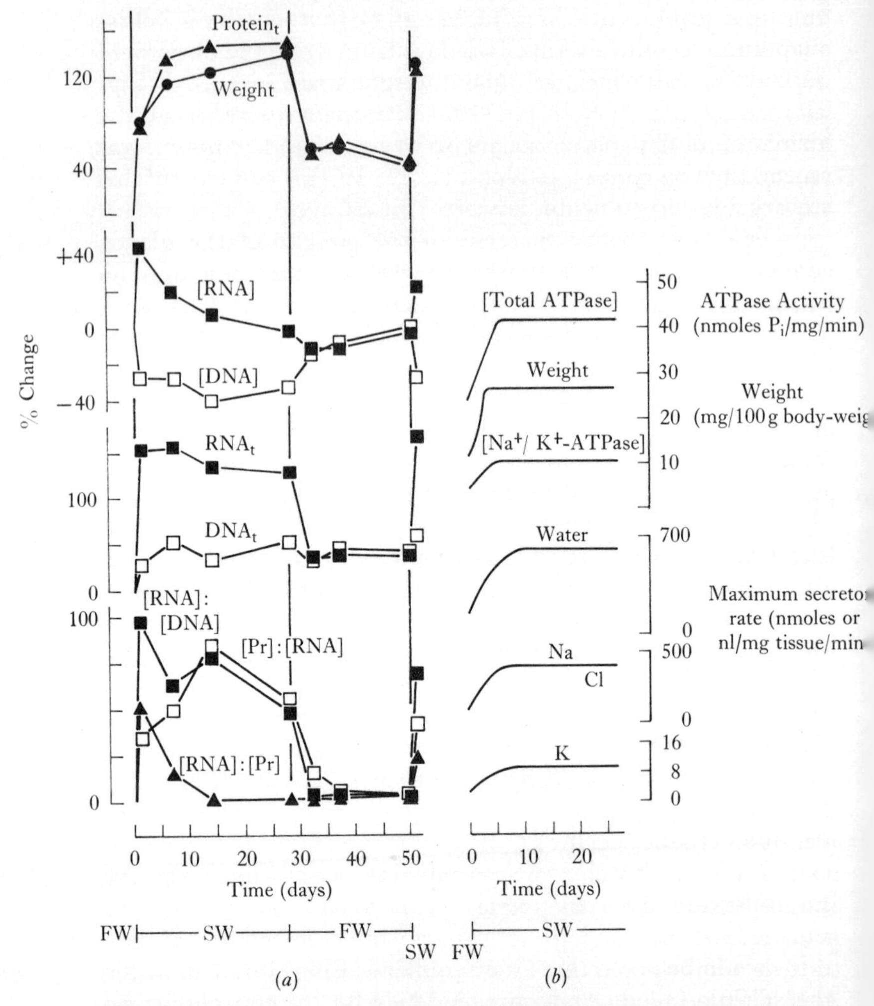

Fig. 9.2. (*a*) Changes in weight, protein, RNA and DNA content of duck salt glands during adaptation to salt water, 'de-adaptation' and 'readaptation'. Mean values shown and expressed as a percentage change from the corresponding value for birds on fresh water only (redrawn from Holmes & Stewart, 1968).

(*b*) Changes in weight, maximum secretory ability and ATPases in duck salt glands during adaptation to salt water. The line which best describes the mean changes is shown, and the time-scale has been drawn as in (*a*) (modified and redrawn from Fletcher *et al.*, 1967).

In both series of experiments the hypertonic drinking water contained 284 mmoles/l sodium, and 6 mmoles/l potassium. Concentrations are shown in square brackets, total gland content with subscript t. SW = salt water, FW = fresh water.

that the ability to secrete first develops three days before hatching (in ducks after 28 days incubation); Lunn & Hally (1967) therefore injected hypertonic sodium chloride into the albumen sac of duck eggs on the fourteen or twentieth day of incubation. The secretory capacity was then determined by injecting hypertonic sodium chloride intravenously into the embryo on day 24. In both groups of birds given sodium chloride previously, the rate of secretion (1.41 ± 0.41 S.E. μmoles Cl/kg embryo/min for the 14 day and 1.48 ± 0.26 for the 20 day) was significantly higher than that of a control group (0.65 ± 0.12). These results clearly indicate that 'a prolonged functional demand accelerates the development of the prefunctional salt gland'. The mechanism by which this accelerated growth is mediated at this very early stage, when, according to Marples (1932) and Zaks & Sokolova (1961), the degree of development of the gland is very meagre (p. 27) is unknown. Another question is whether the embryo is normally 'informed' that it will be hatched in marine or fresh-water conditions so that the gland is prepared accordingly. It could be argued that the mother could include some 'information' on the salinity of the environment in the egg at the time of its formation. In its most simple form this information could be the concentration of sodium chloride in the fluids of the egg, which unfortunately has not been measured. This suggestion is of course speculation but we should point out that Zaks & Sokolova (1961) found that in 1–2 day old Herring Gulls (*Larus argentatus*) from a marine colony the sodium concentration and the rate of secretion were already as high as in older birds; this might suggest some degree of pre-adaptation in the egg.

It is with this background on the gross changes occurring during adaptation that we consider the cellular changes within the salt glands which must ultimately be responsible for the altered efficiency of the secretory process.

Nucleic acids, proteins, cell size and turnover

In this section we must draw a distinction between the total amount of substances in the two glands of a bird (corrected for body-weight if necessary) and its concentration per unit weight of glandular tissue. For the former we are using the subscript t, e.g. RNA_t, and for the latter square brackets to denote concentration, e.g. [RNA].

In the experiments of Holmes & Stewart (1968) it is quite clear that the first signs of a change in the gland were an increase in [RNA] and a decrease in [DNA] during the first 24 hours on salt water. Since the weight of the gland also increased this was reflected in an increased RNA_t, and since the increase in weight was greater than the decrease in [DNA], DNA_t was also raised. RNA_t, DNA_t, and $protein_t$ (which paralleled the increase in weight) remained high while the birds were kept on salt water. Since [DNA] remained lower than the level in birds on fresh water it can be inferred that the size of the cells increased but since DNA_t increased it must also be inferred that the total number of cells in the gland also increased. In other words this evidence (Fig. 9.2) would suggest that the increase in the weight of the glands is due to both hypertrophy and hyperplasia. Holmes & Stewart (1968) calculated from their data that the number of cells increased by 42 per cent and the volume of each cell by approximately 60 per cent. The latter figure was calculated from the gland volume per unit of DNA and could be an overestimate because the volume of the extracellular space in the gland could also have been increased, but must be substantially correct.

The increases in RNA_t and [RNA] were rapid; [RNA] reached a maximum in 24 hours, and then declined to reach the pre-adaptation level after 25 days on salt water. As in other biological systems the maximum [RNA] coincided with the maximum rate of protein synthesis, i.e. during the first 24 hours, although growth continued at a slower rate when [RNA] was declining.

It is clear that in terms of current concepts of genetic expression, RNA synthesis, in some way triggered by the ingestion of salt water, leads to an increased rate of protein synthesis, changes in enzyme levels (see below), ultrastructure (see below) and hypertrophy of existing cells. This evidence would also suggest that either the rate of cell division increases or the rate of cell breakdown decreases, the overall effect being an increase in the number of cells.

The view that both hypertrophy and hyperplasia occur during adaptation is supported by the cytological findings of

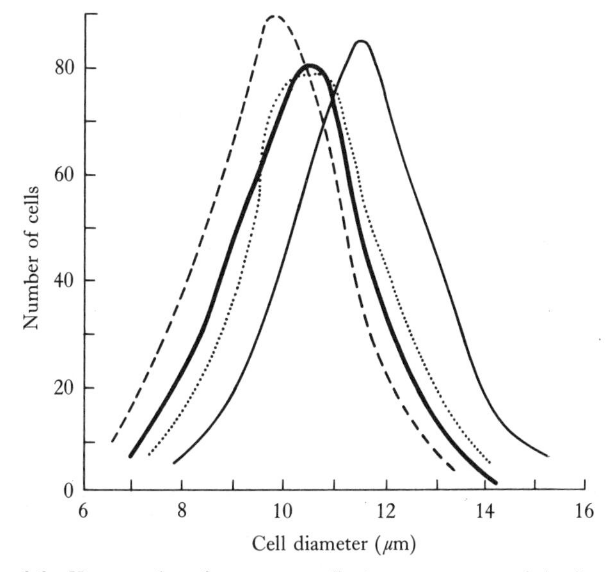

Fig. 9.3. Hypertrophy of secretory cells in response to salt-loading in ducks. Frequency distribution curves show number of cells of a given diameter after 1 hr (– – –), 1 day (· · ·), 4 days (—); fresh water controls (—) (from Ballantyne & Wood, 1969).

Ellis *et al.* (1963). They observed an increase in the size of the secretory cells and also in the number of cells in a tubule, and argued that the rate of mitosis must increase to account for this. Ellis (1965) showed that the site of cell division in the tubule is in the zone of cells lying at the periphery of a lobe i.e. at the blind end of the tubules (p. 15) and therefore it can be argued that any increase in mitotic activity would be expected to occur there. However, there are dissenters from the view that there is hyperplasia as well as hypertrophy. Although Benson & Phillips (1964) and Ballantyne & Wood (1969) found that the size of the secretory cells increases (Fig. 9.3) the latter authors could find no increase in the number of cells in the gland as measured by DNA_t, even after four days on a salt-water regime, or an increase in the incidence of mitosis; they did however detect increases in gland weight, RNA_t, and [RNA], the latter by both chemical and histochemical methods.

In view of the difference of opinion on whether hyperplasia occurs during adaptation more work must be done. It would seem possible or even logical that the number of cells in the gland would increase, but the rapidity with which DNA increased in the experiments of Holmes & Stewart (1968) would necessitate a very great and sudden increase in the rate of mitosis. In comparison Ellis (1965) showed by labelling with tritiated thymidine that the turnover of cells in a secretory tubule is rather slow (six weeks) even in birds on salt water (see p. 16). Nevertheless these doubts do not mean to say that the effect of salt-water ingestion on DNA_t was not a reflection of a genuine increase in the number of cells. Holmes & Stewart (1968) discussed the possibility that the marked rise in the first day on salt water could be due to the presence of a larger number of erythrocytes which, in birds, are nucleated. There is a striking increase in blood flow during secretion and the blood content of the glands increases (Chapter 5). However DNA_t did not decrease when the birds were moved back to fresh water after 28 days and salt-gland secretory activity ceased, and this finding would certainly argue against the whole of the increase in DNA_t being due to an increased vascularity (Holmes & Stewart, 1968).

Whether or not there is hyperplasia, it is still clear that existing cells hypertrophy and, as will become apparent, undergo considerable differentiation when a bird first ingests salt water. It seems that the secretory cells are in a relatively unspecialized state in birds on fresh water and saline conditions either induce or permit full differentiation. Ellis *et al.* (1963) considered that the genome of the domestic duck contains the necessary information for the development of full secretory activity but full expression of the genetic potential is realized only under saline conditions.

The [RNA] in resting and active glands is relatively low, although similar to that of tissues like the kidney which are highly active but which do not synthesize proteins for secretion. In addition to the rise in [RNA] during the first stages of adaptation to salt water, Stewart & Holmes (1970) found that the proportion increased markedly during the first day and remained high during the 14 days these ducks were given hypertonic saline to drink. Like the rise in [RNA] this aggregation preceded the development of full secretory capacity, and Stewart & Holmes (1970) suggested that 'such an accumulation [of larger sized ribosomal aggregates] is indicative of increased m[messenger]RNA synthesis and perhaps increased mRNA stability'. In addition to the change in aggregation of free ribosomes, the proportion of 'bound' ribosomes in the tissue increased as the glands increased in size, but the site of this binding was not determined.

During 'de-adaptation' i.e. when the birds on salt water were returned to fresh water, Holmes & Stewart (1968) found that most but not all the changes associated with adaptation were reversed (Fig. 9.2). The number of aggregated ribosomes decreased, as did [RNA], RNA_t, $protein_t$ and weight; [DNA] increased but, as noted above, DNA_t did not change. As was mentioned in the preceding section, secretory ability again declined. When the time-course of some of these changes was followed by Holmes & Stewart (1968) they found that the changes in weight, $protein_t$ and RNA_t occurred within four days and then levelled off at values somewhat higher than those obtained in birds on fresh water. This, together with evidence from histology, electron microscopy and enzyme levels (see later) indicates that the cells decrease in size and dedifferentiate, and that continuous, or at least continual ingestion of salt water is necessary to keep the cells in their maximum state of development. This rapid dedifferentiation when the stimulus is withdrawn might also suggest that adaptation involves a process of induction rather than the removal of an inhibition on the full expression of the genome: if the latter was the case it would seem likely that once complete differentiation had been initiated, the cells would remain in that condition until replaced by new cells, regardless of subsequent changes in the salinity of the drinking water.

Enzymes and other substances

Adenosinetriphosphatases

In Chapter 6 we discussed the presence, site and role of ion-activated ATPases in relation to the secretory mechanism and stressed the difficulties encountered in trying to extrapolate from ATP hydrolysis measured *in vitro* to the energy requirements for active ion transport *in vivo.* Although these qualms on the quantitative aspects of ATPase measurements exist there is no doubt that the study of these enzymes during adaptation to salt water has provided interesting and important information on the processes occurring in the gland.

Apart from earlier workers who compared Na^+/K^+-ATPase activities in the glands of birds living on salt water or fresh water (p. 101), the first investigators to study the time-course of the changes were Ernst *et al.* (1967). They found increases in both Mg^{++}- and Na^+/K^+-activated ATPases when ducklings were given food containing one per cent sodium chloride and 0.17 M sodium chloride to drink for 12 hours each day starting three days after hatching. No such increases were observed in ducklings given fresh water throughout. The changes in ATPase levels were evident two days after the start of the experiment, and they continued to rise, reaching a maximum at nine days. At this stage, when the plateau was reached, Mg^{++}-ATPase had increased by a factor of 2.5 and Na^+/K^+-ATPase by 5.5. When the birds were returned to fresh water Na^+/K^+-ATPase levels decreased exponentially with a $t_{1/2}$ of five days (Fig. 9.4). In contrast Mg^{++}-ATPase declined only slowly for the first five days but then more rapidly.

In addition to measuring ATPases in homogenates, Fletcher *et al.* (1967) also assessed the changes in secretory ability of the glands during adaptation and 'de-adaptation'. It should be emphasized that they used two enzyme preparations which gave different quantitative results; one had six times the activity of the other in terms of Na^+/K^+-ATPase. However it is clear that the changes in activity they observed do indicate relative levels, and it is evident from Fig. 9.2 that the increase in both 'total' and Na^+/K^+-ATPase (i.e. ouabain-sensitive) paralleled the changes in secretory activity, but reached maximum levels later than the gland weight. An even greater lag between maximum weight and maximum ATPase levels was found by Ernst *et al.* (1967) in ducklings (Figs. 9.1 and 9.4).

Using an 'improved' enzyme preparation, i.e. one which yielded more inorganic phosphate from ATP, Fletcher *et al.* (1967) found that after 30 days on salt-water, 'total' ATPase was increased approximately three-fold, Mg^{++}-ATPase two-

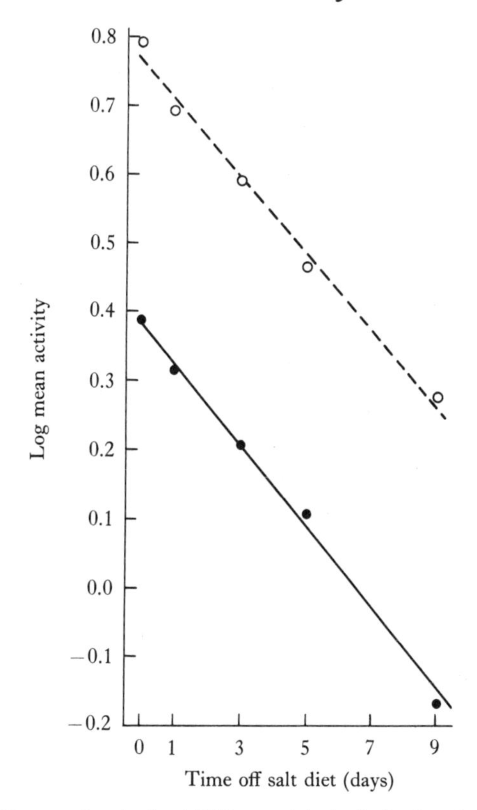

Fig. 9.4. Decrease in salt-gland ATPase activity in ducks after changing from salt water to fresh water: •, log of mean Na^+/K^+-ATPase activity (mmoles P_i per ml homogenate per hr); o, log of mean specific activity (mmoles P_i per mg protein per hr) (from Ernst *et al.*, 1967).

fold, and Na^+/K^+-ATPase four-fold; these changes were reversed when the birds had been back on fresh water for 21 days (Table 9.1). The relative increase in Na^+/K^+-ATPase was almost identical to the increase in sodium secretory ability (Fig. 7.2), so that under all conditions of salinity, the amount of sodium transported per ATP utilized was approximately the same (Table 9.1), about four moles of sodium being transported per mole of ATP hydrolysed.

Ballantyne & Wood (1970*b*) observed a similar pattern although their absolute ATPase activities and the percentage of Na^+/K^+-activated activity were very much lower than those of

TABLE 9.1. *Moles of sodium transported per mole of ATP hydrolysed in ducks maintained on various fresh-water and/or saline regimes* (from Fletcher *et al.*, 1967)

(The ratios were derived from the individual values of the sodium excretory rates and the rates of ATP hydrolysis. All values are expressed as means ±S.E.)

Group	No. of birds	Na transport (moles) per ATP hydrolysed (moles) Total	Ouabain insensitive	Ouabain sensitive
30 days fresh water	5	1.50 ±0.26	2.35 ±0.49	4.48 ±0.65
30 days saline	5	1.93 ±0.16	3.77 ±0.43	4.03 ±0.26
51 days fresh water	5	1.59 ±0.22	2.78 ±0.37	4.17 ±0.74
30 days saline+21 days fresh water	5	2.21 ±0.26	3.93* ±0.18	5.45 ±0.82
58 days fresh water	5	2.41 ±0.31	5.18 ±0.91	4.66 ±0.37
30 days saline+21 days fresh water+7 days saline	6	2.21 ±0.15	4.38 ±0.66	4.70 ±0.16

* $P < 0.05$ with respect to corresponding value for birds maintained on fresh water.

other workers. They detected a rise in 'total' and ouabain-sensitive activity even one hour after salt-loading and by four days, the Na^+/K^+ activity had more than doubled. In contrast to other workers, an even greater rise in 'total' activity was obtained. In addition to these chemical determinations Ballantyne & Wood (1970*b*) also demonstrated that the increase could be detected by the Wachstein-Meisel histochemical technique; this aspect of their work is considered in the chapter on the secretory mechanism (p. 102).

Sulphatides and other lipids

It has been suggested that sulphatides, anionic sphingolipids sulphated in the galactose moiety, could act as a carrier for cations and thus play a role in ion transport. Karlsson,

Samuelsson & Steen (1969; 1971) found that when ducks were given salt water to drink, the relative increase in the concentration of sulphatides in the gland was 2.9, which was almost identical to the relative increase of 2.8 for Na^+/K^+-ATPase obtained in the same experiments. None of the phospholipid components studied increased to this extent and in view of the correlation between the presence of sulphatides and ion transport in a number of tissues, these authors consider that these substances may well have a role in active ion transport.

Glycolytic enzymes

Stainer *et al.* (1970) have measured the activity of a number of glycolytic enzymes during the adaptation of ducks to 200 mM sodium chloride drinking water. When expressed per mg tissue protein, the concentrations of the following enzymes were found to increase over a 14-day period: glutamic-oxaloacetic-transaminase (GOT), phosphofructokinase (PFK), hexokinase (HK) and glucose-6-phosphate dehydrogenase (G-6-PDH). However the time-course of the changes showed that the main rise in HK and GOT levels occurred at 14 days while PFK and G-6-PDH were increased after four, but not two, days on salt water (Fig. 9.5). It has been suggested that PFK may well be the rate-limiting enzyme for glycolysis and the marked increase early in adaptation supports this view. The increase in PFK was similar to that of Na^+/K^+-ATPase and sodium-secreting ability, indicating, as the authors suggest, that the ATPase is at least partly, if not wholly, dependent on glycolysis and associated cycles for ATP.

Histochemical evidence also shows that enzymes associated with energy metabolism increase during adaptation, namely, succinic dehydrogenase and cytochrome oxidase (Natochin & Krestinskaya, 1961; Ellis *et al.* 1963; Spannhof & Jürss, 1967). All these findings are to be expected since anatomical studies clearly show an increased number of mitochondria in the secretory cells of birds kept on salt water.

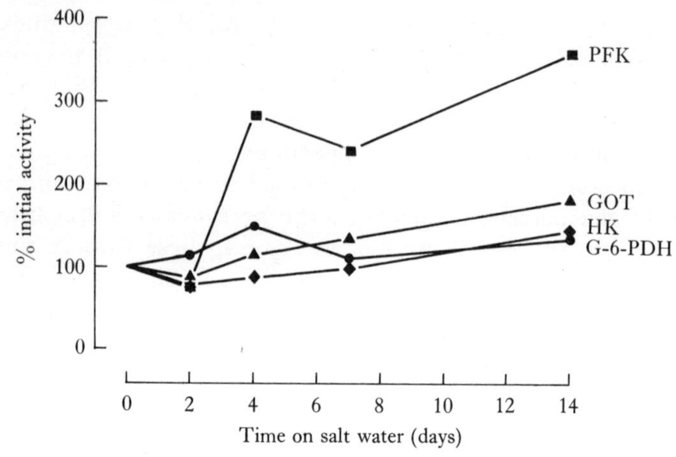

Fig. 9.5. Activities of four enzymes in salt gland homogenates showing a significant change during adaptation of ducks to salt water (0.2 M sodium chloride). The activities (per mg protein) are expressed as a percentage of that obtaining in ducks on fresh water. G-6-PDH, glucose-6-phosphate dehydrogenase; HK, hexokinase; GOT, glutamic oxaloacetic transaminase; PFK, phosphofructokinase. The activities of isocitric dehydrogenase, malic enzyme and lactic dehydrogenase were not significantly altered and are not shown (calculated from the mean data of Stainer *et al.*, 1970).

Carbonic anhydrase

Although Bonting *et al.* (1964) showed that the salt gland of the Herring Gull (*Larus argentatus*) has a high level of carbonic anhydrase, which was not changed when birds from marine conditions were given fresh water for seven weeks, Spannhof & Jürss (1967) showed that the appearance of carbonic anhydrase in the gland (as observed by a histochemical method) was accelerated when young Common Gulls (*Larus canus*) were given salt water. In birds maintained on fresh water the adult distribution of the enzyme was reached at 22 days of age whereas in those given salt the same distribution was reached in 16 days.

β-glucuronidase

We have already discussed the occurrence of β-glucuronidase in the salt gland and the role suggested for it by Ballantyne and his co-workers (p. 142). Wood & Ballantyne (1968) have shown, using a chemical method, that the concentration of enzyme in the glands of ducks increases by 52 per cent after salt-loading, and by 87 per cent after four days on a salt-water regime; the increased levels were observed histochemically in the secretory cells.

Butyrylcholinesterase

As we have pointed out on page 37, butyrylcholinesterase is not confined to nervous tissues in the salt gland but is also present in secretory cells. There have been several suggestions that this enzyme may have some role in ion transport and Ballantyne & Fourman (1967) and Fourman (1969) showed that there is an increase in its concentration in the salt gland of ducks given salt water for several days. As might be expected, such an increase was not apparent after an acute salt-load (Smith *et al.*, 1971 *a, b*). It is possible, as these authors point out, that the enzyme might be associated with the basal cell membrane and that the marked increase in the quantity of membrane per cell, which occurs during adaptation, may be responsible for the rise in butyrylcholinesterase activity observed. Its role however remains unknown.

Histology and ultrastructure

The early reports on the histology of salt glands in ducks during adaptation (e.g. Ellis, DeLellis & Kablotsky, 1962; Ellis *et al.*, 1963; Benson & Phillips, 1964) which suggested hypertrophy of the secretory cells, and increases in cell membrane and in the number of mitochondria, were followed by an excellent electron microscopical study (Ernst & Ellis, 1969; see also Ellis & Ernst, 1967). It will be recalled that several authors had shown that the basal and lateral membrane of the secretory cells from marine birds are extensively infolded and that mitochondria are abundant (p. 22). One might therefore have expected that the main cytological changes during adaptation would involve chiefly these features. Ernst & Ellis (1969) found this to be the case.

The regime of adaptation Ernst & Ellis (1969) employed was similar to that used in their other studies (0.17 M sodium chloride available from 10.00 hours to 22.00 hours, fresh water for the remainder of each day), and this was begun four days after hatching i.e. day 0 of adaptation. In birds on fresh water and at all stages of adaptation, a zone of peripheral cells, which is thought to be the site of cell division (p. 15), was observed at the blind end of the tubules. In ducklings kept on fresh water only one type of cell – partially specialized – was present in the remainder of the tubule. This type has short projections along the lateral membrane but the basal membrane is almost flat and the mitochondria are no more abundant than in peripheral cells. In growing ducklings kept on salt water, the tubules elongated and the partially-specialized cells were only present towards the periphery of the lobe where they formed an ill-defined zone. Towards the centre of the lobe, fully-specialized cells were present in two stages of development. In early adaptation (day 2), the stage I type was present. They have more extensive lateral infolding, basal infolding and more mitochondria. By day 11, both stages of the fully-specialized type were present, stage II as a zone adjacent to the central canal. In these the basal and lateral infoldings are even more extensive and many more mitochondria are present, particularly in the basal region (Plate 7.1). At this stage therefore the cells resemble those of truly marine adult birds. Only slight modification of the apical cell membrane was apparent during adaptation, the fully-specialized type of cell having more microvilli than those less-fully developed.

Ernst & Ellis (1969) have argued that as the cells migrate along the tubule from the periphery of the lobe to the centre they become progressively more specialized when salt water is being ingested and the glands are active. The time-course of the changes indicate that although the cells grow early in adaptation, the full development of Na^+/K^+-ATPase and (extrapolating from the work of Fletcher *et al.* (1967) in older birds) secretory ability is only realized when the fully-specialized stage II cells are developed. Thus it would appear that these cells are more capable of pumping sodium at a high concentration than those towards the periphery of the lobe – an aspect considered in the chapter on the determination of the concentration of nasal fluid (p. 125).

Although these studies were carried out in young ducklings, similar changes probably occur when older birds are first given

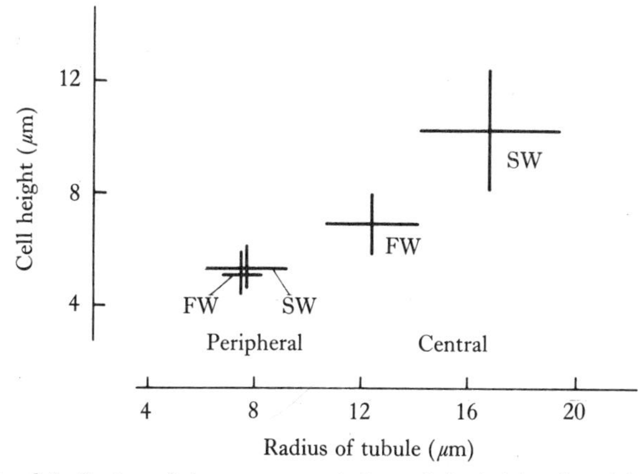

Fig. 9.6. Radius of the secretory tubules and the height of peripheral and central cells in Herring Gulls (*Larus argentatus*) kept on salt water (SW) or fresh water (FW) (mean±S.E.) (redrawn from Komnick & Kniprath, 1970).

salt water to drink. For example, Holmes *et al.* (1969) have published electron micrographs by Cronshaw & Holmes and these show that the degree of basal infolding is greater in ducks on salt water. Although adult geese which have never been given salt water have some fully-specialized stage I cells in the tubule near the central canal (M. Peaker & N. Ackland, unpublished), there is no reason to doubt that during adaptation further development occurs in this species also.

Komnick & Kniprath (1970) have recently applied morphometric methods to their micrographs of the salt gland of the Herring Gull (*Larus argentatus*) given salt water or fresh water to drink for two months. Their results showed clearly that the size of the more central secretory cells is greater in salt-adapted birds (Fig. 9.6). With salt water, the volume of the intercellular space between the lateral membranes, the number of mitochondria and the surface area of the basal and lateral cell membranes are all increased by a factor of approximately two, compared with birds kept on fresh-water (Table 9.2). In salt-adapted birds it was calculated that mitochondria occupy approximately 50 per cent of the volume of the cytoplasm and the surface area of the baso-lateral membrane is about 1000 times larger than that of the apical membrane.

TABLE 9.2. *Morphometric estimations of the intercellular volume and proportion of cytoplasm occupied by mitochondria in Herring Gulls* (Larus argentatus) *given either fresh water or salt water to drink for two months* (modified from Komnick & Kniprath, 1970)

	Intercellular volume as % epithelial volume	Volume of mitochondria as % cytoplasmic volume
Salt water	22.2	49.6
Fresh water	9.8	26.3

No work appears to have been done on the ultrastructure of the duct system during adaptation.

Cell membranes

The site of assembly of cell membrane constituents in tissues is of fundamental interest, and Levine *et al.* (1972) realized that a study of the salt glands during adaptation, when a great increase in the formation of basal and lateral membranes occurs, could provide important evidence on this problem. Acyltransferases are involved in membrane phospholipid synthesis, catalysing the transfer of a long-chain fatty acid from acyl-CoA to α-glycerophosphate; this reaction releases free CoA. The activity of acyltransferase per unit weight of gland increased markedly when ducklings were given salt water to drink, even after one day. The rise continued for six days although at a lower rate, and then declined. The total activity per gland however remained constant after this time which suggests that the amount per cell also remained unchanged. These workers have also developed a histochemical technique to detect the site at which this occurs so that the reaction product is indicative of acyltransferase activity.

Acyltransferases were found in the Golgi apparatus, suggesting that this is the site of at least the first stages in membrane lipid formation. The Golgi apparatus was most prominent in the partially-specialized secretory cells of birds drinking sea water and of course it is these cells, which also showed the greatest acyltransferase activity. Levine *et al.* (1972) considered that the membrane-bound vesicles, which Komnick (1965) has also described, serve to transport new membrane to the baso-lateral region.

It is relevant that in the mammary gland, where apical cell membrane is continually lost as fat globules are pinched off covered by cell membrane, that the apical membrane is replenished as vesicles, derived from the Golgi apparatus and containing casein, fuse with the apical plasmalemma and eject their contents (see Linzell & Peaker, 1971). For the salt gland, Levine *et al.* (1972) concluded, 'Proteins are synthesized in the ribosomes, either free or bound, and are moved to the Golgi apparatus, where the phospholipid is synthesized. The protein and phospholipid are assembled to produce either plasma membranes or precursors of the plasma membrane. These move in the form of vesicles from the Golgi region to fuse with the existing plasma membrane'.

The site of ion transport, ion-activated ATPases and other enzymes associated with cell membranes are considered in Chapter 6.

Control of adaptive changes in the salt gland

The work discussed above provides little information beyond the basic fact that environmentally-induced hypertrophy of the salt gland involves a complex series of intracellular processes. This does not mean that information on the phenomenon is not valuable, indeed an accurate and quantitative description of a phenomenon normally precedes knowledge of the controlling mechanism. Now that the 'natural history' of adaptive hypertrophy has been so well documented by the two groups of workers in the USA, perhaps the time is ripe for interesting investigations to be made on the physiological control of the process.

The rise in [RNA] early in adaptation can be taken to indicate an increase in transcription, but to consider the biochemical mechanisms by which this is brought about in the nucleus of the secretory cell would involve delving deeply into the realms of molecular biology without the support of any experimental evidence. However it is clear that in order that further differentiation may be initiated, some extracellular factor is responsible for transmitting to the cells of the salt gland the information that the bird is ingesting excess salt. It is the identity of this factor (or factors) that is of immediate physiological interest. It would seem that, since the rise in [RNA] is very rapid, the trigger for the process must operate shortly after the bird first drinks salt water. Furthermore, since the adaptive changes are quickly reversed by returning birds to fresh water we can infer that a factor is not only required to induce adaptation but also to maintain the cells in the highly specialized state. In other words it must act continuously, or at least continually, on the gland. The following might be responsible for adaptive hypertrophy, acting alone or in combination:

1. Acetylcholine released by the cholinergic nerves.
2. A raised plasma tonicity acting directly on the secretory cells.
3. Hormones.

Acetylcholine. In an attempt to investigate whether nerves are important, Hanwell & Peaker (1973) denervated one gland of geese by removing the secretory nerve ganglion, a procedure which blocks acetylcholine release and thus secretion (p. 52). In birds kept on fresh water, killed 8 days after the operation, no effect of denervation on [RNA] or [RNA]:[DNA] was apparent. However in birds given an oral salt-load and then 0.2 M sodium chloride to drink for 24 hours, starting 7 days after the operation, [RNA] and [RNA]:[DNA] increased in the intact but not in the denervated gland (Fig. 9.7). In other words denervation blocks both secretion and adaptation. These findings clearly indicate that under normal circumstances nervous impulses to the gland are necessary for the early stages of adaptation.

Similar evidence has also been obtained independently by Pittard & Hally (1973) who used salt-gland weight as an indicator of adaptation. They denervated one gland in three-day old ducklings and to one group gave an intravenous salt-load every day for ten days. Denervation prevented the increase in salt-gland weight in birds on a salt-water regime. However the

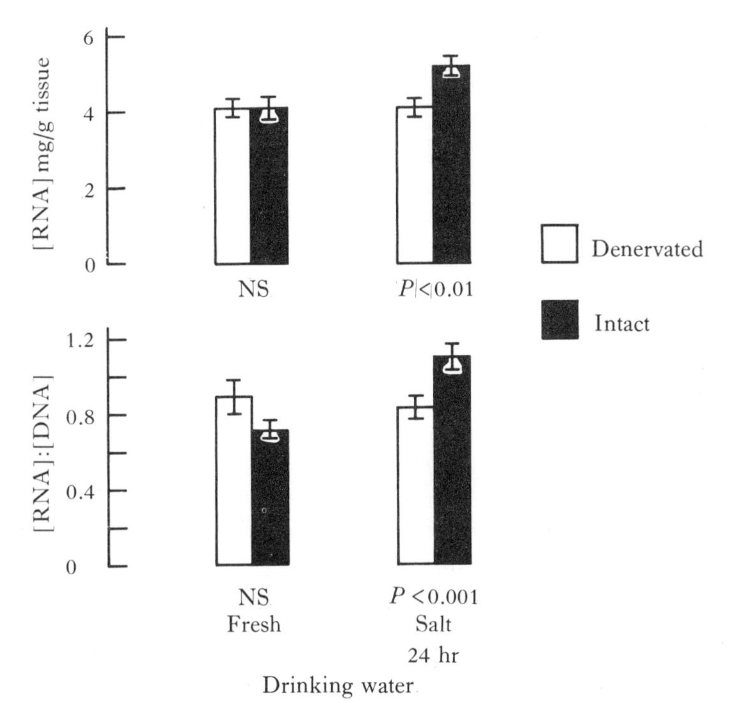

Fig. 9.7. Effect of unilateral postganglionic denervation on salt gland [RNA] and [RNA]: [DNA] in geese drinking fresh water or given salt water (100 ml sea water orally and then 0.2 M sodium chloride to drink for 24 hr) Mean ± S.E., comparison between denervated and intact glands made by paired *t*-test, NS = not significant (drawn from the data of Hanwell & Peaker, 1973).

denervated glands grew as body-weight increased in these young birds so these authors concluded that while genotypic growth is largely independent of nerve supply, growth in response to functional demand is wholly dependent on an intact nerve supply.

Hypertonicity. The experiments of Pittard & Hally (1973) and Hanwell & Peaker (1973) also show that the raised blood tonicity does not cause hypertrophy by a direct action on the secretory cells because unilateral denervation blocked the response, even though the gland was perfused by hypertonic blood.

Adaptation of the gland

Hormones. Bellamy & Phillips (1966) showed that duck salt glands have a remarkable affinity for corticosterone, and that secreting glands extract more of the hormone than inactive glands. They suggested that since this hormone is extracted from the blood at a higher rate than most tissues, then the rate of blood flow could be rate limiting. During secretion the rate of blood flow increases approximately 14-fold (Hanwell, Linzell & Peaker, 1971*a*) and since the salt gland does appear to take up more corticosterone from the blood then the possibility must be admitted that this hormone could be responsible for the adaptive changes even though its concentration in the circulation does not apparently change (p. 142). Since the increase in blood flow in the active gland appears to be under nervous control (p. 74), this would also offer an explanation for the findings of Hanwell & Peaker (1973) and Pittard & Hally (1973).

At this stage one cannot decide whether nerves or hormones are the initiators of adaptation. The simplest explanation that acetylcholine is responsible has some attractions. According to this explanation secretion and the onset of hypertrophy would begin at the same time; both processes require continuous (in the case of secretion) or continual stimulation, and the simple reflex arc could be the only means of control. However even if the apparently somewhat strange mechanism acting via corticosterone is operative, the control is still mediated by the secretory reflex arc. On the other hand if it does not work in this way but by the simple, wholly-nervous mechanism suggested above then we still have to explain why the gland has such an affinity for corticosterone. Of course it can also be implied that unless given in very large amounts to compensate for the low resting blood flow, corticosterone would not induce adaptation in the gland in the absence of nervous stimulation, i.e. in birds kept on fresh water. However even in experiments in which large doses of corticosterone have been given all the signs of increased secretory ability were not apparent. Although the output of nasal fluid increased, the concentration of sodium did not (Holmes, Phillips & Butler, 1961) and since there is the possibility that the output was influenced by an indirect effect on plasma composition (Peaker, Peaker, Phillips & Wright, 1971), the view that adrenocorticosteroids are involved has not yet been proven. So far no experiments have been done in which the reliable signs of adaptation (i.e. changes in weight, in nucleic acids, ATPases and other enzymes, and ultrastructure) have been studied following the administration of hormones, or for that matter cholinomimetics, and this clearly is an important field for future work. Corticosterone is, of course, not the only hormone that is possibly involved (Chapter 8).

Whatever the mechanism, it must be triggered by the ingestion of salt. This is most likely to be the raised plasma tonicity acting on the osmoreceptors in the region of the heart (Chapter 4). Goertemiller & Ellis (1966) and Ernst *et al.* (1967) found that other salts, apart from sodium chloride (potassium chloride, calcium chloride and magnesium chloride) in the drinking water caused some increase in weight and in tissue Na^+/K^+-ATPase.

Compensatory hypertrophy

The mechanism by which a paired organ grows following the removal of its partner on the opposite side of the body has excited the curiosity of cell biologists for many years. One theory to explain this compensatory hypertrophy is that the increased functional demand is responsible but that of Weiss (1952) is that all tissues produce specific inhibitors of their own growth so that a reduction in tissue mass lowers circulating inhibitor. Jephcott & Hally (1970) claimed to have distinguished which of these explanations is responsible for compensatory hypertrophy in the salt gland of ducks. They removed one gland from each bird and found no change in the remaining gland in birds kept on fresh water. These findings, the authors argued, indicate that compensatory hypertrophy depends on functional demand rather than on tissue mass. However we cannot agree that they found compensatory hypertrophy at all – the gland of the birds kept on sea water would have increased in weight anyway, and what the authors required was two groups of birds kept on salt water, one with both glands present and the other with one gland removed. A difference in weight between the gland of the birds in which one had been removed and the corresponding gland in intact controls would indicate that compensatory hypertrophy had occurred and that it results from functional demand.

10 INTEGRATION BETWEEN THE SALT GLAND AND OTHER ORGANS

The title of this chapter is somewhat misleading because it is one which we hope will apply in the future. At present little is known about integration between the organs involved in adaptation to marine conditions and all we can really do is to indicate the nature of the changes in different systems, the complications of interpretation and the reasons why 'more work must be done'.

It is true to say that in most experiments on conscious birds it is impossible to study salt-gland function in isolation. Any factor which alters renal excretion, intestinal absorption or the passage of solutes and water between the different fluid compartments of the body can indirectly influence salt-gland secretion by altering the tonicity of the plasma. It is for this reason that most workers collect the fluid voided from the cloaca in order to assess whether major alterations in salt and water balance have occurred during an experiment. This does not mean to say that the results of such routine collections can be used to explain the renal mechanisms involved, for example, and how control is integrated.

The factor which most complicates studies of salt and water metabolism in birds is that urine and faeces are voided from a common orifice. Because the ureters as well as the rectum open into the cloaca, the separate collection of urine and faeces cannot easily be achieved. While surgical procedures have been used to overcome this problem it is difficult to be certain that normal function is maintained. For example, it is known that alterations in sodium balance can occur in the domestic fowl and that such procedures as touching the ureters and even handling birds can markedly affect urine flow (see Hart & Essex, 1942). A further problem is the role of the cloaca and rectum in modifying the composition of the urine before it is voided (p. 188). This means that one should not use the terms 'renal excretion' or 'urine' for fluid collected from the cloaca. Instead, terms like 'cloacal discharge', 'cloacal flow' etc. have been used and, although cumbersome, are more accurate. (A few unpublished attempts have been made by workers in this field to devise a word which describes the mixture of faeces and urine voided by these non-mammalian vertebrates; most are derivatives of Anglo-Saxon words which are not suitable for publication.)

The results of integration

This has been studied in short-term experiments following salt-loading, and in long-term experiments in birds adapted to salt water. The first method was used by Schmidt-Nielsen, Jörgensen & Osaki (1958) in cormorants, as well as by later workers, and these studies clearly indicate that in many birds the major route for the elimination of sodium chloride is via the salt glands (Table 10.1). Staaland (1968) has found that although many wading-birds have functional salt glands, the cloacal route is the major route of sodium chloride excretion. However he used small oral loads of concentrated salt, and the possibility exists that this procedure may have inhibited secretion (see p. 30 and 135). Quantitative information is best obtained in long-term experiments at steady state. In short-term studies the amount of salt given is usually greater than would normally be ingested and in excess of that which a bird could cope with for an extended period; in fact such birds often enter a state of negative water balance. There is the additional problem that if an oral salt-load is given at least some of the solutes collected from the cloaca may have passed through the gut unabsorbed (p. 135).

The best studies have been carried out by W. N. Holmes and his collaborators at the University of California at Santa Barbara, in long-term experiments on the domestic duck (Fletcher & Holmes, 1968; Holmes, Fletcher & Stewart, 1968; Stewart, Holmes & Fletcher, 1969). They gave groups of birds either fresh water or salt water (284 mM sodium, 6 mM potassium), approximately equal to 60 per cent sea water, for drinking. After fourteen days they measured the output of ions and nitrogenous excretory products from the cloaca. Food was

TABLE 10.1. *Partition of electrolyte elimination between cloacal excretion and nasal secretion in the cormorant,* Phalacrocorax auritus. *Sea water was given by stomach tube and the collections were for 8 hr* (modified from Schmidt-Nielsen *et al.*, 1958)

	Na		K		Cl		Water	
	(mmoles)	(%*)	(mmoles)	(%*)	(mmoles)	(%*)	(ml)	(%*)
Total given	54	—	4.0	—	54	—	50	—
Cloacal excretion	25.6	52	2.66	90	27.5	51	108.9	68
Nasal excretion	23.8	48	0.31	10	26.1	49	51.4	32
Total excretion	49.4	—	2.97	—	53.5	—	160.3	—

* % of total excretion.

withheld for a day before these collections were made so that faecal contamination of the urine was thought to be negligible. Knowing the cloacal losses and assuming a figure for evaporative water loss obtained by other workers, they were able to calculate that under these steady-state conditions the salt glands were excreting 90 per cent of the sodium, whereas the kidney removed 70 per cent of the potassium excreted, as well as other excretory products. Their calculations also show that the salt glands need only excrete at 20 per cent of their maximum daily rate to achieve this output.

In short-term studies, Holmes *et al.* (1972) have applied the concept of 'clearance', beloved by renal physiologists, to estimate the volume of osmotically-free water obtained by activity of the salt glands. Following a salt-load it was calculated that salt-gland secretion would have provided 35–40 ml of free water per kg body weight in 440 minutes. On a daily basis this means the provision of a volume of water amounting to 50 per cent of the cloacal output. In providing this water 75–87 per cent of the total osmotic load was eliminated, and these figures illustrate the vital importance of salt glands in ducks living on brackish water. Comparable data and calculations for a truly marine bird have not yet been obtained but they would clearly be of interest.

Renal mechanisms and their control

This is not intended to be a treatise on renal physiology but we must indicate the way in which the kidney responds to saline conditions and how any control might be mediated. Holmes and his group in the work referred to in the preceding section were probably wise to study cloacal output at steady state. In this way the immediate effects of a salt-load, which could be mediated by alterations in the cardiovascular system for example, could be avoided and a true assessment of renal plus cloacal function obtained. In addition to collecting 'urine' they also carried out classical renal physiology in order to determine the site at which any changes occurred. It must be made clear that these workers believe their experiments reflect renal rather than renal plus cloacal function.

The main difference in renal excretion in birds kept on salt water, as opposed to fresh water, was an increase in urine osmolality to hypertonicity. This was due to increases in sodium and chloride concentrations and was reflected in the daily output of these two ions since the volume of urine produced did not change. Apart from a rise in calcium excretion, the output and concentration of other substances was not changed. There were no changes in glomerular filtration rate (GFR) or in renal plasma flow (RPF) and it was therefore deduced that any changes are the result of alterations in tubular function.

The significant feature of renal physiology is that the sodium concentration in the urine of salt-adapted birds is lower than that of the plasma. This means that the other constituents are excreted at a high concentration and the sodium which is reabsorbed from the tubules is then excreted as a hypertonic solution by the salt glands, or in other words, sodium occupies relatively little of the 'available osmotic space' of the urine as Holmes *et al.* (1968) have described it. The relatively small changes might imply that tubular reabsorption of sodium is not really controlled and is always very high. However, Holmes & Wright (1969) considered that the extent of sodium reabsorption is under some control in the duck kept on salt water. The evidence is that while tubular water reabsorption is similar in birds kept either on fresh water or salt water, sodium and water reabsorption are closely correlated in birds kept on salt water but not in those on fresh water. This, they suggested, implies that different control mechanisms are at work in the two states. Furthermore transient increases in cloacal sodium output have been reported immediately after salt-loads have been administered, which would imply that when a steady state is reached some control over sodium movements may be exerted in salt-adapted birds, and a different form of control over water movements. The case they quote however is after an oral load of very strong salt when it is very likely that the urine was seriously contaminated by the passage of unabsorbed salt through the gut (p. 135). Nevertheless Douglas (1970) has shown by collecting urine directly from the ureters that such transient increases do occur, although even then the sodium concentration does not reach that of the plasma.

It would appear from these considerations that sodium excretion by the kidneys is controlled in salt-adapted birds, and that the effect is exerted on the kidney tubules. However sodium appears not to be reabsorbed in exchange for potassium or ammonium, the latter incidentally accounting for approximately 30 per cent of the nitrogen excreted in these supposedly uricotelic animals (Holmes *et al.*, 1968; Stewart *et al.*, 1969).

By mammalian standards the relatively low sodium concentration in the urine of birds drinking 60 per cent sea water seems strange and it would clearly be unwise to attempt to apply mammalian control mechanisms. One might think that aldosterone could be responsible but its secretion is not normally associated with an excessive intake of sodium, and moreover, Holmes & Adams (1963) showed that exogenous aldosterone acts on the kidney of ducks to reduce both sodium and potassium concentrations. Holmes, Fletcher & Stewart (1968) have suggested that tubular water and sodium reabsorption may be under the influence of arginine-vasotocin, which is thought to be the antidiuretic hormone of birds (p. 154). The action of this hormone would explain the retention of sodium as well as water because in the domestic fowl and in frogs there is clear evidence that exogenous arginine-vasotocin reduces the renal output of water, sodium and chloride, increases urine osmolality but has no effect on GFR. It would also seem likely that arginine-vasotocin is released as a result of increases in plasma tonicity, as suggested by Phillips, Holmes & Butler (1961), and there is now evidence from examination of hypothalamic nuclei that this could be the case (Rhees, Abel & Frame, 1972); the possibility that this hormone might be involved in modifying salt-gland secretion is discussed in Chapter 8. Nevertheless it must be pointed out that cloacal output fell in neurohypophysectomized ducks after salt-loading, and the concentration of sodium was less than in intact control birds (Wright *et al.*, 1967; M. Peaker & A. Wright, unpublished). This might imply that other factors control renal function in salt-loaded birds and Holmes & Wright (1969) have suggested that a glucocorticoid could be responsible because Holmes & Adams (1963) found that low doses of corticosterone and cortisol induced changes in urine output and concentration which are very similar to those seen in birds kept on salt water. Another possible candidate for the factor controlling renal function is prolactin. Lockett & Nail (1965) and Lockett (1965) have shown in rats and cats that prolactin causes a decrease in renal sodium and water excretion, and to a much lesser extent in potassium excretion; the action seems to be direct since the effects were obtained in a heart–lung–kidney perfusion preparation in cats. Since GFR and RPF were also unchanged and the pattern of changes in urine concentration and production was similar to that observed in the salt-adapted duck the possibility that prolactin may have a role must be admitted especially when the interesting changes in pituitary prolactin content at this time (Chapter 8) are also considered.

The question remains; is there integration between salt-gland secretion and kidney function? It is obvious that a common osmotic stimulus could induce both salt-gland secretion and the release of hormones which could then act on the kidneys and also on the salt glands. This might be the sole means of control, each system being affected by the results of the action of both on plasma composition. Another view is that, in addition to such a common mechanism, there is closer integration between the two excretory systems, and that 'information' on what is trying to be achieved in maintaining homeostasis is being passed from one to the other. In other words does the kidney know when the salt glands are switched on and vice versa?

There is a little evidence that when the salt gland cannot cope with excess salt kidney function is markedly different from that in birds which are maintaining homeostasis on salt water. For example Ensor & Phillips (1972*b*) found that 'urine' production decreased markedly in gulls given unnaturally strong sodium chloride solutions to drink which resulted in a marked increase in plasma sodium, an increase in the sodium concentration of the cloacal fluid and even inhibition of salt-gland secretion. Hughes (1972*b*) observed a rise in the sodium chloride concentration of cloacal fluid, but no change in volume, in Kittiwakes (*Rissa tridactyla*) from which the salt glands had been removed. Normal plasma sodium and chloride concentrations were maintained in these birds which were given fish as their only source of food and water. These findings indicate that the kidneys act in a different manner as far as sodium and chloride excretion is concerned in the absence of a normally-functioning salt gland, and the results, particularly those of Hughes (1972*b*), could suggest some close integration between salt-gland secretion and the control of renal excretion. Is it possible that there might be a humoral substance produced by the active salt glands that could influence renal excretion?

An interesting difference in kidney weight between species of birds with and those without an active nasal gland has been found by Hughes (1970*d*); birds with salt glands have larger kidneys for their body-weight so that the total amount of excretory and osmoregulatory tissue is greater.

Role of the cloaca and rectum

After many years of uncertainty there is now no doubt that the cloaca and rectum of birds and reptiles are actively involved in modifying the composition of the urine which flows from the ureters; radiographic evidence has also been obtained for the passage of urine back into the rectum (see Shoemaker, 1972). The active and passive mechanisms by which the urine is modified have been the subject of a number of studies. In general sodium and water are withdrawn from the urine and the relatively insoluble uric acid is precipitated, which in turn leaves more water to be reabsorbed. This is a clear adaptation to uricotelism with its advantages for water conservation. In the context of this chapter our concern is whether cloacal–rectal reabsorption is more or less important to animals with salt glands. However definitive studies have only been carried out on such terrestrial birds as the domestic fowl and the information available on marine birds is sparse.

Schmidt-Nielsen *et al.* (1963) suggested that extra-renal excretion of salts in birds and reptiles may be required for advantage to be taken of water reabsorption in the cloacal region. They argued that if water withdrawal is linked to sodium transport then as well as conserving water, sodium would be present in the body in excessive amounts as a result of this mechanism and that extra-renal excretion of a sodium chloride solution at high concentrations would be a prerequisite for the efficient reabsorption of water in the cloaca. However it is now clear that birds without functional salt glands utilize this mechanism by what appears to be the simple expedient of excreting more sodium through the kidneys so that this can be reabsorbed with water in the cloaca and rectum. Indeed it has long been known that if ureteral urine is diverted to the exterior by surgical means domestic fowls become sodium deficient (Hart & Essex, 1942). The whole mechanism serves to precipitate uric acid at this site rather than in the kidneys, which clearly would be disastrous, as sufferers from gout would no doubt testify.

Clearly the process could occur in marine birds. Hughes (1970*a*) found that fluid which remained in the cloaca until voluntarily voided by gulls had a lower sodium chloride concentration than samples removed after a shorter time in the cloaca. She also calculated that the amount of sodium reabsorbed is the same as in the domestic fowl (Skadhauge, 1968) if it was assumed that the reabsorption of water was the same in both cases. The concentrations of sodium, potassium and chloride in fluid taken from the cloaca at short intervals were all low which indicated either that the movements of ions across the wall are very rapid or that the fluid entering from the ureters is relatively low in concentration. The latter is the more likely explanation because Douglas (1966*a*; 1968; 1970) succeeded in collecting ureteral urine from salt-loaded gulls and found the sodium concentration to be about half that of plasma.

Holmes *et al.* (1968) have suggested that the shorter the time hypertonic urine of salt-adapted birds is in contact with the cloacal epithelium the less chance it would have of being diluted by the osmotic passage of water from the blood. They found that voiding occurred more frequently in birds kept on salt water and suggested that the cloacal epithelium may be sensitive to the more hypertonic urine. Certainly this form of behavioural control could be an important means of determining the extent to which urine is altered in the cloaca and rectum but until it is known whether any advantage could be obtained in terms of water reabsorption from hypertonic urine at this site or whether it would be a disadvantage, as these authors suggest, it is difficult to speculate further.

Until studies on cloacal and rectal sodium and water reabsorption have been carried out in marine birds in different physiological states, as Skadhauge (1967) and Bindslev & Skadhauge (1971*a*, *b*) have done in the domestic fowl it is not possible to assess whether this mechanism could play a major role in salt and water metabolism. The situation in the fowl, particularly in terms of water movements, is complex and affected not only by the osmolarity of the urine entering from the ureters but also by its rate of flow. The only attempt to assess cloacal reabsorption of water after salt-loading was made by Peaker, Wright, Peaker & Phillips (1968). In these experiments the passage of fluid from the cloaca into the rectum was prevented by means either of an inflated Foley catheter or by ligating the gut at the recto–cloacal junction because at that time it had not been conclusively demonstrated that the rectum plays a part in the modification of urine in the fowl. Hypotonic solutions (to match the sodium concentration of urine collected one hour after salt-loading) introduced into the cloaca disappeared after about 30 minutes, and tritiated water included in the solution appeared in the blood, but it was not possible to detect any differences between salt-loaded, water-loaded and water-deprived birds.

Integration between the salt gland and other organs

Intestinal absorption

This is an important but neglected aspect of the physiology of marine birds. In order to gain free water in a marine habitat, sodium chloride and water must be absorbed from the intestine and the sodium chloride then excreted at a higher concentration than the water ingested. In birds, as indeed in all animals, it is difficult to obtain quantitative information on the actual mechanisms involved from studies *in vivo* but there are added complications which apply to birds. If urine mixes with the contents of the rectum then the absorption or other events occurring there will be determined by the flow and concentration of urine as well as by the flow and composition of the digesta. Indeed the resulting absorption will be determined by both urine and rectal contents rather than one or the other, and studies in which only one variable is studied could well fail to reflect the normal situation.

It appears that marine birds resemble marine teleost fish in terms of intestinal absorption, and indeed to a great extent, kidney function. In other words sodium and chloride and water are absorbed from the intestine leaving magnesium and sulphate to be voided in a more concentrated solution. Douglas (1970) gave a stomach load of sea water containing polyethylene glycol to gulls. When polyethylene glycol appeared in the cloacal fluid, the sodium concentration was low which implies that absorption of sodium from the intestine is rapid.

The concentration of sodium chloride in the absorbate is probably isosmotic with the fluid in the gut but hypertonic to the plasma, as in marine teleosts. Water therefore moves against an apparent osmotic gradient but coupled to sodium transport, an aspect of intestinal absorption that has received a great deal of attention. The highest concentration of fluid in the lumen at which this movement across the epithelium can occur is not known but Douglas (1970) suggested that the limit must be more similar to that found in the eel in sea water than in the domestic chicken. In the former of course the situation closely resembles that in marine birds, sodium chloride being excreted extra-renally via the gills in order to gain free water. This limit is an important aspect of the physiology of marine birds because, together with the concentrating and secretory abilities of the salt glands, it must be a main determinant of the salinity of drinking water to which a bird can adapt successfully.

The fact that the absorbate can be hypertonic to plasma does not preclude the possibility that water can or does move into the gut lumen down an osmotic gradient. Douglas (1970) found that when polyethylene glycol first appeared in the cloacal fluid after it had been mixed with a sea-water load, it was very dilute, implying that it had mixed with a large volume of water either already in the gut or drawn in osmotically by the sea water. Similarly Hanwell *et al.* (1972) found that soon after sea water was given to geese by stomach tube, haematocrit increased and blood volume decreased. This implies that water passed from the extracellular fluid into the intestinal lumen. Eventually these changes were reversed and the haematocrit decreased presumably as absorption began and as water passed from the tissues into the hypertonic extracellular fluid. It is difficult to believe that absorption of water does not occur until the contents of the gut are reduced markedly in osmotic concentration and the fact that absorption of water and sodium can occur from hypertonic solutions in other marine vertebrates would certainly suggest that this situation prevails in birds as well.

The only definitive study on intestinal absorption *in vitro* in marine birds has been made by Crocker & Holmes (1971*a*, *b*). They kept five- to ten-day old ducklings on salt water for ten hours and then fresh water for 14 hours in each day. Intestinal absorption was then studied *in vitro* using the everted sac technique, with Krebs-bicarbonate medium on both the mucosal and serosal sides. In birds on the salt water regime for four days, the initial rate of fluid uptake by the sacs was much greater than in ducklings given fresh water (Fig. 10.1). Five equal segments of gut were taken starting 10 cm behind the entry of the bile duct and ending at the junction with the caecal pouches. Adaptation i.e. greater rate of absorption, was most rapid in the anterior segments (Fig. 10.2), a difference which could reflect differences in epithelial cell turnover along the length of the small intestine. The fluid absorbed in these in-vitro experiments had the same concentration as that of the medium; higher concentrations were not studied.

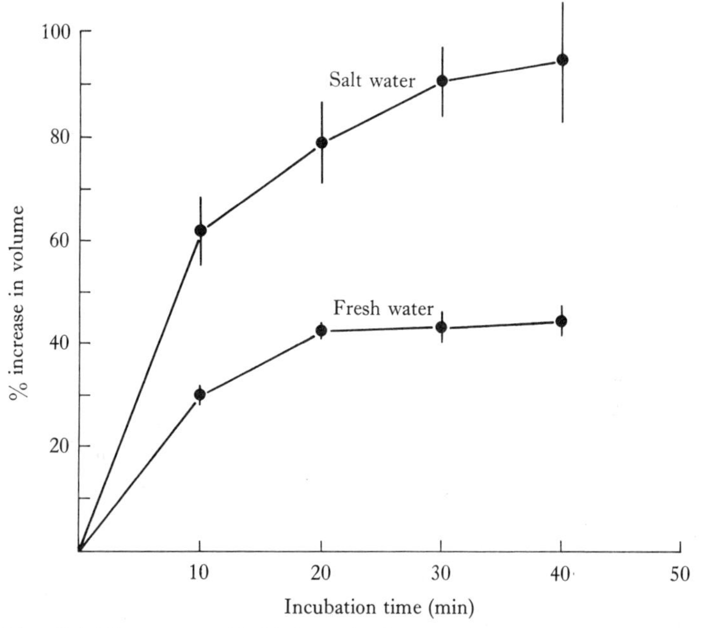

Fig. 10.1. Time-course of fluid absorption by everted intestinal sacs, prepared from ducklings kept on fresh water or salt water (mean±s.e.) (from Crocker & Holmes, 1971*b*).

The changes in intestinal absorption preceded adaptation of the salt gland (see Chapter 9) which implies that maximum absorption is required for the full development of the salt glands. When the aldosterone antagonist spironolactone was administered before the ducklings were given salt water, salt-gland secretion was not observed and intestinal absorption remained at the same level as in birds kept on fresh water. This evidence certainly suggests that a mineralocorticoid may be responsible for adaptation of the intestine to absorb more salt and water under these conditions. However there is no indication that aldosterone is secreted in these circumstances (Chapter 8), but corticosterone, which has also been suggested as a candidate for the induction of adaptive changes in the salt gland, could well be involved at this site. Cortisol appears to be responsible for similar changes in euryhaline fish (Hirano & Utida, 1968).

Crocker & Holmes concluded, 'Clearly the effective absorp-

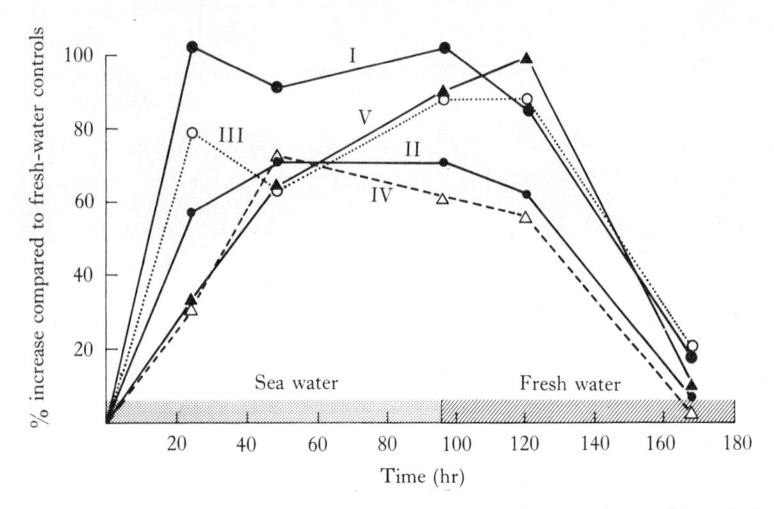

Fig. 10.2. Time-course of adaptation of the gut to salt water. Everted intestinal sacs were prepared at various times after transfer of ducklings to hypertonic saline drinking water and back to fresh water. The intestinal segments I–V are as described in the text; I is the most anterior (mean±S.E.) (from Crocker & Holmes, 1971*b*).

tion of water and ions by the intestinal mucosa is fundamental to the initiation and continuation of the homeostatic mechanism in marine birds and factors affecting absorption will seriously affect their survival. In this regard, an examination of the effects of ingested crude oil and oil dispersant on intestinal absorption mechanisms of marine birds may yield useful insights into the fundamental nature of the physiological hazards associated with oil pollution.' All we can add is a plea for more work to be done both on the fundamental nature of the mechanisms for absorption using techniques which have been so successfully applied in other animals, and the overall role of the alimentary tract as a factor in maintaining homeostasis in marine birds.

Passage between fluid compartments

Once sodium chloride has been absorbed from the intestine other extra-renal factors could affect the activity of the salt glands. That is the rate of passage of ions and water between the intracellular and extracellular fluid compartments. While there is some suggestion that sodium may pass into the cells of the body after salt-loading (M. Peaker, unpublished) there is no information whether such processes occur normally or whether such processes change upon adaptation to salt water.

Harderian gland

An interesting suggestion, that the Harderian glands of ducks can act as potassium-secreting salt glands, has been made by Hughes & Ruch (1968; 1969). They noticed that in ducks kept on sea water the feathers between the eye and the beak were wet and indeed contained depressions which were filled with a clear fluid. They therefore examined the concentrations of ions in the tears of birds kept on different concentrations of sea water or equivalent concentrations of sodium chloride. The results clearly show that while sodium is at approximately the same concentration as in plasma, the potassium concentration was higher and increased with the salinity of the drinking water, the main rise occurring once the salinity exceeded that of the plasma (Fig. 10.3). On full-strength sea water a tear:plasma ratio of 39:1 was obtained in one bird, although the mean ratio was about 17:1. It was also found that the Harderian glands were slightly but significantly larger in the birds kept on 0.42 M sodium chloride. It seems likely that it is the Harderian glands that are responsible for the secretion because the lachrymal glands are very small, only one three-hundredth of the size of the Harderian glands. This evidence therefore suggests that adaptation can occur in the Harderian glands as well as in the nasal salt glands of this species.

The demonstration that tears are rich in potassium does not of course indicate that this is an important route for the elimination of potassium because the rate of secretion must also be taken into account. Hughes & Ruch (1969) determined this in ducks kept on fresh water by wiping the eye and recording the time taken to fill a micropipette from the inner canthus of the eye. This rate they point out may be an underestimate of the rate in birds kept on salt water. When calculations were made it was clear that this route could be as important as cloacal

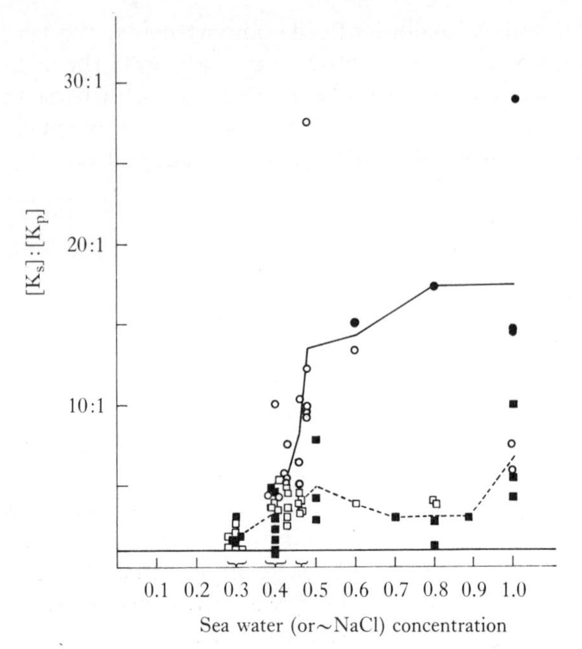

Fig. 10.3. Changes in potassium concentration in salt-gland secretion and tears of ducks drinking different concentrations of sea water. The ratio of potassium in the secretions relative to plasma ([K_s]/[K_p]) is plotted against the salinity of the drinking water (expressed as a fraction of full-strength sea water). The ○–● line passes through the mean tear:plasma ratios and the □–■ line passes through the mean salt-gland secretion:plasma ratios; the solid horizontal bar indicates a ratio of 1:1. For the eye ○, NaCl; ●, sea water; for the salt gland □, NaCl; ■, sea water (from Hughes & Ruch, 1969).

excretion for the elimination of both potassium and sodium since in addition to the relatively high potassium concentration the sodium level was similar to that in plasma (Table 10.2). Although no data for cloacal and salt-gland excretion were obtained in these experiments and figures had to be taken from the literature it is clear that, if the rate of tear production is maintained the Harderian glands must be considered as an organ for extra-renal excretion in the duck. However Hughes & Ruch (1969) do make it clear that the amount of fluid which actually spills from the eye to the exterior is not known and some or even most could pass back into the body along the naso-lachrymal duct. They also found a band of smooth muscle around the opening of this duct which they suggest could

TABLE 10.2. *Estimation of sodium and potassium excretion in cloacal fluid, salt-gland secretion and tears of ducks acclimated to drinking salt water* (from Hughes & Ruch, 1969)

	Sodium		Potassium	
	mM	mmoles/ day/kg	mM	mmoles/ day/kg
Cloacal fluid*	76.3±2.9	4.37±0.86	45.5±6.9	2.18±0.37
Salt-gland secretion†	461.8±24.0	11.08±0.58	16.0±2.2	0.38±0.05
Tears‡	142.3±7.7	4.10±0.22	57.8±8.3	1.66±0.24

* Data from Holmes *et al.* (1968). Ducks drinking salt water (284 mM NaCl, 6 mM KCl) for 14 days.

† Secretion rate estimated to be 24 ml/day/kg (based on data from Schmidt-Nielsen & Kim, 1964, and from Holmes, Phillips & Butler, 1961).

‡ Secretion rate assumed to be 20 μl/min (measured in ducks on fresh water, therefore possibly an underestimate).

act as a sphincter to regulate the passage of tears from the eye.

Hughes (1969) then went on to examine the tears of the Glaucous-winged Gull (*Larus glaucescens*) and found in contrast to the duck that the potassium concentration is very low (1–2 mM). The concentration moreover was not affected by the salinity of the drinking water and there were no patches of dried tears on the feathers as in ducks. Hughes therefore suggested that since this species has a greater ability to adapt to sea water than the duck, which at these concentrations is in a precarious state of water balance, the ancillary mechanism of the Harderian gland is not required. She concluded, 'too little is known about the eye fluids of birds to permit more than speculation upon the meaning of the differences noted between duck and gull tears, but these differences would seem to be sufficiently great to warrant further investigation of other species to assess any functional significance these differences might have'.

From the point of view of the physiology of the Harderian glands it must be inferred that secretion occurs in the long-term because we have not observed tears in ducks or geese up to five hours after a large salt-load. Harderian-gland blood flow was not affected by salt-loading in geese (Hanwell *et al.*, 1971*a*) although the rate of sodium and potassium secretion quoted by Hughes & Ruch (1969) could be supported by the rate of blood flow (2 ml/g/min) found in geese.

Growth and metabolism

When birds with salt glands are given salt water to drink there is interference with metabolism and growth. This has been observed in young ducks and gulls when fed *ad libitum* and it seems likely that this is due, at least in part, to a decrease in food intake (Krista *et al.*, 1961; Holmes, Butler & Phillips, 1961; Ellis *et al.*, 1963) (see p. 111). In adults the adrenals and liver increase in weight and the liver glycogen content increases (Holmes *et al.*, 1963). These changes would appear to be manifestations of an increased rate of gluconeogenesis. The cause of the decreased food intake is not known but Stewart *et al.* (1969) have suggested that '...the adaptation of migratory birds to a marine habitat is characterized by inanition, the attendant increase in adrenal size resulting in an increased gluconeogenesis and an increase in the excretion of nitrogen'.

Since corticosterone has been suggested as a factor which acts alone or in concert with other hormones in the adaptive changes that occur in the salt glands, intestine and kidney, the increased rate of secretion of this hormone could perhaps be induced initially by a decrease in food intake and temporary inanition, although it must be pointed out, as discussed in Chapter 8, that apart from the increase in adrenal weight such birds show no signs of a raised concentration of corticosterone in plasma. One might then predict that once a bird is fully adapted to deal with marine conditions, food intake recovers and metabolism reverts to normal. However this does not account for the larger adrenals of truly marine birds (Fig. 8.2, p. 138) but it should be noted that in general these birds are carnivorous and the possibility remains that the increased adrenal weight and the presumed high rate of gluconeogenesis is as much an adaptation to a high protein diet as it is to marine conditions.

The data available at present suggest that saline conditions only affect growth under conditions of *ad libitum* feeding. Ensor

& Phillips (1972*a*) for example found no significant difference in weight between young gulls from an inland and a marine colony. Similarly when ducks kept on fresh water were given the same amount of food as that voluntarily consumed by ducks on salt water, the body-weights of the two groups were the same (Stewart *et al.*, 1969). Thus the effect seems to be on the amount of food consumed which leads to maximal growth, and that other factors like the availability of food, have more effect on growth rate under natural conditions.

11 COMPARATIVE AND APPLIED PHYSIOLOGY

In this chapter, we try to deal in general with the ecological importance and comparative aspects of salt-gland function but we must apologise for using this chapter as a repository for information, much of it interesting, that will not fit conveniently into other chapters.

Marine birds

Comparative aspects of salt-gland function which show that the more pelagic, invertebrate-eating birds secrete nasal fluid of a higher concentration, have been dealt with in Chapter 7. Similarly aspects of integration with other organs like the kidney and gut are discussed in Chapter 10. In this section we have to consider the size of the glands in different birds and a collection of topics that can be labelled 'ecophysiology'.

Nasal gland weight

Technau (1936) compared the weight of the nasal gland with that of the lens of the eye in 83 species of bird. While a clear relationship was found between this index and habitat by Holmes, Phillips & Butler (1961) (Table 11.1) in that the ratio was higher in marine birds, Technau's data have only limited usefulness. This is because the weight of the lens is also a variable which reflects the mode of life. Thus owls and nightjars, for example, which are nocturnal, have large eyes and Technau's index can obviously be affected as much by this variable as by the size of the salt glands. Therefore we have collected data from the literature on nasal gland weights and body-weights (Fig. 11.1). Unfortunately, little information has been published on non-marine birds but it is clear from the figure that marine birds have larger nasal glands than other

Table 11.1. *Nasal-gland index for fresh water, brackish water and sea water-maintained birds.* (Technau's data adapted by Holmes, Butler & Phillips, 1961)

Habitat	Nasal gland index (Lens weight: nasal gland weight)
Fresh water	1:0.193
Brackish water	1:0.890
Marine	1:4.14

species. In marine forms the exponent for body-weight, on a log–log basis, is 0.92, which indicates that the weight of the salt glands is almost in direct proportion to body-weight rather than to metabolic body size. It should also be noted that the weight of the glands can be increased by adaptation to salt water, a facet of salt-gland function discussed more fully in Chapter 9. From the figure it is also clear that one can predict from body-weight the minimum weight of salt-gland tissue in a bird adapted to marine conditions. Thus a figure of 35 mg/100 g body-weight in a bird weighing 4000 g to 45 mg/100 g in a 40-g bird, would be indicative that a bird was marine at the time of death.

Salt glands and survival

Until recently no direct evidence had been obtained for the necessity of the salt glands in birds drinking salt water. Bradley & Holmes (1972*a*) showed that ducks in which the nasal glands had been excised could not survive on hypertonic saline as their only source of drinking water on which intact ducks will survive indefinitely. Such birds lost weight while plasma osmolality, sodium, potassium and chloride concentrations increased markedly; these changes were reversed when the birds were returned to fresh water (Fig. 11.2). There was also a marked reduction in urine production when the birds without a salt gland were given salt water. Since the sum of sodium, potassium and chloride concentrations in plasma was very much less than the measured osmolality, Bradley & Holmes suggested that osmotically-active nitrogenous

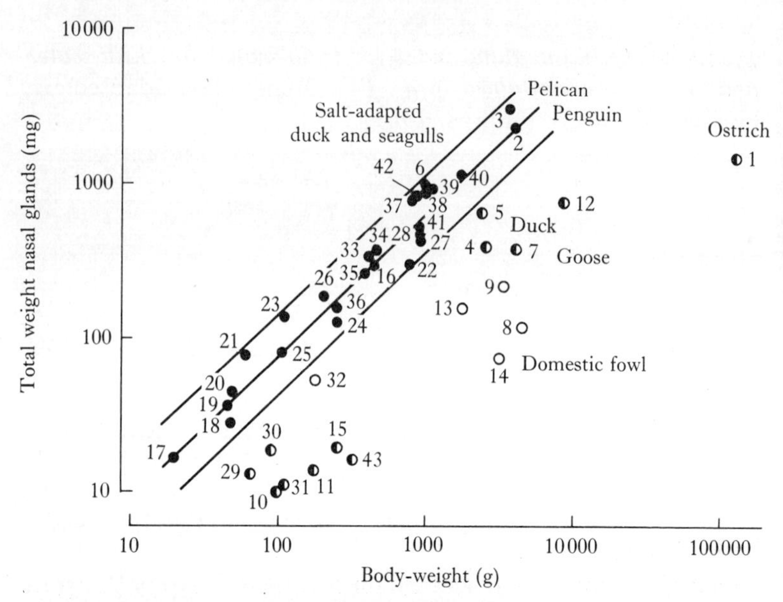

Fig. 11.1. Relationship between body-weight and nasal gland weight in different birds: •, truly marine birds; ○, birds habitually drinking fresh water (those tested incapable of secreting salt); ◐, terrestrial and brackish-water birds capable of secreting. The regression line and 95 per cent confidence limits are for the marine birds. The equation of the line is $y = 0.09 + x^{0.92}$ where y = weight of nasal glands (mg) and x = body-weight (g).

Key to species

No.	Species		Habitat	Ref.
1	*Struthio camelus*	Ostrich	T	b
2	*Pygoscelis adeliae*	Adélie Penguin	M	k
3	*Pelecanus occidentalis*	Brown Pelican	M	f
4	*Anas platyrhynchos*	Domestic duck (Mallard) on fresh water	FW–M	i
5	*Anas platyrhynchos*	on salt water		i
6	*Anas platyrhynchos*	on salt water		l
7	*Anser anser*	Domestic Goose	FW–B	j
8	*Balearica pavonina*	Crowned Crane	T	b
9	*Ciconia ciconia*	White Stork	T	b
10	*Poliohierax semitorquatus*	Pigmy Falcon	T	g
11	*Micronisus gabar*	Gabar Goshawk	T	g
12	*Torgos tracheliotus*	Lappet-faced Vulture	T	g
13	*Cathartes aura*	Turkey Vulture	T	g
14	*Gallus domesticus*	Domestic fowl	T	h
15	*Rallus owstoni*	Guam Rail	FW–B	c
16	*Fulica americana*	American Coot	FW–M	c
17	*Calidris minuta*	Little Stint	M	a
18	*Actitis hypoleucos*	Common Sandpiper	M	a
19	*Charadrius hiaticula*	Ringed Plover	M	a

materials, perhaps urea and ammonium, were accumulating in the body, but they could not decide whether death was due to dehydration or to the accumulation of nitrogenous excretory products.

Hughes (1972*b*) has also studied the effects of removing the salt glands. When Kittiwakes (*Rissa tridactyla*) without nasal glands were given only fish to eat as a source of water and food, no changes in plasma sodium, potassium and chloride concentrations were apparent although the output of sodium and potassium in cloacal fluid increased. However her balance studies suggested that some sodium chloride was still being excreted extra-renally. Evidence in support of this deduction was that the plastic sheets lining the cages, and onto which nasal

Key to Fig 11.1 (*cont.*)

20	*Calidris alpina*	Dunlin	M	a
21	*Crocethia alba*	Sanderling	M	a
22	*Alca torda*	Razorbill	M	a
23	*Calidris canuta*	Knot	M	a
24	*Limosa lapponica*	Bar-tailed Godwit	M	a
25	*Plautus alle*	Little Auk	M	b
26	*Plautus alle*	Little Auk	M	a
27	*Uria aalge*	Guillemot	M	a
28	*Uria lomvia*	Brünnich's Guillemot	M	a
29	*Tringa glareola*	Wood Sandpiper	FW–B	a
30	*Capella gallinago*	Snipe	T–FW	a
31	*Tringa ochropus*	Green Sandpiper	T–FW	a
32	*Pluvialis apricaria*	Golden Plover	FW(B?)	a
33	*Cepphus grylle*	Black Guillemot	M	a
34	*Fratercula arctica*	Puffin	M	a, o
35	*Rissa tridactyla*	Kittiwake	M	a
36	*Larus ribibundus*	Black-headed Gull	M	a
37	*Larus argentatus*	Herring Gull	M	d
38	*Larus argentatus*	Herring Gull	M	a
39	*Larus argentatus*	Herring Gull	M	e
40	*Larus hyperboreus*	Glaucous Gull	M	a
41	*Larus glaucescens*	Glaucous-winged Gull	M	
		on fresh water		m
42	*Larus glaucescens*	on salt water		m
43	*Geococcyx californianus*	Roadrunner	T	n

Habitats: M – marine, B – brackish-water, FW – fresh-water, T – terrestrial.

References: (a) Staaland, 1967*b*, (b) Technau, 1936 & Hughes, 1970*d*, (c) Carpenter & Stafford, 1970, (d) Fänge, Schmidt-Nielsen & Osaki, 1958, (e) Bonting *et al.*, 1964, (f) Schmidt-Nielsen & Fänge, 1958*a*, (g) Cade & Greenwald, 1966, (h) McLelland & Pickering, 1969, (i) Fletcher *et al.*, 1967, (j) Hanwell, Linzell & Peaker, unpublished, (k) Douglas, 1964, (l) Schmidt-Nielsen & Kim, 1964, (m) Holmes, Butler & Phillips, 1961, (n) Ohmart, 1972, (o) Hughes, 1970*c*.

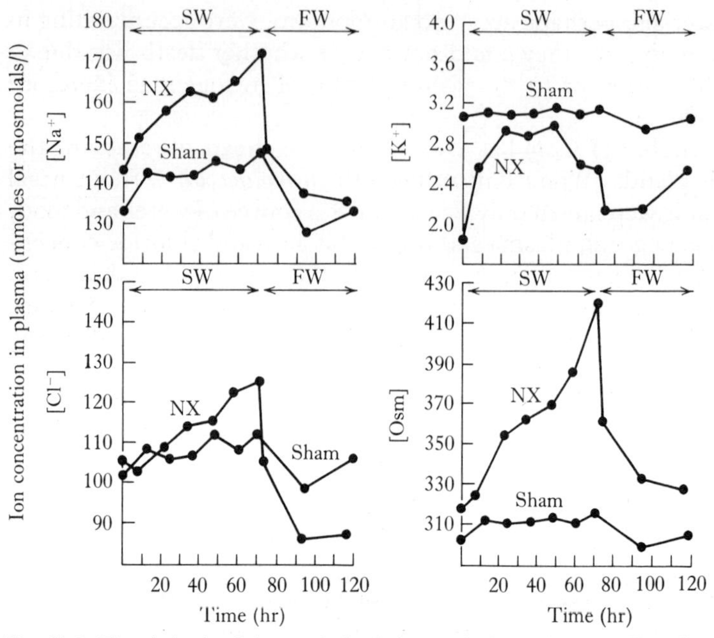

Fig. 11.2. The vital role of the nasal glands for survival on salt-water. The effect of giving hypertonic saline (284 mM sodium chloride, 6 mM potassium chloride) drinking water to ducks from which the nasal glands had been removed (NX) or subjected to sham operation (Sham) one week before the beginning of the experiment, on plasma composition. The birds were maintained on hypertonic saline (SW) or fresh water (FW) during the periods indicated by the arrows. Each point is the mean from five birds (from Bradley & Holmes, 1972*a*).

fluid is normally thrown by the bird, still had salt on them. Nevertheless, nasal salt secretion was not observed when the birds were given hypertonic sodium chloride intravenously although it was noted that the tears from one of the birds were somewhat hypertonic (sodium 255 mM, chloride 179 mM). Thus it is possible that orbital glands could have been acting as salt glands (see p. 195) in these circumstances. Of course these experiments did not constitute such a severe test for survival under marine conditions as those of Bradley & Holmes (1972*a*); further studies are awaited.

The effect of removal of the salt glands of gulls on their survival under natural conditions has been studied by Schwarz and his co-workers (Schwarz & Spannhof, 1961; Schwarz, 1962; Schwarz & Nehls, 1967). They found that the gull, *Larus canus*, survived in the wild after the salt glands had been removed and concluded that extra-renal excretion is an accessory mechanism for survival. There is no doubt that this is true in birds which live near or on the shore since fresh water is usually near to hand; the situation may well be different in pelagic, invertebrate-eating, tropical birds. They also found that secretion appeared in the nostrils after extirpation of the nasal glands in gulls (*Larus argentatus*, *L. canus* and *L. ribibundus*). However, these experiments are, as Schmidt-Nielsen (1965) has remarked '...difficult to interpret because one lacks assurance that there was complete removal of all salt-gland tissue, some regeneration might have taken place, and studies of total salt balance were not executed'.

There is no clear evidence on the extent to which marine birds rely on sea water for drinking nor consequently on the extent to which salt-gland secretion occurs in the wild. Many sea birds range over land where fresh water is available and some gulls probably return to drink fresh water after relatively short periods at sea. Some authors have claimed that marine birds do drink sea water (e.g. Murphy, 1936) while others, including Homer Smith (1953), have maintained that they can rely wholly on the free water obtained in their prey and upon metabolic water from the utilization of food. These arguments were made before extra-renal excretion was discovered and there is little doubt that marine birds can drink sea water if necessary. In fact several factors probably determine the extent to which the salt gland is used, for example the type of food consumed (p. 115) and environmental temperature, evaporative water loss being greater at high temperatures.

Changes in salt-gland function with age

Hughes (1968) has argued that in nestlings exposed to the sun on open beaches or rock ledges, water intake must be low and evaporative water loss high and, therefore, that extra-renal excretion may be more important at this stage than in older birds. Two nestling Common Terns (*Sterna hirundo*), when given a standard intravenous salt-load, did show a decline in nasal secretion with age (although two others did not). However, changes in renal function may well have been responsible since the data show clearly that the cloacal output of water and sodium increased with age. Thus it could be argued that in very young birds the kidney would be removing less sodium which would leave more to be excreted by the salt glands. Certainly the capacity for salt-gland secretion appears to change little with age since the salt glands were 0.1 per cent of body-weight from 8–100 g in this species. In the Adélie Penguin (*Pygoscelis adeliae*) on the other hand, there is certainly evidence that the capacity for secretion decreases with age. The slope of the log–log plot between body-weight and salt-gland weight (calculated from the data of Douglas, 1964) is 0.63 which implies that extra-renal excretion could be more important in the young than in the adult. The relative decrease in salt-gland weight is only counteracted to a small extent by an increase in the rate of secretion (measured as chloride output) relative to weight, the slope relating these two variables being 1.12 (Douglas, 1964).

Diurnal variation in secretion

By giving a salt-load at different times of the day Ensor & Phillips (1970*b*) found a circadian pattern of salt-gland secretion. The peak of activity was two hours after food and water were given. The prolactin content of the pituitary mirrored this pattern (see Chapter 8) suggesting that prolactin is involved. In the wild, secretion is also associated with feeding, as might be expected. Blösch (1966) found by observing Herring Gulls (*Larus argentatus*) that secretion starts while they search the mudflats for food and continues for two to three hours after feeding. Again this is presumably related to the ingestion of their food, marine invertebrates rich in sodium chloride and which Blösch has also shown are a more potent stimulus for secretion than fish (Fig. 11.3). Thus these studies do not decide whether there is also an inherent circadian rhythm, independent of the time of feeding.

Salt tolerance

In contrast to studies on fish and aquatic invertebrates very little work has been done on the ability of different species to

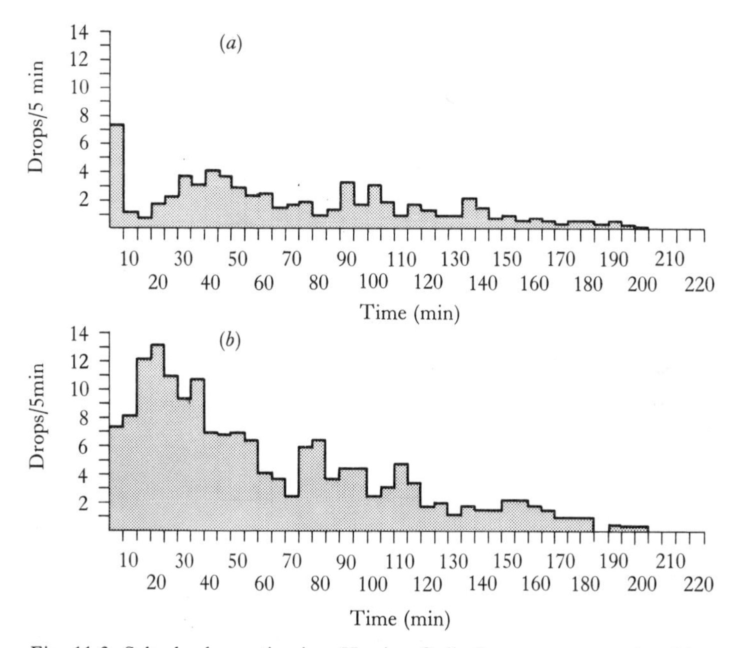

Fig. 11.3. Salt-gland secretion in a Herring Gull, *Larus argentatus*, after (*a*) a meal of fish, (*b*) mussels (which contain more salt) (from Blösch, 1966).

survive on different concentrations of salt water. There is no doubt that such differences do exist but no systematic study has been made. This is a pity because by testing the tolerance of a species the genetic potential for adaptation to a marine environment becomes apparent. For such studies, birds could be slowly accustomed to more hypertonic solutions so that adaptation can occur gradually. This procedure, of course, measures something quite different from a sudden change which would give more of an indication of the ability of a bird to fly from fresh water to sea water and to remain there indefinitely. A useful discussion on the design of experiments for such research is given by Gordon (1968).

It has been found that the adult Laughing Gull (*Larus atricilla*) can tolerate sea water whereas the Herring Gull can live on water of only half this strength (Harriman & Kare, 1966*a*; Harriman, 1967). This, these authors suggested, could reflect the different habits of the two species, the Laughing Gull

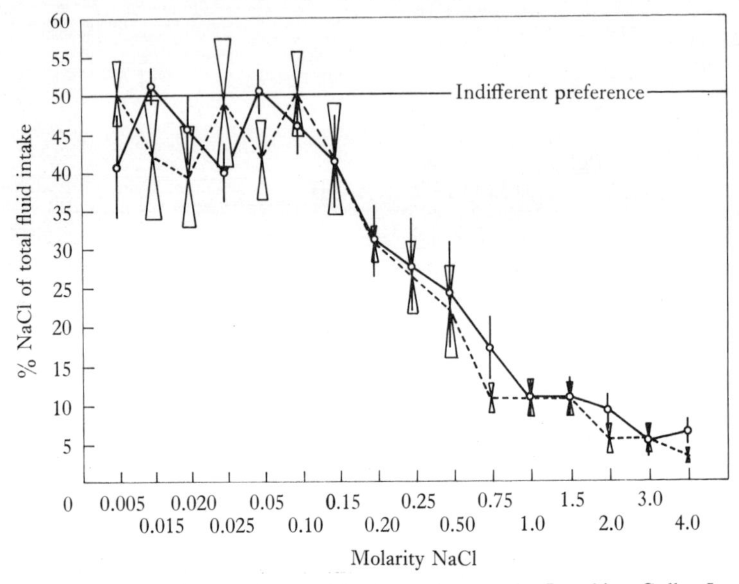

Fig. 11.4. The voluntary intake of fresh or salt water by Laughing Gulls, *Larus atricilla*. The birds were offered fresh water and salt water of increasing strength. Above 0.1 M sodium chloride, the birds increasingly selected fresh water. Circles and midpoints show means; vertical lines and outlines, standard errors; dashed line, adults; solid line, chicks (from Harriman, 1967).

being more pelagic. Schmidt-Nielsen & Kim (1964) succeeded in acclimating domestic ducks to three per cent sodium chloride (514 mM) by slowly raising the concentration, although it should be noted that some birds died (see also Hughes & Ruch, 1969). Wild mallards were less tolerant of hypertonic drinking water than domestic ducks – a somewhat surprising finding. On the other hand, Fletcher & Holmes (1968) found that domestic ducks could not tolerate 470 mM sodium chloride and 10 mM potassium chloride even if they were previously given 284 mM sodium chloride and 6 mM potassium chloride for three weeks. These results indicate that the duck, *Anas platyrhynchos* cannot tolerate truly marine conditions for an extended period.

Salt intake

Apart from an unsubstantiated report that the Kittiwake (*Rissa tridactyla*), which lives on the open sea, drinks sea water in preference to fresh, there is no doubt that all the birds that have been studied prefer fresh water when given the choice (Drost, 1931; Schwarz, 1966; Harriman & Kare, 1966*b*; Harriman, 1967) (Fig. 11.4); once the strength of the drinking water exceeds the osmolarity of blood plasma there is a marked aversion to it. Harriman & Kare (1966*b*) found no difference in preference between Herring Gulls and two terrestrial birds which lack salt glands.

When offered only one of a number of sodium chloride solutions of different strength, Harriman & Kare (1966*a*) found that the fluid intake was less in the gulls given the stronger solutions. Fletcher & Holmes (1968) observed the same effect in domestic ducks but made the important and extremely interesting finding that once the concentration of the drinking water exceeded that of plasma, the daily intake of sodium remained constant while fluid intake declined (Fig. 11.5). This implies that ducks appreciate in some way, and can regulate their sodium intake, perhaps so as not to exceed their ability for excretion. This obviously implies that at concentrations of sodium chloride in the drinking water above that with which renal and extra-renal excretion can cope, the fluid intake would be so low that dehydration could result from the low fluid intake rather than from the inability to excrete all the ingested salt. Such a mechanism controlling fluid intake when the birds are kept on salt water might also ensure that they do not drink large quantities of sea water merely because they are thirsty and it is clear that further work on the physiological and psychological mechanisms controlling fluid intake in marine birds would be of great interest. Comparative studies are also required to investigate the possibility that the control may vary in species which are more, or less, tolerant to salt.

Dehydration and salt-gland function

The absence of sufficient water in the diet to balance renal, faecal and evaporative losses is a similar environmental stress,

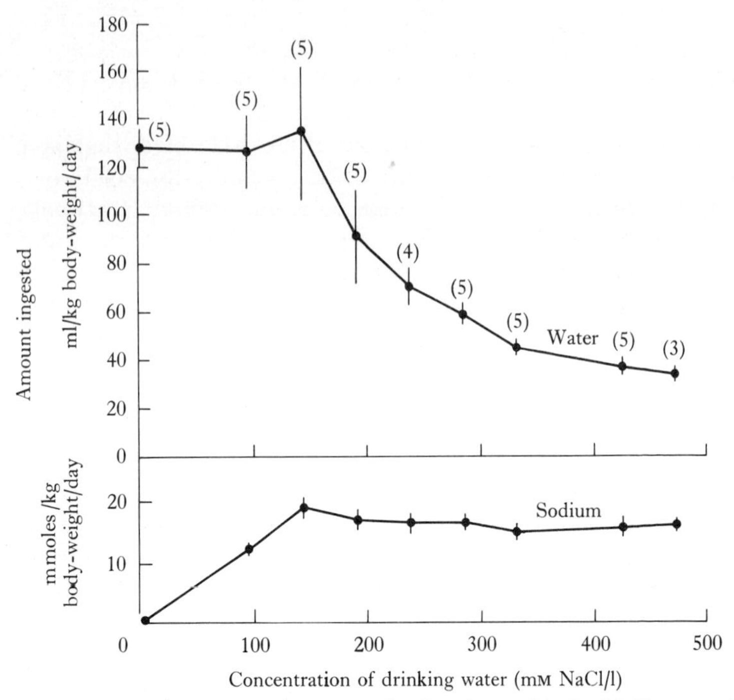

Fig. 11.5. Volume of water and amount of sodium ingested in the drinking water by ducks given various concentrations of sodium chloride as their sole source of drinking water. Daily food and water intakes were measured in ducks on fresh-water for a period of 2–3 weeks. The ducks were then given 284 mM sodium chloride as their sole source of drinking water, and daily food and water intakes were again observed for an additional three-week period. Following this the ducks were given various concentrations of sodium chloride to drink, and the daily intakes measured for periods of 5–10 days. Mean±S.E. (number of determinations in parentheses) (from Fletcher & Holmes, 1968).

at least in the short term, to that of drinking hypertonic saline; in both there is a lack of osmotically-free water. Therefore it might be thought that the salt glands could assist in such circumstances by removing sodium chloride which accumulates in the extracellular fluid as a result of the loss of water by evaporation. Although one of us has observed secretion in the nostrils of domestic geese deprived of water for 24 hours, Stewart (1972) has provided the only published evidence that it occurs under these circumstances. He kept ducks without drinking water and found that secretion started after about twelve hours. After 33 hours without water, the salt glands accounted for approximately 75 per cent of the total quantity of sodium excreted but only two per cent of the potassium. These data are similar to those of birds kept on hypertonic saline (Chapter 10). It would seem that in the wild, the salt gland would be of value in maintaining homeostasis during similar periods of water shortage, for example, while nesting, sheltering from storms, etc.

There is obviously a limit to the time salt-gland secretion can offset dehydration because renal excretion and insensible losses must continue. Ensor & Phillips (1972*b*) studied the effect of long-term dehydration on salt-gland function in juvenile gulls (*Larus argentatus, L. fuscus*). They did this either by keeping them at an environmental temperature of 30 °C for five days with fresh water only available for twelve hours each day; by depriving them of drinking water for four days; or by giving them hypertonic saline to drink (80 per cent sea water or 800 mM sodium chloride) in excess of the concentration on which they can maintain homeostasis. In all cases the signs of dehydration were that the water content in both plasma and muscle was decreased; in the birds given the strong salt solutions plasma sodium concentration was also raised markedly. The effect of these treatments on salt-gland secretion was investigated by giving 500 mM sodium chloride and 10 mM potassium chloride orally before and after the period of dehydration; in all cases, the secretion of nasal fluid was depressed. The mechanism by which this reduction in salt-gland secretion was brought about was not investigated but Douglas & Neely (1969), in a short abstract, found that the effect of water deprivation on salt-gland secretion in response to an oral salt-load and which was evident as early as twenty hours, could be overcome by treating the birds with the α-adrenergic blocking agent, phenoxybenzamine. This drug had no effect on salt-gland secretion in normal birds and it was suggested that '...dehydration diminishes the ability of the gull to eliminate a salt load extra-renally through vasoconstriction of the salt gland's arterioles'.

It appears then that the bird short of free water attempts to maintain its extracellular fluid volume rather than to osmoregulate against overwhelming odds since Ensor & Phillips

(1972*b*) also found that urine production was markedly reduced. These authors were surprised that their gulls could not adapt to the environmental conditions they imposed on the group deprived of water since, in the wild, birds face water restriction and high environmental temperatures while nesting. However, these birds can clearly tolerate such conditions and it would seem that adaptation is concerned with saving fluid, by reducing both renal and extra-renal excretion to a minimum, obtaining metabolic water from the breakdown of stored fat and possibly protein. These measures would act to preserve the volume of extracellular fluid and blood pressure. In other words the birds become osmoconformers rather than osmoregulators under such circumstances. However it is odd that the cloacal (? renal) output of sodium increased during water deprivation and that salt-gland secretion, which is the more economical route for sodium chloride excretion and water conservation, was depressed. Ensor & Phillips (1972*b*) suggested that since dehydration results in a decrease in the prolactin content of the pituitary it is possible that in the wild prolactin is released '...at specific times of the life cycle as part of the regulation of reproduction and this enhanced release has a secondary benefit, the maintenance of salt gland activity during the isolation of the animal from sources of drinking water and when heat exposure can be considerable' (see Chapter 8).

It should be pointed out that, the experiments of both Douglas & Neely (1969) and Ensor & Phillips (1972*b*) to test salt-gland function were done by giving the birds an oral salt-load after dehydration, and it may be that the vascular bed of the alimentary canal was constricted and that the salt was not absorbed to the normal extent in the dehydrated birds. Before it can be decided with certainty that dehydration affects the salt gland *per se* it would seem desirable that concentrated salt should be given intravenously to overcome any effects on the gut, a point considered in Chapter 3.

Ecological clues

Several authors have pointed out that the state of development of the salt glands is a useful variable to study in investigations into ecology, migration and population. For example, Anderson & Warner (1969) found by weighing the salt glands that, during migration, populations of Lesser Scaup (*Aythya affinis*), which had spent the winter either on salt water or fresh water, converged into one population. Le Maire (1971) has also pointed out the usefulness of the salt gland in studies on migration, and has suggested that measurement of ATPase activity in the gland may be valuable. This may well be better than just measuring salt gland weight because the ATPase activity per unit weight of tissue increases markedly in birds kept on salt water (Chapter 9), and allowance would not then have to be made for body-weight.

In ecological studies it is important that the birds should be killed soon after capture (or in work on migration, as soon as they arrive at the new site), since adaptation to sea water or de-adaptation to fresh water occurs within a few days. Ensor & Phillips (1972*a*) have checked that the differences in salt-gland function observed when captive birds are kept either on fresh water or salt water do actually exist in wild populations. They collected young gulls (*Larus argentatus, L. fuscus*) from an island breeding site, thirty miles from the sea, and from a marine colony. The salt glands were very much larger in birds from the latter group and both the rate and concentration of the secretion obtained in response to an oral salt load were higher (Table 11.2). Such differences were not apparent after both groups of birds had been kept in captivity (on fresh water) from the time of collection at four to five weeks old until they were 15 to 16 weeks of age. However their results indicate that the birds from the inland colony had caught up with those from the marine site and that salt water '...accelerates the onset of the full secretory capacity of the gland but that in the absence of this environmental stimulus such capacity is acquired with time...'. It is, of course, possible that the weight of the glands may have been even greater if the older birds had been kept on salt water.

Ecological studies of the type undertaken by Ensor & Phillips are useful because they indicate the importance of

TABLE 11.2. *Salt-gland function in young gulls* (Larus argentatus *and* L. fuscus) *from an inland colony (Tan Hill) and a marine colony (Walney Island). From 4–5 to 15–16 weeks of age the birds were kept in captivity with fresh water for drinking. The birds 4–5 weeks of age were studied three days after capture. Salt-gland secretion was induced by an oral load of 500 mM NaCl, 10.5 mM KCl (20 ml/kg body-weight). Mean±S.E. (six birds in each group)* (modified from Ensor & Phillips, 1972 *a*)

Age (weeks)	Colony	Time in captivity	Body-weight (g)	Salt-gland weight (mg)	Salt-gland weight (mg/100 g)	Time to onset of secretion (min)	Secretion: Rate (ml/h/kg)	Secretion: Osmolality (mosmole/kg)	Secretion: Na (mM)	Secretion: K (mM)
4–5	Inland	3 days	741±79	409±6	57.1±6.1	10.5±1.8	1.5±0.67	1268±124	380±18	25.5±3.1
4–5	Marine	3 days	590±53	519±8*	93.5±6.8*	9.5±0.6	4.2±0.43*	1639±106*	689±53*	44.8±3.9*
15–16	Inland	3 months (on fresh water)	818±34	623±5	76.5±3.0	6.5±0.5	4.1±1.5	1572±110	690±15	32.4±2.0
15–16	Marine	3 months (on fresh water)	827±19	619±8	74.4±2.7	7.8±0.3	6.6±0.7	1695±107	702±10	31.4±4.4

* $P < 0.01$ compared with 4–5 week-old birds from inland colony.

effects observed in captive or domestic birds. For example, in gulls and ducks continuous maintenance on salt water impairs growth (p. 198). In the two groups of young birds collected by Ensor & Phillips (1972*a*), there was no significant difference in body-weight and they point out that other factors, like competition for food, probably regulate growth in the wild.

The use of salt gland size as an aid to taxonomy (Watson & Dovoky, 1971) is clearly unwise since its size is so variable within a species and depends to a great extent on habitat. In contrast, reports of salt secretion being observed under natural conditions clearly indicates that the salt gland is important (see Russell, 1958; Matthews, 1959; Meischke, 1967). Another useful feature for recognizing salt-gland secretion is that the glands swell when secretion starts and protrude visibly above the eye, and Owen & Kear (1972) have noted this in swans when they feed on inter-tidal vegetation.

Applied physiology

There are two fields in which a knowledge of salt-gland physiology is of importance to practical problems. The first is the maintenance of marine birds in zoos or for research purposes; the second is the treatment and rehabilitation of sea birds damaged by oil pollution.

It has been known for many years that some marine birds, which are often difficult to keep in captivity for any length of time, require sea water in order to survive (see Allen, 1925). Soon after the discovery of salt glands in birds, Frings & Frings (1959) found that 'stress', for example handling and excitement, induced salt-gland secretion in captive albatrosses (*Diomedea immutabilis, D. nigripes*) at times when extra-renal excretion would not be expected to occur. They argued that such a response could result in a fall in plasma sodium and chloride concentrations in birds given fresh water to drink, and that this might be of a sufficient magnitude to cause death. In an attempt to maintain these albatrosses in captivity (see Frings, Anthony & Schein, 1958), Frings & Frings (1959) gave salt water to drink and avoided stress as much as possible, for example, by not handling the birds. However, insufficient sea water was available and in addition to a bucket for drinking purposes, the birds were given a tank of fresh water for bathing. Under these conditions one bird died and others were declining but it was noticed that they were drinking the fresh water in the tank rather than the sea water. The sick birds were then given an injection of hypertonic sodium chloride and recovery was rapid. This provided direct evidence that they were hyponatraemic and, therefore, salt (in gelatine capsules or as tablets) was inserted into the fish given for food, in sufficient quantities to induce salt-gland secretion in the absence of apparent stress. In addition, sea water was available for both drinking and bathing and the birds were in fact seen to drink it, even when fresh water was also available after experimental salt-loads.

Many zoos now give marine birds salt in their food, especially when sea water is not available (see the various volumes of *International Zoo Yearbook*, Zoological Society of London, Academic Press). Not all marine birds require such treatment but two interesting physiological problems are raised. The first is the mechanism by which stress induces secretion to such an extent that hyponatraemia supervenes. The second is the control of salt intake because one might have expected that the sodium-deficient birds may have preferred to drink the sea water rather than the fresh water. It is possible, as Frings & Frings (1959) suggested, that they cannot discriminate between the two, but, of course, it is difficult to be certain because the two types of drinking water were not given in identical containers. The fact that the birds drank fresh water from the bathing pool when clearly short of salt, and from a tub of sea water after a salt-load, when both types of water were available, may suggest that they drink from the larger container. If they cannot discriminate, then this implies a major difference in their control of salt intake between these pelagic birds and the gulls studied by other workers (see above, p. 209).

Another possible benefit of giving sea water to marine birds is that it is said to reduce their susceptibility to aspergillosis, to which they are particularly prone (J. J. Yealland, cited by Clark & Kennedy, 1968). The reasons for the effect of sea water are not known but a possible explanation is that the hypertonic secretion from the salt glands washes, or perhaps kills, the fungal spores from the nostrils before they enter the lower respiratory tract.

A good deal of emphasis has been placed on the care and rehabilitation of birds damaged by pollution of the sea with crude oil. Clark & Kennedy (1968) and Peaker (1971*d*) have emphasized that salt and water metabolism is an important aspect to be investigated since at present little information exists on salt-gland activity in oiled birds on which to base practical recommendations.

The immediate care of sea birds, including removing oil, feeding etc., has been dealt with by a number of authors and animal welfare organisations. However it is true to say that the provision of salt or the supply of drinking water has received little serious attention. There are really three courses of action in terms of the provision of salt. The first is to give sea water to drink or to add considerable amounts of salt to the food. This is the procedure recommended by the Royal Society for the Prevention of Cruelty to Animals in its *Amended Procedure for Cleansing and Rehabilitation of Oiled Seabirds* (cited by Clark & Kennedy, 1968), and clearly has a sound basis if, but only if, oiled birds behave like the albatrosses studied by Frings & Frings (1959) and secrete during stress. On the other hand, stress (p. 128), dehydration (p. 209) and loss of blood can lead to inhibition of salt-gland activity, and the provision of excess salt would exacerbate the situation. Crocker & Holmes (1971*b*) have pointed out that intestinal absorption of salt is necessary for salt-gland secretion and for obtaining free water, and that any factor, like oil in the gut, which impaired absorption could seriously affect survival.

The second option is to give fresh water to drink but, if stress does induce secretion, it could result in hyponatraemia as Frings & Frings (1959) have shown. We have several times in this chapter urged that more research should be carried out on the regulation of salt and water intake in marine birds of different species. If it were to be found that the preference for saline drinking water is related to the level of sodium chloride in the body, or, in other words, that hyponatraemic birds would drink sea water in preference to fresh water, and vice versa, a sensible regime, the third option, would be to provide a choice. If controlled experiments do not show this and the birds show no discrimination or prefer to drink one or the other in a manner independent of the state of their internal environ-

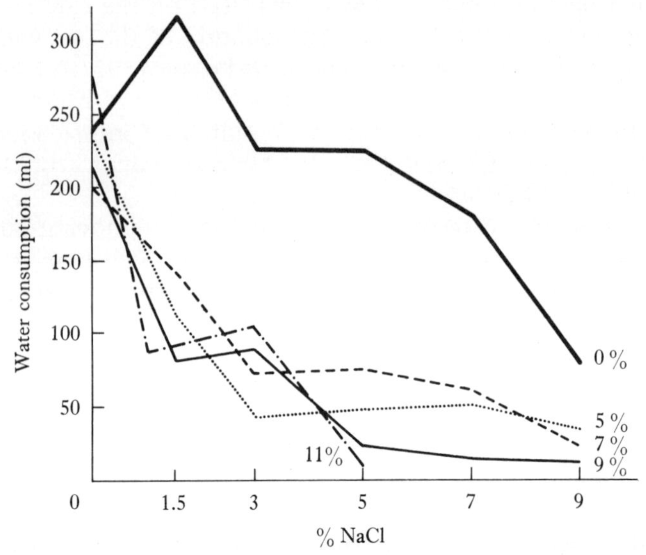

Fig. 11.6. Experiment showing that the voluntary intake of saline solutions is affected by the osmotic state of the animal. The consumption of drinking water of different concentrations of sodium chloride in Herring Gulls (*Larus argentatus*) loaded intraperitoneally with 0–11 per cent sodium chloride (30 ml/kg) in a multiple-choice experiment (from Schwarz, 1966).

ment, this treatment could not be applied. In such cases successful care could well depend upon measurements of plasma electrolyte concentrations in individual birds because there may be differences not only between species but also between individuals depending upon the amount of damage they have suffered as a result of the oil. In fact there is some evidence that preference can be affected by the internal environment. Schwarz (1966) found, by loading gulls with distilled water or with hypertonic sodium chloride solutions, that the preference for drinking water of different salinities was altered (Fig. 11.6); the birds given distilled water consumed much larger quantities of the hypertonic solutions than salt-loaded birds, which clearly showed a preference for distilled water.

Information on salt balance in oiled sea birds during their rehabilitation is scanty. C. J. F. Coombs (unpublished, in Clark & Kennedy, undated) has measured plasma sodium and chloride concentrations in guillemots during recuperation.

Sodium and chloride levels were somewhat elevated, but not excessively high in three birds (164–168 mM and 123–130 mM respectively) but in a fourth, sodium was 189 mM and chloride 153 mM. If these birds were being given salt in their food or sea water to drink, as seems likely, it would suggest that the salt gland was not able to cope, and could have been inhibited by some form of stress.

Clark & Kennedy have carried out a survey on the treatment of oiled sea birds by various people, and the following extract from their booklet *How Oiled Seabirds are Cleaned* serves to illustrate how little is known about these problems.

> Fresh water, sea water and artificial sea water are commonly provided for drinking, but one report stated that auks had been observed to drink both sea water and fresh water, while another (Dunstone) that guillemots and razorbills never drank water of any kind. Thomas concluded that sea water is most suitable for auks, but that if this was not available they would take fresh water in preference to artificial sea water. This observation is generally confirmed by other respondents. Although the Mousehole workers provided all three kinds of drinking water, they found no deterioration of condition among birds that had access only to fresh water. Carradice on the other hand stated that the provision of sea water for drinking was essential. It may be that these discrepancies arise from the different nature of the foods provided to different birds, but there is no evidence of this. Another factor which may influence these observations is the quantity of liquid supplied with the food.

It is for such reasons that Peaker (1971*d*) has urged that physiological investigations on salt balance in oiled birds should be carried out.

A further problem to be considered is the rehabilitation of birds for release at sea. Ringing studies have indicated that very few birds survive after release even though they were in apparently good health immediately before. In fact a number of articles appeared in newspapers questioning whether the effort and expense of rescuing oiled seabirds is worthwhile. The state of adaptation of the salt glands could be an important factor after release, particularly in the birds, most often affected by oil, that live on the open sea. If, during the period of rehabilitation, fresh water is given for drinking and there is insufficient salt in the diet it is extremely likely that de-adaptation of the salt glands, as well as other organs like the intestine, would occur; the bird would then be unable to cope with a marine existence when released (Peaker, 1971*d*). In fact, in ducks the $T_{1/2}$ of decline in Na^+/K^+-ATPase activity in the salt glands when transferred from salt water to fresh water is only five days (Ernst *et al.* 1967). It would therefore seem essential to prepare the birds for release by giving them sea water to drink for some time before, in order to bring the osmoregulatory system into the state necessary for survival; if necessary this could be done by gradual adaptation.

Another marine pollutant that could affect the salt glands is the insecticide DDT. Janicki & Kinter (1971) have found that this substance affects salt movements in the intestines of eels adapted to sea water and there is the possibility that it affects the sodium pump. In birds at the top of a food chain, DDT could well act at this site in many organs including the salt gland. In fact, recently, Friend, Haegele & Wilson (1973) have found that DDE (dichlorodiphenyldichloroethylene) in the diet has an inhibitory effect on salt-gland secretion in response to a salt-load in ducks kept on fresh water but, strangely, not in those kept on salt water.

Non-marine birds

The marine environment is not the only habitat in which there is a shortage of free water. Some inland lakes are rich in salt, and, of course, arid conditions are a severe test for water-conserving mechanisms in most animals. However, few birds that live in deserts have salt glands and it should not be imagined that their presence and ability to secrete hypertonic salt solutions is necessary for survival under such conditions. The ability to conserve water can in part be attributed to uricotelism, a preadaptation to desert life, as a number of authors have pointed out (see Serventy, 1971). Even so a number of birds without salt glands can drink salt water, even sea water, without ill effect and in such circumstances the ionic concentration of the urine is high (chloride concentration up to 960 mM) (see Schmidt-Nielsen, 1964; Serventy, 1971). While there is evidence that loops of Henle are present in the kidneys of some birds (Poulson, 1965), the renal and cloacal mechanisms involved are by no means clearly understood.

In the 'higher' orders of birds most species do not have salt glands (Table 14.2), even though many of them live in arid habitats. In fact one passeriform, *Cinclodes nigrofumosus*, lives on the coast of Peru and Chile and feeds amongst the rocks and sand. The coast is in fact the edge of a desert so the only source of drinking water is the sea. Paynter (1971) has found marine invertebrates in the stomach of these birds but measured the nasal glands and found them to be no larger than in other terrestrial passeriforms (the weight of the glands is unfortunately not stated). Therefore it would appear from this morphological evidence that the nasal gland does not function as a salt gland in this species although it must be pointed out that the glands are not large in some other terrestrial birds in which salt secretion has been observed. Perhaps it would be safer to conclude from Paynter's morphological study that the nasal glands are not a major avenue for extra-renal excretion in this interesting South American bird.

Salt glands in terrestrial birds

The ostrich (*Struthio camelus*), which lives in the arid parts of Africa and Arabia, was found by Technau (1936) to have large nasal glands with one duct entering the nostril from each side (Fig. 11.7). Schmidt-Nielsen *et al.* (1963) discovered nasal salt secretion in this species, the first report of a salt gland in terrestrial birds, when they exposed one to high ambient temperatures and also deprived it of drinking water. The fluid they collected from the nostrils was found to contain high concentrations of potassium, sodium, calcium and chloride. Considerable variation in composition was apparent: in some samples sodium and potassium concentrations were similar but in others potassium was five to ten times higher than sodium. A fascinating field for ion transport studies is thus exposed. It is tempting to consider that the relative concentrations of sodium and potassium in the fluid may be related to dietary intake, as appears to be the case in lizards (Chapter 13). The secretion of both cations would seem to be necessary in view of the high potassium content of the diet and the possibility that these birds may drink water from soda or salt lakes (see Schmidt-Nielsen, 1964). Cloudsley-Thompson & Mohamed (1967) also observed nasal secretion in the ostrich but did not report the composition of the fluid.

Schmidt-Nielsen *et al.* (1963) also observed nasal secretion in

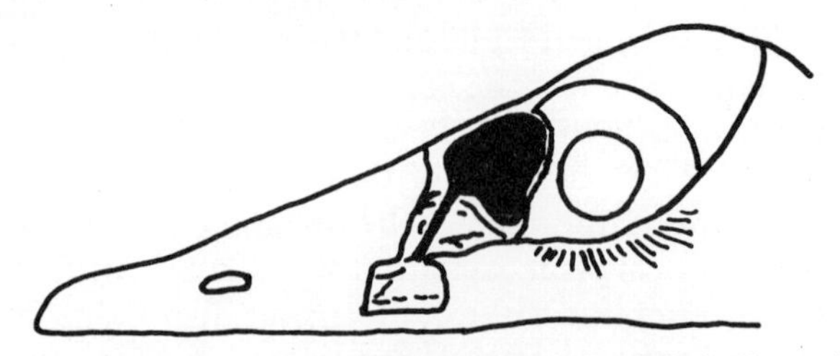

Fig. 11.7. The position of the salt gland (shown in black) in the ostrich (*Struthio camelus*) (adapted from Technau, 1936).

a galliform, the partridge *Ammoperdix heyi*, from North Africa and the Middle East. This species, like the ostrich, was found to secrete in response to high ambient temperatures and water deprivation. In addition, secretion rich in sodium chloride but low in potassium, was obtained after loading with hypertonic sodium chloride; salt-loading was not attempted in the ostrich because of the value of the birds studied!

Nasal salt secretion has also been observed in Falconiformes (birds-of-prey) by Cade & Greenwald (1966). They noticed that many of these birds secrete a clear fluid from the nostrils while eating. The concentrations of both sodium and chloride were high (400–2400 mM). However the amounts of fluid secreted were so small that it is difficult to believe that the salt glands are an important route for excretion. In fact balance studies in the Red-tailed Hawk (*Buteo jamaicensis*) have shown that the nasal gland only accounts for about three per cent of the sodium excreted (Johnson, 1969). However Cade & Greenwald (1966) point out that the nasal glands may well be more important in the nestling than in the adult since its only source of water is in the food brought to it by the parents. In hot regions, like the South African parks where this work was carried out, evaporative water loss would presumably be high, and Cade & Greenwald actually observed the onset of secretion in a four-day old Bateleur Eagle (*Terathopius ecaudatus*) when exposed to sunlight in its nest. In the adult birds studied, sodium and chloride were the predominent ions but the concentration of potassium was higher in many cases than in most marine birds.

The North American Roadrunner (*Geococcyx californianus*), a cuculiform which lives in arid regions, has recently

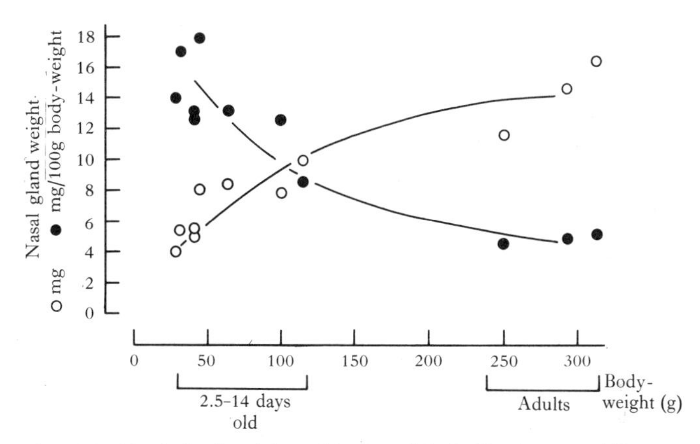

Fig. 11.8. Nasal gland weight and body-weight in Roadrunners (*Geococcyx californianus*) of different ages (drawn from the data of Ohmart, 1972).

been found to have a nasal salt gland by Ohmart (1972). It appears to be relatively more important in the nestling than in the adult because as the birds grow, the relative weight of the salt gland decreases (Fig. 11.8). Secretion was observed after an intravenous salt-load and the mean composition of the fluid was 776 mM sodium, 794 mM chloride and 69 mM potassium; osmolality was 1567 mosmoles/l. Ohmart concluded that nasal salt secretion appears to have its greatest survival value in nestlings. The nests are constructed as platforms in cactus plants and the young are exposed to solar radiation when the adults have to be away from the nest in order to catch lizards during their activity period. Salt-gland secretion therefore appears to counterbalance the water lost by gular fluttering and cutaneous evaporation while the adults are away from the nest and cannot shade their young. It is likely that, when the adults return, the young receive water in addition to that in their food because the adults have been observed to regurgitate a clear liquid and give it to the young with each feed. Ohmart does not believe that the salt glands are used to any great extent by free-ranging adults.

Nothing is known about the control of salt-gland secretion in terrestrial birds but there is no reason to suppose that in those which secrete hypertonic sodium chloride, an osmoreceptor is not responsible as in marine birds, although an association between secretion and feeding seems possible in birds-of-prey.

A species which, for its size, has large nasal glands is the dipper (*Cinclus cinclus*) (Technau, 1936). This bird would appear not to be short of water because they live by fast-flowing, fresh-water streams and dive in to catch their prey – insect larvae, molluscs, worms etc. – and the function of the glands is unknown. However, recently Vader (1971) has found that these dippers in coastal regions feed on marine invertebrates and has reviewed earlier ornithological findings of this species being sighted near the sea. Therefore it is possible that the specimen studied by Technau (1936) came from the coast. If the dipper does have a functional nasal gland it would be the first passeriform to be discovered with this ability. In the only histological study, Goodge (1961) noted that the cells of the nasal gland are columnar and that mucus is not present in the American species, *C. mexicanus.*

The question of whether the nasal gland of the domestic fowl functions as a salt gland has received some attention. While the gland in this species has a structure basically similar to that in marine birds it appears less active in that fewer mitochondria could be seen in the light microscope by McLelland *et al.* (1968). Certainly salt-loads that are effective in other domestic birds, ducks and geese, have been found not to evoke nasal secretion in fowls (M. Peaker & J. L. Linzell, unpublished). McLelland & Pickering (1969) attempted to determine whether the chronic exposure to salt water induces glandular activity. They gave 0.6 per cent sodium chloride drinking water, instead of tap water, to chickens for nine weeks. In weeks six to nine, 1.5–3.0 M sodium chloride was injected daily by the intraperitoneal route. Only in the last weeks was plasma osmolality increased. Nasal secretion was not observed during this period nor did the nasal glands increase in weight. However the saline given for drinking was hypotonic to plasma and since the fowl can produce hypertonic urine (see for example, Skadhauge & Schmidt-Nielsen, 1967), no marked osmotic stimulus would have been given during this period. Moreover the intraperitoneal injections of salt were clearly stressful since one bird died and the others ceased laying eggs, the comb and wattles became pale and 'the group lost its previously healthy

TABLE 11.3. *Composition of inland bodies of water. Part A is from Gordon (1968), B are the data of Kalmbach & Gunderson (1934) recalculated and C is from* Encyclopaedia Britannica. *A dash means that a figure is not given*

	Osmolality (mosmoles/1)	Salinity (g/1)	Na (mM)	K (mM)	Ca (mM)	Mg (mM)	Cl (mM)	SO_4 (mM)	HCO_3 (mM)	Br (mM)
erage' river water	1	—	0.08	0.01	0.3	0.09	0.05	0.08	—	—
e Manitou, Canada	2000	—	780	28	14	500	660	540	—	—
at Salt Lake, Utah	6000	—	3000	90	9	230	3100	150	—	—
ious lakes in	—	0.64	5.7	—	1.6	1.3	6.0	0.5	2.3	—
rth America	—	103.0	1453	—	25.0	135.0	1605.0	79.0	3.9	—
	—	0.26	0.1	—	1.02	0.5	0.3	0.1	2.7	—
	—	0.48	5.1	0.7	0.7	0.8	0.6	0.4	3.5	—
	—	2.71	31.4	—	0.5	1.7	1.3	7.1	16.0	—
	—	1.69	12.9	—	0.7	5.3	0.8	8.1	6.6	—
	—	38.3	399.0	—	16.9	69.7	71.0	241.0	17.5	—
	—	1.05	81.2	—	13.6	13.0	11.4	0.3	122.3	—
	—	0.85	6.8	—	1.3	0.7	1.0	0.9	8.2	—
	—	213.7	3539.0	—	1.0	1.0	1494.0	221.0	860.0	—
d Sea	—	192	972.0	91.0	227.0	1056.0	3549.0	—	—	57

appearance’. It seems to us that perhaps more work is required before the domestic fowl is finally condemned by the salt-gland physiologist as a ‘non-secretor’.

Inland saline waters

Saline waters which may also be alkaline occur inland particularly in arid regions (Table 11.3). In East Africa the soda lakes contain carbonate and bicarbonate and it is in these conditions that flamingos live and breed. Although they obtain their invertebrate food by a highly-efficient filtering process in the beak, it is difficult to see how they can avoid ingesting some of the salt water and an interesting problem to be answered is whether the salt glands of these birds can secrete bicarbonate. They can certainly secrete sodium chloride at high concentrations when given a sodium chloride load (McFarland, 1959) but loading with sodium bicarbonate appears not to have been tried.

A great deal of interest has centred on the waterfowl which inhabit the alkaline, saline lakes of North America because, in some years, many of them die from a paralysis known as 'duck disease'. The first work on this problem was done by the noted ornithologist Alexander Wetmore in the early part of this century and his findings were published as two bulletins of the United States Department of Agriculture in 1915 and 1918. Because many of the lakes and creeks the ducks inhabited were so alkaline and salty, Wetmore thought that the disease was due to 'alkali poisoning', and his was the generally accepted theory for the cause of the disease for a number of years. Many of the inland waters are certainly salty; the total salinity can even be slightly higher than that of the Dead Sea (Table 11.3). Wetmore noted in his reports that loading birds with 'alkali' induced secretion from the nasal region and mistakenly attributed this to a response of the Harderian glands. What a pity it was he did not taste or analyse the secretion for he only just missed discovering salt glands!

Wetmore's theory for the aetiology of duck disease fell into disfavour when Gunnison & Coleman (1932) and Kalmbach & Gunderson (1934) showed that the disease was associated with the ingestion of the toxin of *Clostridium botulinum* Type C, to which birds are particularly susceptible. However Wetmore clearly showed, and this was confirmed by Kalmbach & Gunderson (1934) that many birds recovered when given fresh water to drink and this led F. G. Cooch (1961; 1964) of the Canadian Wildlife Service to suggest that there may be a link between 'salt-water intoxication' and avian botulism.

Following the discovery of salt glands in marine birds Cooch quickly realized that ducks inhabiting salt lakes would need such an excretory mechanism and that any agent which blocks cholinergic transmission could inhibit secretion by the glands and lead to a fatal accumulation of salt in the body. He then showed that the ducks of these regions do have highly active salt glands and that botulinus toxin Type C had a dose-dependent inhibitory effect on secretion. Moreover the lethal dose of toxin was considerably lower in birds given hypertonic saline. Cooch therefore concluded that ducks may die of botulism in two ways: (i) from a massive dose of toxin, ingested while eating insect larval cases in which the bacterium is incubated, which is itself lethal, or (ii) from ingestion of sublethal doses plus water or food containing large amounts of salt. In the latter case the provision of fresh drinking water is an effective cure, allowing the kidney to remove the excess salt. Support for Cooch's theory of duck disease (which also occurs in South Africa, Blaker, 1967), but not quoted by him, which links the two previously proposed is that Shaw (1929) found that ducks suffering from the disease have a higher plasma chloride concentration. We have recalculated Shaw's data and the plasma chloride level was 138 mM (range 124–169) in diseased birds and 129.5 (115–141) in apparently healthy birds from the same habitat, a statistically significant difference ($P < 0.02$).

This is how the situation on duck disease now stands and, apart from being an excellent example of salt-gland physiology being applied to an ecological problem, shows clearly the importance of the salt glands in permitting survival in seemingly unfavourable environments. It is difficult to imagine that birds could permanently inhabit waters with salinities as high as some of those shown in Table 11.3 with osmolalities ten times that of blood plasma or three times that of sea water. However Kalmbach & Gunderson (1934) make the point that fresh water is often near and birds may fly there to obtain relief after feeding.

PART 2 · REPTILIAN SALT GLANDS

12 MARINE REPTILES

After the discovery of salt glands in marine birds it was soon realized that marine reptiles face the same problems of disposing of excess salt. Therefore it was gratifying that Schmidt-Nielsen & Fänge (1958*b*) found salt glands to be present in a number of marine reptiles. One of the most interesting aspects of these glands in reptiles is that, in most groups, they are not homologous with the avian nasal gland and current evidence indicates that nasal, orbital and oral glands have all evolved the ability to secrete salt. In fact one of the main problems has been to identify which gland (or glands) is involved in extra-renal excretion.

Chelonians

Salt glands in a marine turtle, the Loggerhead (*Caretta caretta*) were first discovered by Schmidt-Nielsen & Fänge (1958*b*). They found a high sodium chloride concentration in tears and observed that the secretion appeared from a duct which opens into the posterior corner of the eye; this duct was traced to a posterior orbital gland. They also succeeded in cannulating the duct, and, when methacholine was injected, a markedly hypertonic fluid containing 732–878 mM sodium, 810–992 mM chloride and 18–31 mM potassium was secreted. Later workers showed that other turtles, the Green (*Chelonia mydas*) (Holmes & McBean, 1964) and Ridley's (*Lepidochelys olivacea*) (Dunson, 1969*b*) secrete fluid of a similar composition.

Perhaps at this point it would be advisable to state the system of nomenclature we are using for chelonians. In English usage the term turtle is reserved for truly marine forms, terrapin for fresh water species and tortoise for the wholly terrestrial types. Although there is some overlap in habitats these terms are thought by British herpetologists to be better from both ecological and taxonomic viewpoints than the American system which lumps both turtles and terrapins into 'turtles'.

In addition to the truly marine turtles, Schmidt-Nielsen & Fänge (1958*b*) also observed that the Diamond-back Terrapin (*Malaclemys terrapin*) can excrete salt (616–788 mM sodium) in tears when given a salt-load. This species, which the Americans do call a terrapin, belongs to a large family (Emydidae), and is euryhaline, occurring along the eastern seaboard of the USA.

The tears that turtles produce when they clamber onto sandy beaches to lay their eggs have been thought to either keep the surface of the eye wet or to wash sand from the eyes. Schmidt-Nielsen & Fänge (1958*b*) suggested that tear secretion plays an osmoregulatory role and noted that Carr (1952) had observed that the tears fail to wash the eyes and in fact by the time the female began to lay her eggs the '...eyes were closed and plastered over with tear-soaked sand and the effect was doleful in the extreme'.

Identity of the salt gland

Schmidt-Nielsen & Fänge (1958*b*) did not attempt to determine the homology of the turtle salt gland but later workers called it the nasal gland (Holmes & McBean, 1964; Benson, Phillips & Holmes, 1964), which it clearly is not, the Harderian gland (Dunson & Taub, 1967; Dunson, 1969*a*), or the lachrymal gland (Abel & Ellis, 1966). Both the Harderian and lachrymal glands are located in the orbit and it was not until recently that the question was resolved.

Cowan (1967; 1969; 1971) found that both types of orbital gland are present in terrapins but that the Harderian gland empties its secretion into the anterior corner of the eye and is no different in gross or microscopical structure in euryhaline species compared with those that live in fresh water. In contrast, the posterior orbital or lachrymal gland is larger and has a different structure in the euryhaline species (e.g. *Malaclemys*) compared with fresh water species, a difference also noted by Peters (1890). Since this gland discharges its contents into the posterior corner of the eye, where Schmidt-

Nielsen & Fänge (1958*b*) found the salt-gland duct in the Loggerhead, it seems that, in all probability, the lachrymal land functions as a salt gland in marine and euryhaline chelonians.

Structure of the lachrymal salt gland

The gross, histological and ultrastructural morphology of the salt gland has been studied by Schmidt-Nielsen & Fänge (1958*b*), Benson *et al.* (1964), Ellis & Abel (1964*a*, *b*), Abel & Ellis (1966) and Gerzeli (1967*b*) in marine turtles, and by Cowan (1967; 1969; 1971) in *Malaclemys.*

The lachrymal gland lies in the orbit, posterior to the eye and consists of branched lobules. Tightly-packed, branched secretory tubules radiate from the central canal of the lobules and the canals join to form a short, wide main duct which opens into the posterior canthus of the eye. The blood supply is rich, and according to Abel & Ellis (1966) is arranged countercurrent to the flow of secretion down the tubules, as in the nasal glands of marine birds. Cholinesterase-containing nerve fibres form a dense plexus around the secretory tubules. In addition nerve fibres containing monoamine oxidase are seen around the lobules (Abel & Ellis, 1966), suggesting that in addition to a cholinergic innervation there is also an adrenergic sympathetic supply.

The arrangement of cells in the secretory tubules is also similar to that in marine birds. At the blind end of the tubules the lumen is small and the cells are unspecialized with few mitochondria and little histochemical indication of marked metabolic activity. In contrast the principal secretory cells along the rest of the tubule appear highly active with many mitochondria and histochemical reactions for oxidative enzymes. Infolding of the basal membrane is not marked but along the lateral membranes microvilli form an extremely complex intercellular space which gives a most distinctive appearance even in the light microscope (Plate 12.1). This arrangement is unlike that of the nasal gland in birds where cytoplasmic folds intermingle with those of the neighbouring cell in the lateral intercellular space. A typical junctional complex joins the cells at the junction of the apical and lateral cell membranes. Abel &

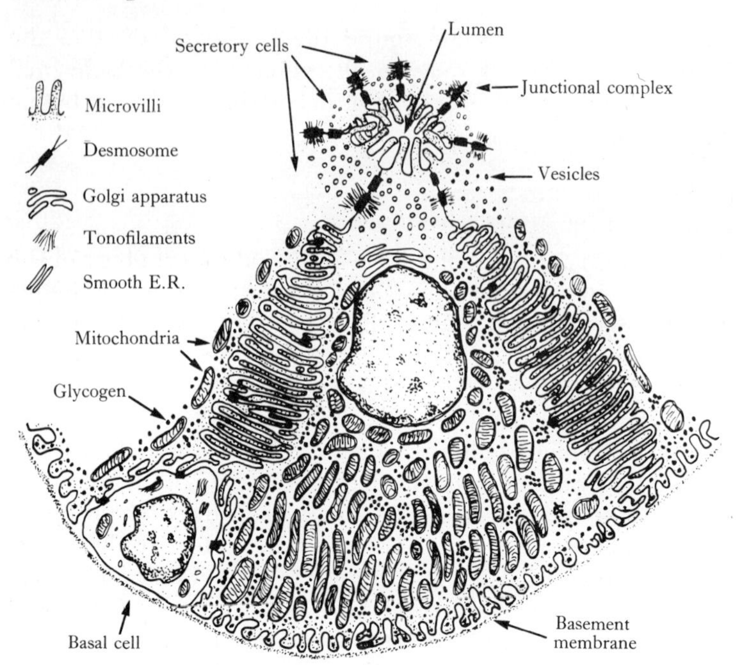

Fig. 12.1. Diagram of the ultrastructure of a secretory cell of the turtle lachrymal salt gland: E.R., endoplasmic reticulum.

Ellis (1966) also found a few microvilli on the apical membrane and observed that the endoplasmic reticulum is of the smooth type. The Golgi apparatus, lying above the nucleus is prominent and glycogen granules are also seen in the secretory cells. A diagram showing the ultrastructure of the secretory cell in a turtle is shown in Fig. 12.1.

A third type of cell is also present in the tubule. These cells lie between the bases of the secretory cells and do not reach the lumen of the tubule. They appear unspecialized with large nuclei and are characterized by the presence of tonofilaments in the cytoplasm. Apart from their position Abel & Ellis (1966) could find no indication that these are myoepithelial cells.

As in birds, the cells lining the central canals of the lobules do not appear to be highly active; they have little cytoplasm and few mitochondria, and are arranged in several layers. However, in addition, mucus-secreting cells also occur in this epithelium and they discharge their contents into the lumen.

Cowan (1971) observed an essentially similar structure of the gland in the euryhaline Diamond-back Terrapin, *Malaclemys*, except that there is more basal infolding. He too found a well-developed Golgi apparatus.

Ellis & Abel (1964*a*, *b*), Abel & Ellis (1966) and Cowan (1971) have found, using histochemical techniques, a large amount of mucopolysaccharide in the lateral intercellular spaces. It seems likely that this could account for the presence of a prominent Golgi apparatus since this organelle is known to be involved in the synthesis of mucopolysaccharides in many different tissues (see Neutra & Leblond, 1969). The presence of vesicles near the lateral cell membranes (Abel & Ellis, 1966) might suggest that material from the Golgi apparatus is released into the lateral intercellular spaces by 'reverse pinocytosis'. Farber (1960) showed that mucopolysaccharides attract cations and function as an ion exchange resin, and Bennett (1963) proposed that their negative charge could attract cations to the absorptive surfaces of cells. As Abel & Ellis (1966) have suggested, this could be their role in the turtle salt gland since large amounts of sodium have to be extracted from the plasma, but an obvious argument against this view is that chloride would be repelled.

Mechanism of secretion

Surprisingly, no work has been done on ion transport in chelonian, or for that matter, in any reptilian salt glands.

Nervous control of secretion

It seems likely that, as in marine birds, secretion is not continuous but occurs in response to the ingestion of hypertonic food or sea water. Holmes & McBean (1964) showed that extra-renal excretion occurred at a greater rate after juvenile Green Turtles (*Chelonia mydas*) had been given food or an intramuscular salt-load (Fig. 12.2). Similarly in salt-adapted *Malaclemys*, salt-loading induced secretion (Dunson, 1970). Since methacholine will induce secretion in marine and euryhaline species (Schmidt-Nielsen & Fänge, 1958*b*; Dunson, 1970) one can only assume that secretion is controlled by a nervous reflex, as in marine birds.

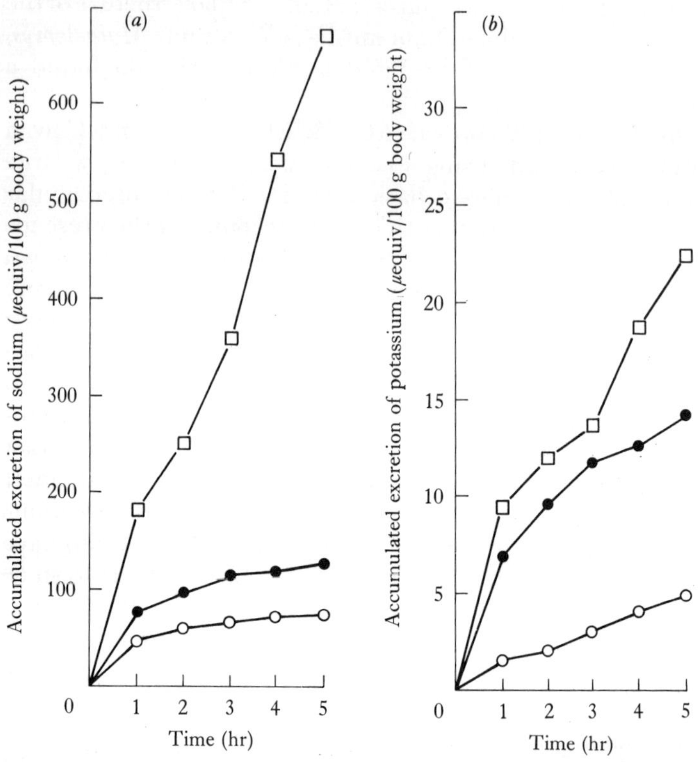

Fig. 12.2. The excretion of (*a*) sodium and (*b*) potassium by fed and unfed green turtles (*Chelonia mydas mydas*) over a 5-hr period. The fed animals ate 2.3 ± 0.25 g shrimp per 100 g body weight, of the following composition: 788.9 ± 5.8 g water per kg wet weight, 54.4 ± 1.3 mmoles sodium per kg and 92.1 ± 3.9 mmoles potassium per kg. Amphenone B was administered as a single intramuscular dose (5 mg) immediately after feeding. □, Fed; ○, unfed; ●, fed+amphenone (from Holmes & McBean, 1964).

Effect of hormones

Holmes & McBean (1964) studied excretion in juvenile Green Turtles by immersing them in de-ionized water and following the time-course of sodium and potassium accumulation in the water. By blocking the cloaca they found that normally the cloacal loss of ions must be negligible. Therefore they concluded that the salt glands are the main route of electrolyte excretion. Amphenone B, which blocks the formation of adrenocorticosteroids, was administered to 'chemically adrenal-

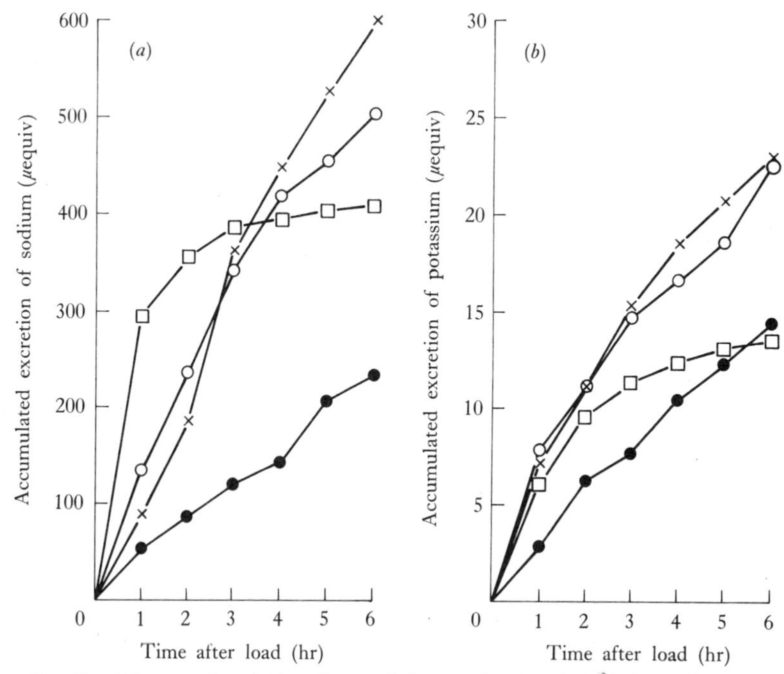

Fig. 12.3. The excretion of (*a*) sodium and (*b*) potassium by salt-loaded juvenile green turtles (*Chelonia mydas mydas*). Each animal received a single intramuscular injection of 500 μmoles sodium chloride; amphenone B and corticosterone (5 and 2.5 mg respectively) were administered at the time of salt-loading. Excretion was measured over a 6 hr period. □, amphenone+corticosterone; ○, controls; ×, tied cloacae; ●, amphenone (from Holmes & McBean, 1964).

ectomize' the turtles; following this treatment the initial rate of secretion after salt-loading was markedly depressed. The administration of corticosterone i.e. replacement therapy, restored the response (Fig. 12.3). Thus, as in birds, it appears that normal circulating concentrations of corticosterone or other adrenocorticosteroids are necessary for salt-gland secretion but it is not known whether these hormones have a controlling function.

Salt-water adaptation in Malaclemys

Cowan (1969) showed that the weight of the lachrymal glands increases when *Malaclemys* is kept in sea water, and Dunson

(1970) showed that the sodium concentration increased from 284 mM when the reptiles were kept in fresh water to 682 mM after several weeks in sea water. Thus the pattern of adaptation is like that in birds. Similarly, the weight of the salt glands decreased when Green Turtles were kept in fresh water (Holmes & McBean, 1964).

The physiological importance of salt-gland secretion

Holmes & McBean (1964) in their experiments on Green Turtles clearly showed that extra-renal excretion is the most important route for sodium and potassium elimination. The concentrations of both these ions were low in urine collected from the bladder, even in animals kept in sea water. They suggested that although the concentration of potassium in tears is low, salt-gland secretion may be the major route for the elimination of this ion, which is ingested in excess quantities of their diet of marine vegetation and animals. They also made the interesting suggestion that '...these animals drink sea water not only to obtain osmotically free water, but also to enable the excretion of the large amounts of potassium ingested as food. That is, sodium from the ingested sea water may act as a "vehicle" for the extra-renal excretion of dietary potassium.'

In *Malaclemys*, the salt gland seems incapable of maintaining homeostasis indefinitely when the animals are in sea water because plasma sodium concentration rises and they lose water. It is probable that migration to less concentrated estuarine water is necessary for these animals to rehydrate (Bentley, Bretz & Schmidt-Nielsen, 1967; Dunson, 1970). In other words the salt glands in this species are not as efficient as in the truly marine forms.

Crocodiles

Of the twenty-five species of crocodilian several are known to occur in, or will tolerate, brackish or sea water. One, the Estuarine Crocodile (*Crocodylus porosus*) from the Far East is known to swim from island to island in the Malay Archipelago and it is usually said that its ability to withstand marine conditions is responsible for its wide distribution throughout South-East Asia, the East Indies, northern Australia and even islands in the Pacific. Dunson (1969*a*; 1970) injected methacholine and collected tears from an American Crocodile (*Crocodylus actus*) (not to be confused with the American or Mississippi Alligator, *Alligator mississippiensis*), which is also known to occur in brackish-water swamps; the sodium concentration in the secretion was 484 mM. In very young Estuarine and American crocodiles, tears could be induced by giving hypertonic sodium chloride or methacholine subcutaneously but the concentration was relatively low (sodium 245 mM, potassium 11.5 mM). Acclimation to sea water raised these concentrations to 343 mM for sodium and 23 mM for potassium.

The identity of the salt gland is not known. Dunson (1970) noted that when nasal secretion was observed, the sodium concentration was low compared with sea water. He suggested (Dunson, 1969*a*) that the pre-orbital Harderian gland is responsible for salt secretion but in a young specimen of the Estuarine Crocodile, *C. porosus*, he found that the gland was small although histologically it bore some resemblance to a typical reptilian salt gland. Thus Dunson (1970) has suggested that the present evidence indicates that extra-renal excretion from orbital glands in these species is minor. However, no really large salt-adapted specimens have been studied and it is hoped that some brave person will try to collect tears from a twenty-foot-long Estuarine Crocodile.

Sea snakes

The Hydrophiidae are a family of approximately fifty species of venomous snakes, almost all of which are marine and occur in the Indian and Pacific Oceans. Our knowledge of salt and water metabolism in these interesting snakes has been gained through the efforts of Dr W. A. Dunson and his associates who have not been deterred by their extremely potent venom.

Identity of the salt gland

Dunson & Taub (1966) found that a hypertonic fluid is excreted from the mouth of sea snakes. They dissected specimens of both *Laticauda semifasciata* and *Pelamis platurus* to search for a gland that could secrete salt and reached the conclusion that neither the nasal nor the Harderian gland could be involved since although *Laticauda* possessed both, *Pelamis* had neither. However in both species, a gland was discovered in the palate which had not been described in any other family of snakes. They suggested that this gland might well be the salt gland because it was found to be larger in *Pelamis*, which is more pelagic, than in *Laticauda.* Taub & Dunson (1966; 1967) therefore called it the 'natrial gland' in view of its supposed function. But the matter did not rest there.

Although Taub & Dunson (1967) showed a drawing of the natrial gland in *Pelamis platurus* in their paper, Dunson (1968; 1969*a*) reported that B. Burns & G. Pickwell had made unpublished observations on the head glands of sea snakes. They examined sections of the head and could find no natrial gland in *Pelamis.* This gland therefore could not be the salt gland of sea snakes. On the other hand, a large 'Harderian gland' was found (cf. Taub & Dunson, 1967) which emptied its secretion into the mouth near Jacobson's organ. Dunson (1968) therefore felt that the Harderian gland might be the salt gland. It should be noted that this orbital gland in snakes is not regarded as being homologous with the Harderian gland of other vertebrates (see Bellairs, 1942).

However, later work by Dunson, Packer & Dunson, much of which was carried out on board the research vessel *Alpha Helix*, clearly showed that none of the glands previously considered was responsible for salt secretion, and that, in fact, the salt gland is the posterior sublingual gland which empties into the tongue sheath. Direct confirmation of the identity of the salt gland was obtained by the hazardous procedure of cannulating the tongue sheath into which only the posterior sublingual gland empties; a hypertonic secretion was then obtained. Moreover the size of the gland in *Pelamis* was 0.043 per cent of the body-weight – a similar figure to that for salt glands in other marine reptiles.

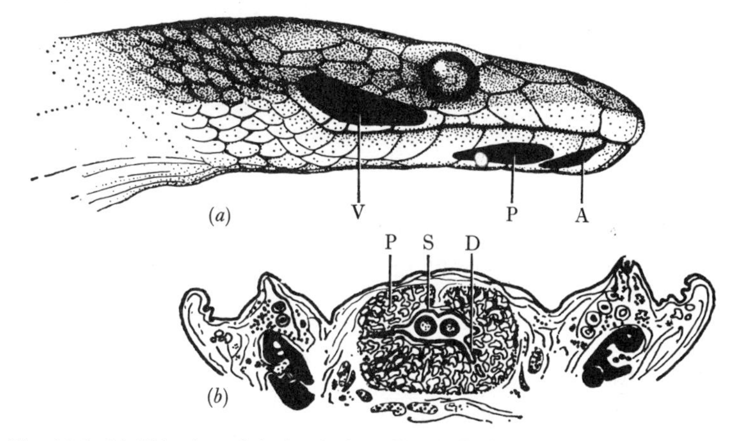

Fig. 12.4. (*a*) Side view of the head of a yellow-bellied sea snake (*Pelamis platurus*) showing the positions of the venom (V), posterior sublingual (P) and anterior sublingual (A) glands. The Harderian gland is medial to the eye.

(*b*) A cross-section through the lower jaw of *Pelamis* showing the large posterior sublingual gland (P) which almost completely surrounds the tongue sheath (S); portions of three ducts (D) are shown entering the sheath (from Dunson *et al.*, 1971, copyright © 1971, American Association for the Advancement of Science).

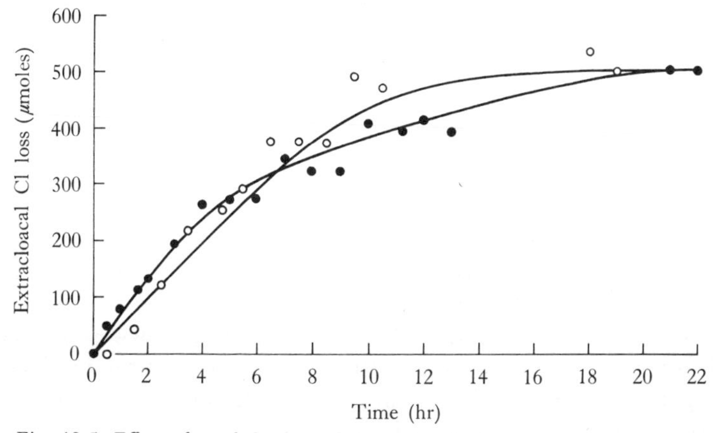

Fig. 12.5. Effect of a salt-load (○, 0.5 mmoles NaCl) and an osmotically equivalent sucrose load (●, 1.0 mmoles sucrose) on the extracloacal chloride loss in a specimen of the sea snake, *Pelamis platurus* (from Dunson, 1968).

Structure of the posterior sublingual gland

The gland is single and almost covers the tongue sheath into which a number of ducts open (Fig. 12.4); the secretory tubules are arranged radially around the sheath. Dunson,

Marine reptiles

Packer & Dunson (1971) examined the secretory tubules in the electron microscope and found only one type of cell, similar in structure to other reptilian salt glands. Mitochondria are numerous and '...numerous villous extensions of the cell surface interdigitate with those of the other cells, and there are extensive intercellular spaces'. The cells also seem to be joined to each other at the apex by zonulae adhaerens and not by zonulae occludentes (tight junctions) which might imply that there is some continuity between the lateral interstitial spaces and the lumen of the tubule. The sheath itself as well as the lower parts of the ducts, are lined by a keratinized, stratified squamous epithelium. It seems likely that the action of the tongue moving in and out of the mouth carries the secretion to the exterior.

Dunson *et al.* (1971) make the interesting point that it would be valuable to examine the structure of the glands in the few fresh-water species which live in lakes in the Phillipine and Solomon islands.

Effects of salt-loading

Dunson & Taub (1967) and Dunson (1968) studied cloacal and extra-renal excretion in the two species they had used for their anatomical studies. The rate of extra-renal excretion was assessed by immersing the snakes in distilled water or 1 M sucrose, and following the rate of sodium, potassium and chloride appearance. Cloacal fluid was collected separately by sewing a catheter into the cloaca and collecting the fluid in a rubber bag. After sodium chloride loads were given the maximum rate of extra-renal excretion was 72 μmoles sodium chloride and 3 μmoles potassium/100 g/hr in *Laticauda*, and 200 μmoles sodium chloride and 1.9 μmoles potassium/100 g/hr in the more pelagic species *Pelamis platurus*. Extra-renal sodium loss was greater than cloacal loss and it seems that, as in other marine reptiles and birds, the salt glands are the major route for the elimination of sodium chloride.

In both species hypertonic sucrose given subcutaneously induced extra-renal excretion, implying that, as in birds, hypertonicity of the blood is the trigger for the onset of salt-gland secretion at a high rate (Fig. 12.5).

The actual collection of secretion from the tongue sheath in the two species by Dunson *et al.* (1971) has permitted a more detailed study to be made of the changes in secretion that occur after salt-loading. A spontaneous secretion was obtained with sodium, potassium and chloride concentrations similar to, and only a little higher than those of blood plasma. After subcutaneous salt-loading both the rate and concentration of the secretion increased markedly (Fig. 12.6). These changes occurred in hours rather than minutes but the rate at which salt passed into the circulation from the site of injection was not known.

The plasma-like concentration of the spontaneous secretion is interesting. Is it possible in view of the apparent lack of tight junctions between the secretory cells that spontaneous secretion is in fact an ultrafiltrate of plasma to which is added, in much greater quantities, a true secretion from the cells in response to salt-loading? Data on inulin or sucrose passage into the secretion would be most welcome.

Other species

A number of snakes apart from hydrophids are known to occur in, or will tolerate, brackish water or even sea water (see Schmidt, 1957 for list). Of these only two have received any attention. Schmidt-Nielsen & Fänge (1958*b*) found that in the Mangrove Water Snake (*Natrix sipedon compressicauda*) salt-loading evoked no secretion from the head region. Dunson & Taub (1967) studied extra-renal excretion in the Elephant-trunk Snake (*Acrochordus javanicus*) from the Far East using the methods they employed for *Laticauda* (see above). Extra-renal sodium losses were small and were not increased by salt-loading. Another species in the same family (*Acrochordidae*), *Chersydrus granulatus*, was said by the late Dr Malcolm Smith, the renowned herpetologist, to be more marine and to occur along the coasts of India to Queensland and the Solomon Islands (see Schmidt, 1957); no study of its salt and water metabolism has been made.

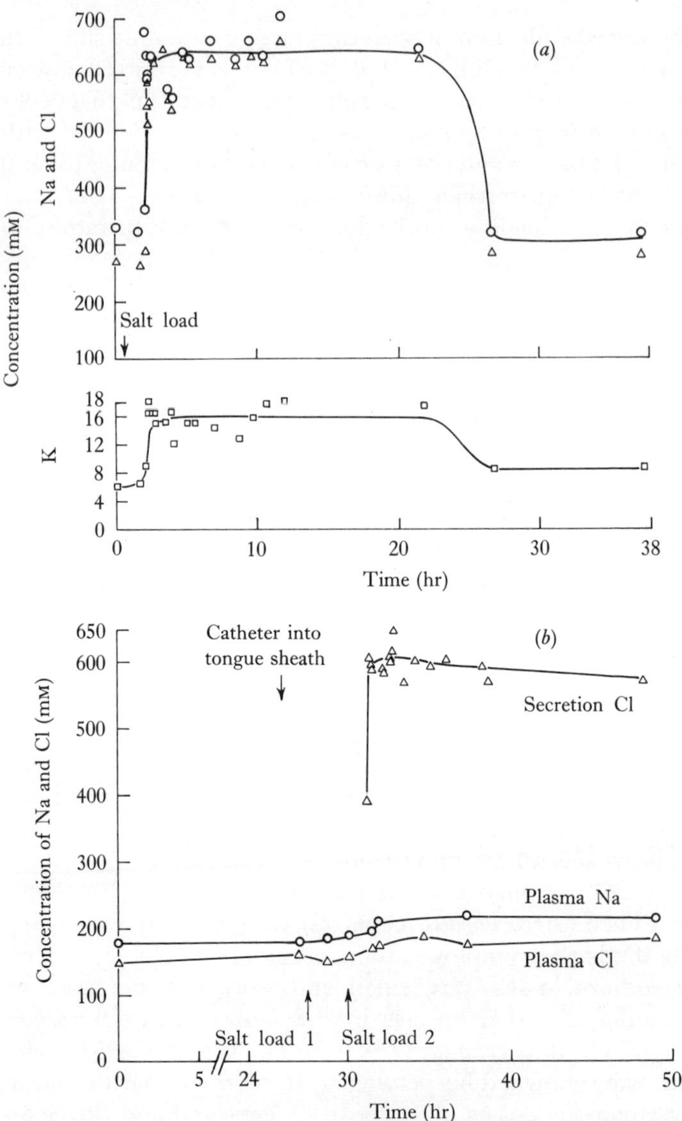

Fig. 12.6. (*a*) The effect of a salt-load (2 mmoles sodium chloride/100 g subcutaneously) on the composition of the fluid secreted by the posterior sublingual gland of a sea snake (*Pelamis*): ○ sodium, △ chloride, □ potassium.

(*b*) The relation between plasma sodium chloride concentration and the time of initiation of secretion of the posterior sublingual gland in *Pelamis*. Salt-load 1 was 0.5 mmoles sodium chloride per 100 g; salt-load 2 was 1.5 mmoles per 100 g (from Dunson *et al.*, 1971, copyright © 1971, American Association for the Advancement of Science).

The Galápagos marine iguana

It is a hideous looking creature, of a dirty black colour, stupid and sluggish in its movements... They do not seem to have any notion of biting; but when much frightened they squirt a drop of fluid from each nostril.

Darwin, 1889

One of their most curious habits was revealed on a later afternoon when I lay flat on the sand watching the ageless surf pounding on the lava boulders. Over the jagged, tortured summits there climbed the largest iguana I saw on the islands. It was a full four feet in length – appearing forty to my lowly viewpoint. His head was clad in rugged scales, black and charred, looking like the clinker piles of the island; along his back extended a line of long spines, as if to the skin of lava he had added a semblance of cactus. He saw me and stopped, looking long and earnestly with curiosity, not fear; then with his smug lizard smile unchanging, he dismissed me with an emotional feat as strange as his appearance; he twice solemnly nodded his whole massive head, he sniffed and sent a thin shower of water vapour into the air through his nostrils and clambered past me on down towards the water. If only a spurt of flame had followed the smoky puff of vapour, we should have had a real old-fashioned dragon! He had come from whatever an Ambly finds to do inland at midday, and was headed seaward at high tide on whatever errand calls such a being into activity at such a time.

Beebe, 1924

The Galápagos islands have many interesting animals. Not least amongst them is the only truly marine lizard, the Marine Iguana (*Amblyrhynchus cristatus*). These lizards, which are up to five feet in length, live on the volcanic rocks lining the shore and enter the sea to eat seaweed which they obtain by diving. They are apparently purely herbivorous and the salt content of their diet is therefore high. It is not surprising that one of the first species to be studied by Schmidt-Nielsen & Fänge (1958*b*) was this lizard. They obtained nasal salt secretion in a captive specimen following salt-loading or the administration of methacholine. They also found that the nasal gland is relatively large (0.06 per cent of body-weight) and resembles in microscopical structure the avian nasal salt gland. A short duct empties into the lateral nasal cavity which can apparently act as a reservoir for the secretion. A ridge in the nostril prevents the fluid flowing back into the posterior part of the nostril from where it could drain into the oesophagus. The normal method of expelling the secretion seems to be by a sudden expiration blowing the fluid out as a shower of fine drops. Dunson (1969*b*) has also studied the anatomy of the gland and Fig. 12.7 is from his paper.

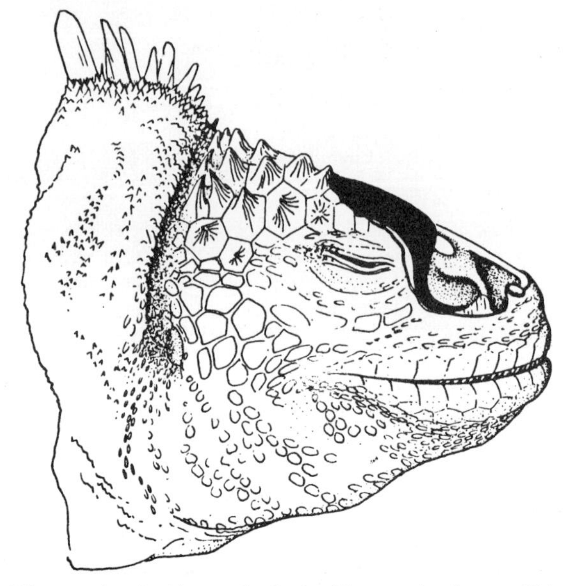

Fig. 12.7. Diagram showing the nasal salt gland in a marine iguana. Side view of a dissection showing the nasal gland (black) lying above the orbit and extending under the nasal passage (drawing by R. D. Chambers, from Dunson, 1969).

Dunson (1969*b*) has also collected secretion from wild specimens and he found a mean concentration of 1434 mM for sodium, 1256 for chloride and 235 for potassium. One of the great difficulties in working on these lizards is that they have been under great pressure from domestic dogs and cats, which were released on the islands by sailors, and their conservation is vital. The number of experiments that can be performed is therefore limited and Dunson, after his studies were completed, released all the iguanas he had obtained at the site of capture. However he did study the rates of extra-renal and cloacal loss using the same method he employed for sea snakes and was able to show that extra-renal losses of sodium, potassium and chloride were very much higher than from the cloaca.

13 TERRESTRIAL REPTILES

Terrestrial reptiles and in particular those that live in the arid regions of the world and eat plants with a high potassium content face the problem of obtaining free water. Some lizards have salt glands, which secrete potassium at a high concentration, but they do not necessarily play a major role in electrolyte excretion because other lizards and all terrestrial snakes appear to lack a salt gland, even those which occur in deserts. One of the most important factors in permitting their survival under such conditions appears to be the ability to excrete sodium and potassium from the cloaca as insoluble urate salts (see Minnich, 1968; 1972). Even in a desert lizard with a salt gland (*Dipsosaurus dorsalis*) living in simulated wild conditions, extra-renal excretion seems relatively unimportant (Minnich, 1970). However, this does not mean that the physiology of the salt glands in lizards is uninteresting or unimportant because, as we shall show, terrestrial lizards possess the ability to vary the secretion of both cations and anions according to the needs of the animal at any one time. In this respect the salt glands act more like a kidney than those of marine birds and reptiles.

Lizards

Nasal salt secretion was discovered, apparently independently, by Schmidt-Nielsen *et al.* (1963), Templeton (1963, 1964*a*, *b*) and by Norris & Dawson (1964). Schmidt-Nielsen *et al.* (1963) observed that a captive specimen of the Green or Linnean Iguana (*Iguana iguana*) was sometimes found to have white incrustations around the nostrils. These deposits were analysed and found to consist mainly of sodium, potassium, chloride and bicarbonate. The potassium content was three times that of sodium, and the ratio of bicarbonate to chloride was two to three. In other words the secretion resembles that of

TABLE 13.1. *Comparison of sodium and potassium concentrations in nasal fluid and plasma of lizards* (modified from Templeton, 1964*b*)

Species	No. of animals		Nasal fluid Mean	Nasal fluid Range	Plasma Mean	Plasma Range	Ratio: (Nasal fluid/ Plasma)
			Concentration (mM)				
Sauromalus	7	K	378.5	268–493	4.87	1.5–9.4	77.2
(Chuckwalla)		Na	120.8	64–176	169.4	161–180	0.67
Ctenosaura	8	K	302.4	124–475	4.41	2.4–6.5	68.7
(False Iguana)		Na	74.9	39–143	170.9	153–182	0.43

the ostrich (p. 221) and differs markedly in composition from that of marine birds and reptiles. Nasal secretion of a similar composition was also observed in the Desert Iguana (*Dipsosaurus dorsalis*) from the arid regions of North America and in a mastigure (*Uromastyx aegyptius*), an agamid from the deserts of North Africa and the Middle East. In both cases secretion was obtained after the intraperitoneal injection of hypertonic sodium chloride.

Norris & Dawson (1964) found that in three species of chuckwalla (*Sauromalus*), the secretion is blown out of the nostrils, as in the Marine Iguana. Templeton (1964*b*) also observed that fluid may be allowed to evaporate in the nostrils leaving dry deposits which then flake off. Norris & Dawson checked that the secretion was in fact from nasal glands by collecting secretion from the eyes and nostrils of chuckwallas after the injection of sodium chloride or methacholine into the lateral accessory lymph sacs (which increase extracellular fluid volume and probably assist in withstanding arid conditions); only nasal fluid was found to be markedly hypertonic. They also analysed the secretion and found twice as much potassium as sodium. Furthermore, the chloride content of the salt deposits was very low and the main anion appeared to be bicarbonate.

Templeton (1964*b*) succeeded in collecting uncontaminated fluid from the salt glands of the False Iguana (*Ctenosaura pectinata*) and a chuckwalla (*Sauromalus obesus*) by drilling holes through the snout and inserting cannulae into the 'posterior nasal tube', into which the ducts from the glands empty. In *Ctenosaura* the pH of the nasal fluid was approximately 8.5, again indicating the presence of bicarbonate. In both species the potassium concentration was higher than that of sodium (Table 13.1), and Templeton also noted that there was considerable variation between animals and between consecutive samples from the same animal.

Murrish & Schmidt-Nielsen (1970) have examined the heat exchange that occurs in the nasal passages of *Dipsosaurus dorsalis.* During inhalation, the air is warmed and humidified by the nasal mucosa and of course, during this process heat is lost from the mucosa. As air passes over the cooled nasal mucosa during exhalation, it loses heat and water and thus leaves the nasal mucosa at a temperature below that of the lungs. The whole system therefore operates as a counter-current heat-exchanger. These authors point out: 'The geometry of the nasal passageways of the desert iguana suggest one very interesting consequence. The distal portion of the passageway forms a slight depression. Fluid secreted from the nasal salt gland accumulates in this depression and contributes to the humidification of the inhaled air. As the fluid gradually becomes more concentrated, the salts crystallize and form incrustations in or at the opening of the nares. Thus the water contained in the secretion from the nasal gland is not a net loss to the body. Since water in any event must be used to humidify the inhaled air the evaporation from the nasal secretion serves to utilize water that has already been "excreted". The net effect on the overall water balance is therefore the same as if the salts were secreted by the nasal gland in the crystalline state.'

Effects of salt loads and environmental changes

Schmidt-Nielsen *et al.* (1963) found that when a specimen of *I. iguana* was given an intraperitoneal load of sodium chloride, secretion appeared in the nostrils in two hours. Some of the fluid was collected, analysed, and found to contain 340–849 (mean 507) mM sodium and 316–862 (mean 497) mM potassium. However, the lizard was still secreting two days after the load was administered but by this time the fluid contained

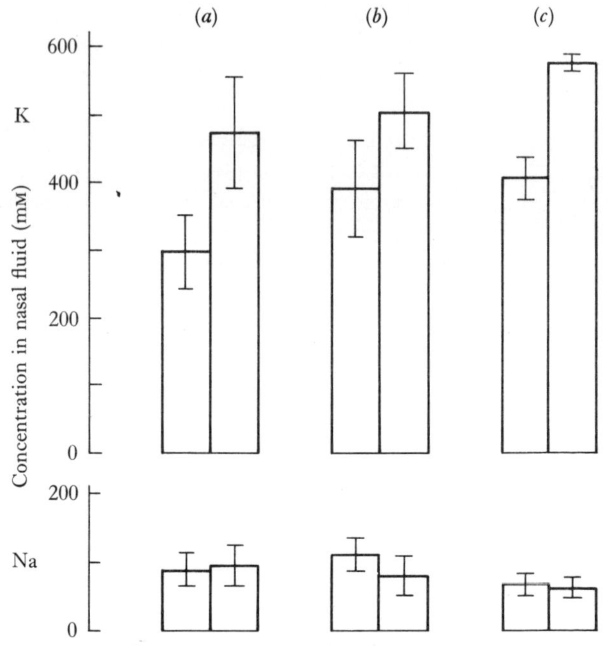

Fig. 13.1. Salt-gland secretion in lizards. Paired data for each of three groups showing cation concentration in nasal fluid before and during induced hypernatraemia (1 ml M sodium chloride/100 g body weight intraperitoneally) in five *Ctenosaura* (*a*), in five *Sauromalus* (*b*), and before and during hyperkalaemia in five *Ctenosaura* (0.5 ml M potassium chloride/100 g) (*c*); mean±S.E. This nasal fluid was collected over hours rather than days as in Fig. 13.2 (drawn from the data of Templeton, 1964*b*).

728 mM sodium and 290 mM potassium. In other words the output of cations was being adjusted to excrete sodium in greater quantities; this effect was also seen in *Dipsosaurus dorsalis.*

Templeton (1964*b*; 1966; 1967) has made a full study of the effects of administering sodium chloride or potassium chloride, both in the long- and short-term, on the output of sodium and potassium in the nasal secretion. In the short-term (i.e. hours) the intraperitoneal injection of sodium chloride or potassium chloride into specimens of *Ctenosaura pectinata* and *Sauromalus obesus* resulted in an increase in the rate of secretion and in the potassium concentration of the nasal fluid; sodium concentration was not affected (Fig. 13.1). In the long-term (i.e. days) in *Dipsosaurus dorsalis* and *Sceloporus cyanogenys,* adaptation occurred in that the quantities of

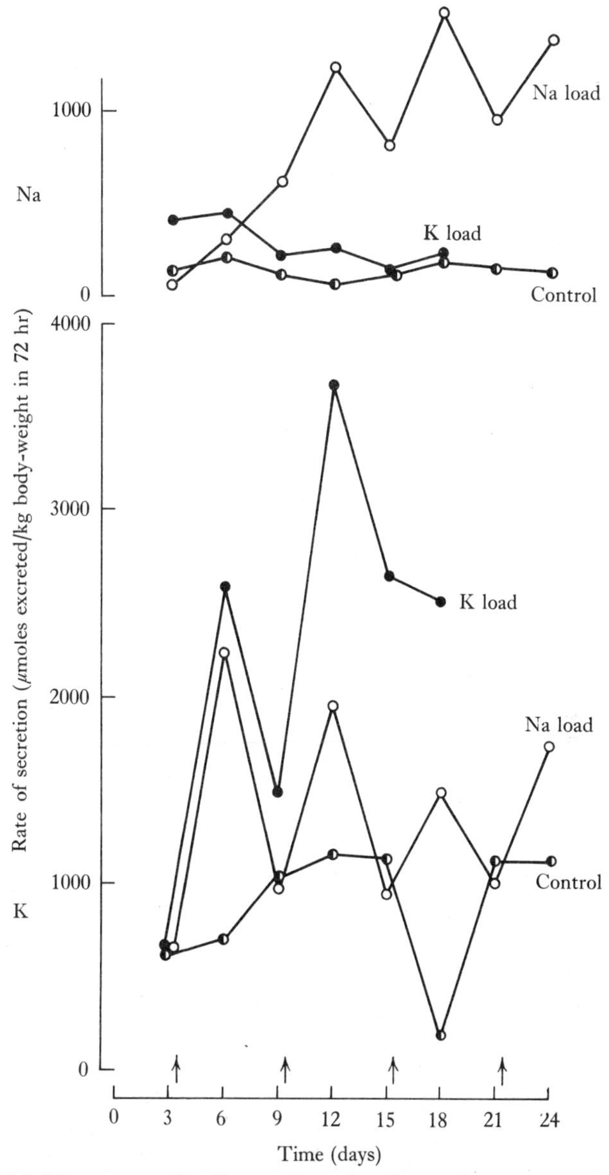

Fig. 13.2. Mean output of sodium and potassium from the salt glands of the desert iguana (*Dipsosaurus dorsalis*). Intraperitoneal injections (1 ml/100 g) were given at the times shown by the arrows; ●, M potassium chloride; ○, M sodium chloride; ◐, 'isotonic lizard saline' (redrawn from Templeton, 1966).

TABLE 13.2. *Composition of nasal fluid of captive false iguanas* (Ctenosaura). *Ion figures are mean*±S.E. (mM) (modified from Templeton, 1967)

	Group A	Group B
Number of animals	4	3
Mean body weight (g)	783	1346
Time in captivity (weeks)	12	1
K^+	253±9.35	527±98.7
Na^+	439±85.7	77.7±17.9
Na^++K^+	692±79.7	605±81.3
K^+:Na^+ ratios	0.58:1	6.78:1
Cl^-	477±78.5	487±102
HCO_3^-	210±33.7	123±16.4
Cl^-+HCO_3^-	688±48.1	610±87.9
Cl^-:HCO_3^-	2.27:1	3.96:1

sodium and potassium secreted were altered. The dried fluid was collected by taping a mask over the nostrils. In both species nasal potassium excretion increased over several days after potassium chloride was injected into the peritoneal cavity; nasal sodium excretion was not affected (Fig. 13.2). When 1 M sodium chloride was injected instead of potassium chloride, sodium output slowly increased in *Dipsosaurus* (Fig. 13.2) but was not sustained in *Sceloporus.* In this latter species the rate of secretion was low and it appears that extra-renal excretion of ions is normally very small.

In *Ctenosaura pectinata,* where nasal fluid was collected from cannulae in the nasal tubes, the daily administration of sodium chloride markedly raised the sodium concentration but lowered that of potassium (Fig. 13.3). Templeton (1967) felt that the ability to vary the output of sodium and potassium in the nasal fluid might be related to possible seasonal or geographical changes in the sodium and potassium content of the diet. He also found that a group of *Ctenosaura pectinata* had a higher sodium and lower potassium concentration in their nasal fluid. These animals had been in captivity for some time and the diet included tinned dog food which contained five hundred times more sodium than the usual food plants; the concentrations of bicarbonate and chloride in the secretion were not affected

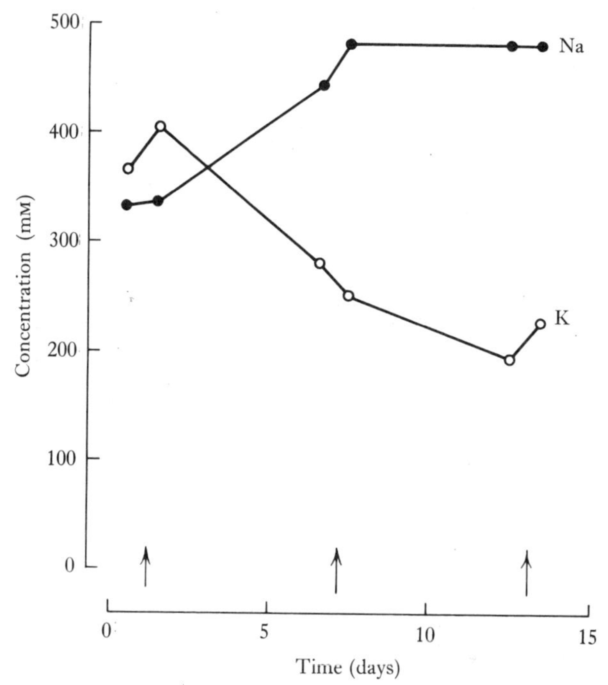

Fig. 13.3. Mean sodium and potassium concentrations in the nasal fluid of the lizard, *Ctenosaura pectinata*, following the intraperitoneal injection of 0.5 ml M sodium chloride/100 g (shown by arrows) (redrawn from Templeton, 1967).

(Table 13.2). Later work by Templeton, Murrish, Randall & Mugaas (1972*a*) has shown that changes in food and water intake can affect salt-gland and renal excretion in *Dipsosaurus dorsalis.*

Braysher (1971) studied nasal secretion in the Stump-tailed Skink (*Trachydosaurus rugosus*). Previous workers had failed to find a salt gland in this Australian species and the reasons for their failure are not clear. As in the iguanids studied by Schmidt-Nielsen *et al.* (1963) and by Templeton (1966; 1967) sodium output increased in the long-term when sodium chloride was injected into the peritoneal cavity every day. However loading with potassium chloride was without significant effect on the sodium:potassium ratio of the secretion. Chloride was the predominant anion and there was no bicarbonate.

However, variations in the proportions of chloride relative to bicarbonate have been observed by Shoemaker, Nagy & Bradshaw (1972) in *Dipsosaurus* who found that if sodium or potassium chloride loads were given, chloride was the main anion. In contrast when potassium salts of organic acids (acetate, bicarbonate or succinate) were given instead, a substantial proportion of the anionic content was bicarbonate. Thus it appears that at least in one species the chloride: bicarbonate ratio can be varied in addition to the sodium: potassium ratio – mechanisms ripe for investigation! Shoemaker *et al.* (1972) make the point that this lizard obtains more sodium and potassium than chloride in its diet, and the versatility in varying chloride and bicarbonate output is necessary if a large quantity of cations is to be excreted by the salt gland.

Structure of the nasal salt gland

The gross and microscopical anatomy of the salt gland in terrestrial lizards has been studied by Templeton (1964*b*), Norris & Dawson (1964), Philpott & Templeton (1964), De Piceis Polver (1968), Crowe, Nagy & Francis (1970), Lemire, Deloince & Grenot (1970), Duvdevani (1972), Van Lennep & Komnick (1970) and by Braysher (1971); all are agreed on the gross anatomy. The nasal glands lie, embedded in cartilage, lateral to the nasal cavity (Fig. 13.4). A short wide duct opens into the overlying vestibule and secretion flows into the principal nasal chamber where it pools, as in the Marine Iguana, just below the nostril. Norris & Dawson (1964) noted that the '...trap-like arrangement just inside the external nares is deep enough that the secretions can block the nasal passage'. This probably necessitates the periodic sneezing of fluids observed in both *Sauromalus varius* and *Sauromalus hispidus* in the wild (see above, p. 248). The duct ramifies into lobules in a similar manner to other salt glands. Unmyelinated nerve fibres and blood vessels are abundant.

The most complete ultrastructural study is that of Van Lennep & Komnick (1970) in Bell's Mastigure (*Uromastyx acanthinurus*). They found that the terminal parts of the secretory tubules contain mucus-secreting cells amongst which are interspersed principal secretory cells. The rest of the

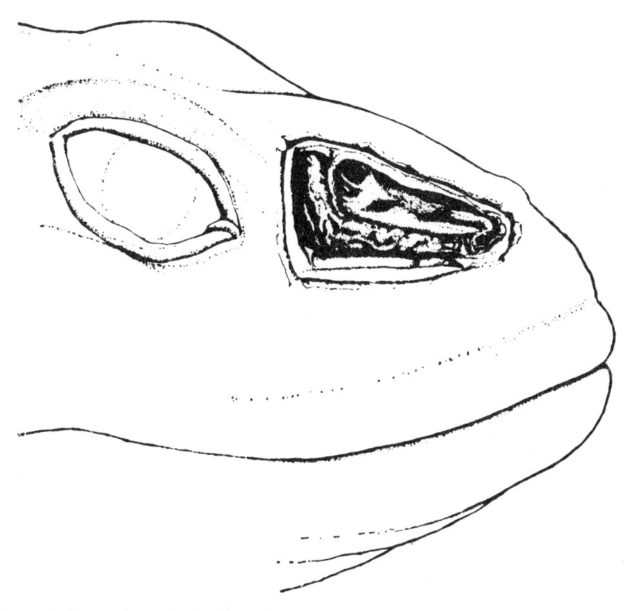

Fig. 13.4. A dissection of the lizard, *Sauromalus hispidus*, showing the lateral nasal gland lying beneath the vestibular portion of the nasal passage (from Norris & Dawson, 1964).

tubules contained principal secretory cells which were called 'light' or 'dark' on the basis of their appearance in the electron microscope. However they felt it was not justified to call them definite cell types because intergradation occurred. It is now well known that such differences in appearance of cells can result from poor fixation and since immersion rather than perfusion fixation was employed it seems possible that the principal secretory cells all have a similar structure.

The apical membranes of these cells have either a few microvilli or a true brush border. The cells are joined at their apices by typical junctional complexes of zonulae occludentes and zonulae adhaerens. Basal cell processes were observed by Van Lennep & Komnick but they found that the basal membranes are not infolded, in contrast to the lateral cell membranes which interdigitate with those of neighbouring cells. Mitochondria are abundant and occur mainly towards the base of the cells. Mucopolysaccharide was detected in the lateral intercellular spaces, as in turtles, and a small Golgi apparatus with vesicles near the lateral membranes; these may be responsible for the secretion of this substance at this site (see p. 235). As in other salt glands small basal cells are seen in the tubules but their function is unknown; Van Lennep & Komnick (1970) thought that some could be myoepithelial in nature.

The major difference between the salt glands of the other birds and reptiles and those of lizards is the presence of cells in the duct system which appear highly active. Apart from a portion of the duct nearest the exterior where the epithelium is keratinized, the rest of the duct system is a pseudostratified epithelium but lining the lumen is a layer of columnar cells, with many mitochondria, similar in structure to the principal secretory cells. This arrangement may well account for the versatility of lizards' salt glands to elaborate fluids of different composition because the ducts could well act to modify the primary secretion, as in many other exocrine glands, by re-absorbing or secreting sodium, potassium, chloride and bicarbonate.

Control of secretion

The secretory mechanism by which lizards can adjust the output of sodium, potassium, chloride and bicarbonate in nasal fluid is of obvious physiological importance. Data do not exist on relationships between secretory rate and composition and so it is impossible to assess to what extent the changes occur as a result of alterations in the rate of secretion. However two papers have appeared which begin to explain some of the controlling mechanisms although it is not yet possible to interpret the effects at the cellular level.

Shoemaker *et al.* (1972) found in small specimens of *Dipsosaurus dorsalis* that while sodium chloride, potassium chloride, potassium succinate, potassium acetate, potassium bicarbonate and rubidium chloride induced nasal secretion when given intraperitoneally, sucrose and mannitol were not effective stimulants. Therefore it was concluded that '...nasal gland secretion in *Dipsosaurus* is elicited by increased concentrations of alkali metals'. This interesting difference from birds and marine reptiles does indeed suggest that specific ion receptors rather than osmoreceptors are involved. Unfortunately, plasma was not analysed and in view of the slow response time (hours or days) it is possible that concentrated sugar solutions, given as a purely osmotic stimulus, were poorly absorbed from the peritoneal cavity. Therefore before the existence of cation receptors can be considered it would be preferable to use intravenous loading and to follow changes in plasma osmolality.

The same authors, as well as Templeton *et al.* (1968; 1972*b*) have found that aldosterone acts on the salt gland of *Dipsosaurus dorsalis* in a similar manner to its action on the mammalian kidney. In other words it favours sodium retention (Fig. 13.5). Shoemaker *et al.* (1972) found that corticosterone, cortisol and mammalian ACTH were less effective in this respect. Thus it is possible that when wild lizards are secreting more potassium than sodium, aldosterone is acting on the gland. This conclusion is compatible with the effects of adrenalectomy which results in an increase in sodium output relative to that of potassium (Templeton *et al.*, 1972*b*). The results are difficult to interpret because only the rates of sodium and potassium excretion were known while the rate of fluid excretion and its concentration were not. However it does seem clear that hormones are probably involved in adapting salt-gland secretion for the excretion of a particular cation but it is to be hoped that experiments will be done where nasal fluid, rather than dried flakes of salt, is collected, as Templeton succeeded in doing in other experiments. Shoemaker *et al.* (1972) failed to induce secretion with adrenocorticosteroids or ACTH whereas Templeton (1964*b*) and Norris & Dawson (1964) observed secretion after the administration of methacholine. (Templeton also noted that secretion is intermittent rather than continuous even though plasma concentrations remain high.) More recently, Templeton (1972) has stated that atropine inhibits secretion. Therefore the available evidence suggests that while hormones may modify secretion, nerves actually initiate it, as is fully documented in birds.

Other species of lizard

A list of species and families in which nasal salt secretion has been observed is shown in Table 13.3. Most terrestrial lizards

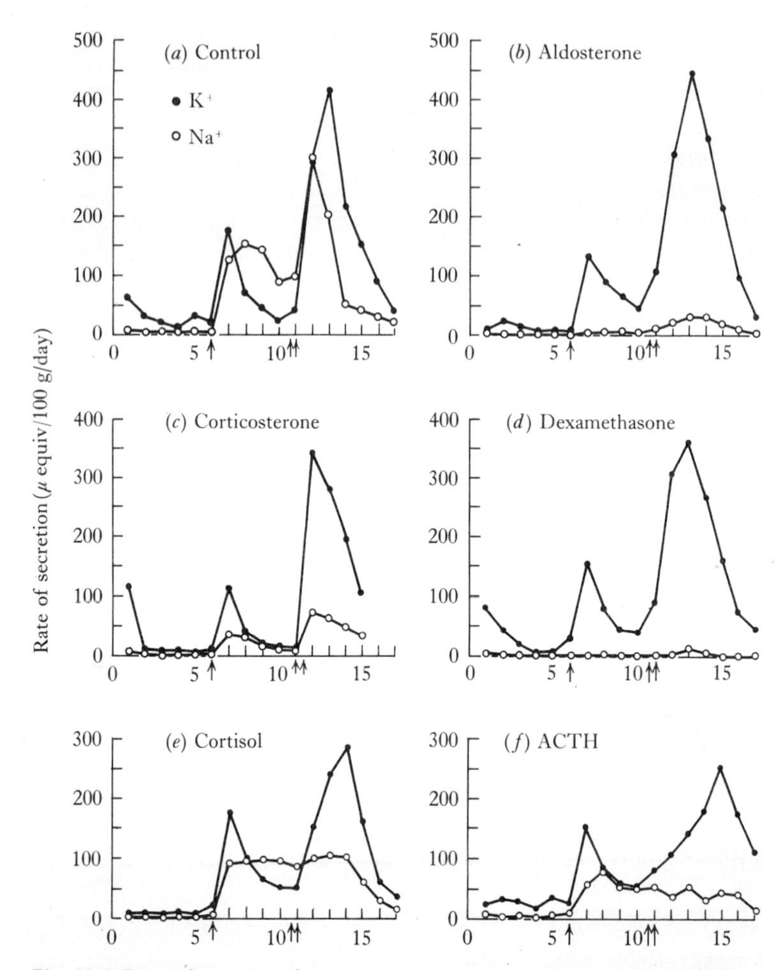

Fig. 13.5. Rates of secretion of potassium (•) and sodium (○) by the salt gland of the desert iguana (*Dipsosaurus dorsalis*). Two mmoles of sodium chloride and potassium chloride were given on day 6 and day 11 respectively. Animals received daily injections of water (*a*) or various hormones (*b–f*) beginning on day 0. Points represent salt collected over the previous 24 hr (from Shoemaker *et al.*, 1972).

secrete more potassium than sodium, at least when taken straight from the wild, but the Galápagos Land Iguana (*Conolophus subcristatus*), whose vegetarian diet is rich in sodium (Dunson, 1969*b*), and a species of monitor lizard (*Varanus*), which is carnivorous (Minnich, unpublished, in Dunson &

TABLE 13.3. *Salt glands in lizards*

Family	Species
	Nasal salt secretion observed
Iguanidae	*Iguana iguana*[1]
	Dipsosaurus dorsalis[1,2,4,5,13]
	Crotaphytus collaris[2]
	Ctenosaura pectinata[2,7]
	Sauromalus obesus[2,12]
	Sauromalus hispidus[12]
	Sauromalus varius[12]
	Sceloporus cyanogenys[3]
	Conolophus subcristatus[8]
	Amblyrhynchus cristatus[8,16] (Marine)
Agamidae	*Uromastyx aegyptius*[1]
	Uromastyx acanthinurus[6]
Scincidae	*Eumeces skiltonianus*[9]
	Trachydosaurus rugosus[10]
Xantusidae	Species not stated[15]
Lacertidae	*Acanthodactylus* sp.[17]
Teiidae	Species not stated[15]
Varanidae	*Varanus gouldii*[11]
	Nasal salt secretion not observed
Gekkonidae	*Coleonyx* sp.[13]
Agamidae	*Amphibolurus ornatus*[14]
Lacertidae	*Lacerta viridus*[18]
Anguidae	*Gerrhonotus* sp.[3]

Other families with representatives lacking salt glands: Anniellidae, Teiidae, Cordylidae, Helodermatidae.[15]

REFERENCES

[1]Schmidt-Nielsen *et al.* (1963), [2]Templeton (1964*b*), [3]Templeton (1966), [4]Shoemaker *et al.* (1972), [5]Templeton *et al.* (1968; 1972*a*, *b*), [6]Grenot (1968), [7]Templeton (1967), [8]Dunson (1969*b*), [9]Minnich, unpublished, in Braysher (1971), [10]Braysher (1971), [11]Green, unpublished, in Braysher (1971), [12]Norris & Dawson (1964), [13]Minnich (1970), [14]Bradshaw & Shoemaker (1967), [15]Minnich, unpublished, in Dunson & Taub (1967), [16]Schmidt-Nielsen & Fänge (1958*b*), [17]Duvdevani (1972), [18]Gerzeli & De Piceis Polver (1970).

Taub, 1967), secrete more sodium than potassium. In view of the ability of terrestrial iguanids to vary the output of sodium and potassium in nasal fluid it is possible that in the two species where sodium is high, environmental rather than genetic influences are involved.

Gabe & Saint Girons (1971) have studied the histology of the salt glands in 36 species of lizard from 16 families. They noted the presence or absence of striated cells which would appear to correspond with the principal secretory cells seen in the electron microscope, the lateral cell membranes giving the striated appearance in the light microscope. In all but one species where striated cells were present they were found at the proximal end of the tubules whereas mucoserous cells were located at the distal (blind) end; this is in agreement with electron microscopical observations (Van Lennep & Komnick, 1970). However in *I. iguana* the two types of cells were found side by side along the tubules. Gabe & Saint Girons concluded that striated cells are found in the species of a family that live in more arid habitats. However, correlation with nasal salt secretion was not made and there are some inconsistencies. For example, in the desert gecko *Coleonyx* striated cells are present whereas Minnich (1970) failed to demonstrate salt secretion in this species.

Although attempts to induce secretion in the Green Lizard (*Lacerta viridis*) by administering salt-loads and cholinomimetics have failed, the glands have been observed to show histochemical changes, which might imply a vestigial adaptive response (Gerzeli, 1967*a*; Gerzeli & De Piceis Polver, 1970).

Finally, Gabe & Saint Girons (1973) have failed to detect cells characteristic of salt glands in the poorly-developed nasal glands of the Tuatara (*Sphenodon punctatus*); this animal, not a lizard, is the only extant member of the Rhynchocephalia.

14 EVOLUTION OF SALT GLANDS

The evolution of salt glands, or indeed of any homeostatic system, is clearly a fascinating subject for speculation. Cranial salt glands are only known to exist in birds and modern reptiles. Therefore it is the reptiles to which we must concentrate most of our attention.

Nasal glands probably first appeared in the Amphibia serving to moisten and cleanse the nasal passages of the adults, which are usually terrestrial (see Bang & Bang, 1959). Judging by extant species the kidneys of reptiles and birds are much less 'efficient', in terms of concentrating ability, than those of mammals. Therefore evolution of amphibian nasal glands into salt-secreting glands might be inferred to have occurred in the early reptiles since, in truly terrestrial vertebrates, two major avenues for ionic and osmotic regulation present in lower vertebrates would have been lost – the gills and the skin. In fact the development of the reptilian skin, which is relatively impermeable, from the amphibian type across which ions and water movements occur and can be controlled, must have meant a major alteration in the control of salt and water metabolism. Therefore any animal with an additional means of excreting ions would be at an advantage. It would enable such creatures to live on plants, which are rich in potassium and/or to return to the sea and eat plants and invertebrates, which have the same ionic content as sea water. Therefore one might argue that the evolution of a salt gland together with all the other adaptations which enabled the initial colonization of land, was an important means of permitting the vast and complex adaptive radiation that occurred to fill all the varied ecological niches.

Sokol (1967) has suggested that in terrestrial lizards the presence of nasal salt glands is associated with a herbivorous diet (rich in potassium) and that the role of the glands may be

TABLE 14.1. *Occurrence of salt glands in birds and reptiles*

	Gland	Type	Main cation in secretion
Birds	Nasal	Nasal	Na^+ – marine Na^+ or K^+ – terrestrial
Lizards	Nasal	Nasal	Na^+ – marine Na^+ or K^+ – terrestrial
Snakes	Posterior sublingual	Oral	Na^+
Chelonians	Lachrymal	Orbital	Na^+
Crocodilians	? (Harderian)	Orbital	Na^+

to regulate the body's sodium and potassium balance rather than to conserve water. This line of reasoning would certainly explain the presence of salt glands in such lizards as the Common Iguana (*I. iguana*) which eat plants but normally have free access to fresh water. Thus it is possible, or even probable, that nasal salt glands were present in early and now extinct reptiles, in the herbivorous dinosaurs for example, even though they may have had free access to water. One might then suggest that the archetypal salt gland was present at an early stage of reptilian evolution serving originally as an ionic rather than an osmotic regulator, and that the primitive type resembled that of modern lizards in being versatile and able to secrete either sodium or potassium together with chloride or bicarbonate in different proportions depending upon which were being ingested in excess. Of course if this line of argument is accepted then the sodium chloride-excreting gland of marine birds and lizards must have been an adaptation to this particular habitat which involved the loss of the ability to secrete potassium and bicarbonate.

An important problem to consider is, if the nasal gland was a functional salt gland in early reptiles, why are other cranial glands adapted to serve the same purpose in modern marine chelonians, snakes and crocodilians (Table 14.1). All the reptiles with a non-nasal salt gland seem to be secondarily marine and may well have passed through a stage in evolution when the nasal gland was lost. Many comparative anatomists believe that snakes evolved from lizards and that they passed through a fossorial stage early in their evolution, emerging later to occupy successfully the niches they fill at the present time (see Bellairs & Underwood, 1951). It is quite possible that the nasal salt gland was lost in the underground habitat of these carnivores and that another gland, the posterior sublingual (p. 240) became modified when the sea snakes evolved. Similarly, the crocodilians are typically fresh-water carnivores and such a mode of life might well be expected to lead to the loss of the primary nasal salt gland. It is certainly difficult to accept that orbital glands were archetypal salt glands in terrestrial forms since the hypertonic salt solutions could not be washed from the eyes. Therefore it would seem that the orbital glands of marine chelonians and crocodilians became modified into salt glands as an adaptation to a secondarily marine mode of life.

We do not know which gland, if any, served as a salt gland in the extinct groups of aquatic and marine reptiles such as the ichthyosaurs and plesiosaurs. We can only hope that the Loch Ness Monster proves to be such an animal – and that it will be considerate enough to respond to a salt-load.

Fig. 14.1 is a 'family-tree' of terrestrial vertebrate evolution and in it we have shown the lines along which salt glands may have evolved. Another point to be considered is whether salt glands were present in the theropsid evolutionary line which led ultimately to mammals. Palaeontologists believe that the theropsid line split from the sauropsid line (which included all the other reptiles, and birds) at a very early stage; some even believe it occurred at an amphibian stage of evolution. The question is whether a salt gland was ever present in mammalian ancestry and was lost when more efficient kidneys evolved. It is interesting to note that recent evidence suggests that the nasal gland of the dog secretes a hypotonic fluid during panting and that this is probably a major source of water for evaporative cooling (Blatt, Taylor & Habal, 1972). To be sure, there is no physiological evidence for salt glands in most mammals, as any ship-wrecked mariner would verify. But are we accepting too blindly that the mammalian kidney can be so efficient that there is no need for extra-renal excretion? In this respect it is interesting that sirenians (manatees and dugongs) have been observed to produce copious tears although they apparently lack both Harderian and lachrymal glands (see Bang

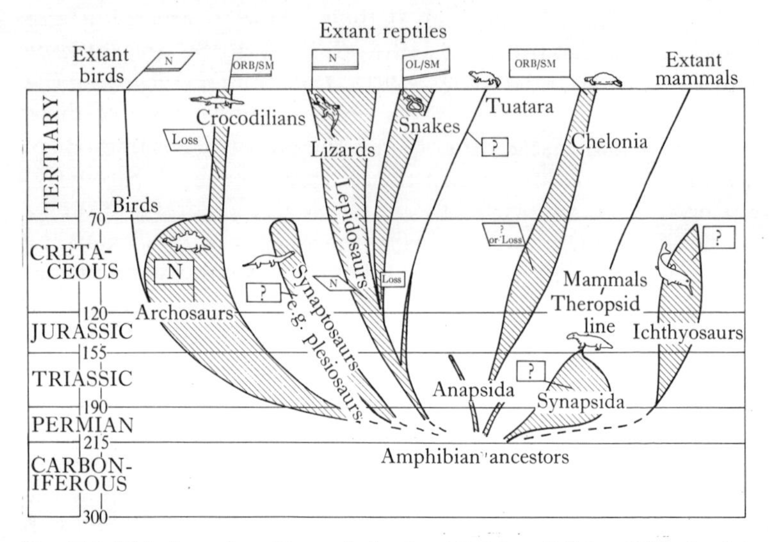

Fig. 14.1. Main lines of reptilian evolution (modified from Bellairs, 1957); the time scale on the left is in millions of years. The type of cranial gland acting or thought to have acted as a salt gland is shown in the flags: N = nasal gland, ORB = orbital gland, OL = oral gland, SM = in secondarily marine forms, ? = not known, Loss = original type lost during evolution.

& Bang, 1959). These mammals, which probably gave rise to the legends of mermaids, are found in coastal waters as well as in large rivers and eat prodigious quantities of seaweed – rich in potassium. Similarly, Gordon (1968) sums up what is known about marine mammals in the following: 'The kidneys presumably are the major routes for elimination of salt loads from marine animals, but this is by no means certain in all cases. The urine of these animals is usually strongly hyperosmotic to the blood and, quite often, to sea water. However, most species are carnivores that produce large amounts of waste urea. The largest part of urinary osmotic concentration often derives from this urea. Urinary sodium chloride concentrations higher than sea water levels have been recorded for a few species, but the difficulties of experimentation with animals such as whales have made our understanding of overall salt-balance regimes in most forms very fragmentary. There is a distinct possibility that extra-renal pathways for salt excretion, perhaps the salivary glands, are important for many of these species.'

Table 14.2. *Orders of birds in which salt-gland secretion has been observed or can be inferred* (the classification is taken from Storer, 1971)

		Evidence for salt glands marked by *			
E	Archaeornithes				
	Neornithes				
E	Super Order	Odontognathae	*†		
		Neognathae			
	Order	Tinamiformes		Gruiformes	*
		Rheiformes		Charadriiformes	*
		Struthioniformes	*	Gaviiformes	*
		Casuariiformes		Columbiformes	
E		Aepyornithiformes		Psittaciformes	
		Dinornithiformes		Cuculiformes	*
		Podicipediformes	*	Strigiformes	
		Sphenisciformes	*	Caprimulgiformes	
		Procellariiformes	*	Apodiformes	
		Pelecaniformes	*	Coliiformes	
		Anseriformes	*	Trogoniformes	
		Phoenicopteriformes	*	Coraciiformes	
		Ciconiiformes	*	Piciformes	
		Falconiformes	*	Passeriformes	
		Galliformes	*		

† from anatomical evidence only E = extinct

As we have mentioned, it is probably reasonable to assume that nasal salt glands were present in the early reptiles and, therefore, in the reptilian ancestors of birds, the archosaurs. Certainly all the groups of birds that diverged from the main stock comparatively early have members with salt glands, for example, the ostrich amongst the ratites, and the penguins. Depressions in the supra-orbital region of the skull of *Hesperornis* and *Ichthyornis* (Marples, 1932) suggest that salt glands were present in these extinct aquatic piscivorous birds. It can also be suggested that the versatile sodium- and potassium-secreting type of gland was archetypal and that this was retained in the terrestrial ostrich and in certain galliforms, while the potassium secreting ability was lost in marine species and some terrestrial birds short of free water so that the gland became more of a mechanism for osmoregulation rather than ionic regulation.

Table 14.2 shows the different orders of birds and whether

salt-gland secretion has been observed or can be inferred from the size of the glands. Since the arrangement of orders in this way is thought to show phylogenetic relationships to some extent, it is clear that the evidence available now indicates that the 'higher' birds do not, in general, have a salt gland even though their presence might be expected in some of the species that live in arid regions (p. 220). It seems likely that the passeriform birds in particular have lost salt glands while evolving in terrestrial regions with water freely available and eating an omnivorous diet. An often repeated (but sometimes contested) maxim is that if the function of an organ is lost during evolution then that same function is not re-acquired by the same organ in later descendants. If this is so in this case then the lack of salt glands in desert finches might be a manifestation of such a phenomenon since they have instead developed an efficient renal excretory mechanism (p. 220).

Possible mechanisms

If, as we have suggested, the versatile type of salt gland developed from an unspecialized nasal gland, and then in marine forms the ability to secrete potassium and bicarbonate was lost, we must consider how such changes might have occurred.

A typical exocrine gland consists of a duct system into which secretory alveoli empty. It seems possible that the original amphibian nasal gland secreted mucus in an isotonic fluid from two main types of cell, one secreting mainly mucus, the other ions and water. Typical exocrine glands also modify the primary secretion from the alveoli by secreting or re-absorbing substances in the duct system (see Botelho, Brooks & Shelley, 1969). Development into a salt gland could occur by an increase in the number of cells secreting ions and water and a decrease in those secreting mucus, together with a modification of the ion transporting cells to form a hypertonic solution. We have noted that in the salt glands of lizards, as opposed to those of the other animals studied, the layer of cells lining the duct system appears metabolically active as in mammalian salivary glands, where they are involved in modifying the secretion from the alveoli. If this is the mechanism by which the composition of the final secretion is achieved then we must

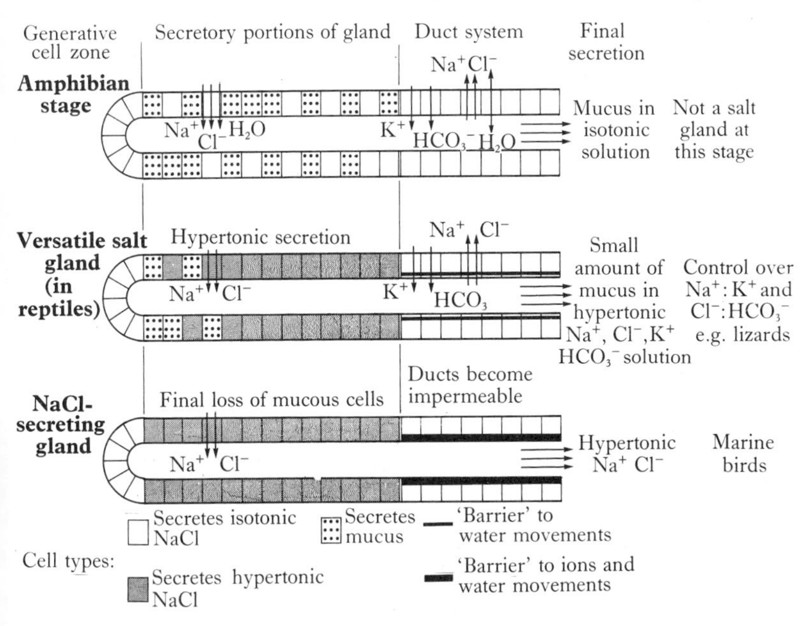

Fig. 14.2. Possible stages in the evolution of the nasal salt gland.

assume that this was the type of salt gland present in extinct reptiles and in early birds (Fig. 14.2).

Since in lizards sodium secretion is depressed by the administration of aldosterone, it is possible that the primary secretion is a sodium chloride solution and that sodium can be re-absorbed in the ducts which would also have the ability to secrete potassium. Similarly the chloride:bicarbonate ratio could also be controlled by the ducts, as in the mammalian pancreas (Fig. 14.2). The loss of the ability to secrete potassium and bicarbonate in marine birds could then simply be achieved by a loss of activity of the ducts and the development instead of a permeability barrier to the movements of ions and water as, indeed seems to be the case. Essentially the same process could have occurred in the other salt glands of marine reptiles, except that more mucus cells seem to have been retained.

The lack of typical secretory alveoli surprises many workers when they see sections of salt gland for the first time. Since salivary glands have different regions of their ducts which

TABLE 14.3. *Comparison between origin, epithelial fine structure, concentration and rate of secretion of the salt glands in a bird, lizard, turtle and elasmobranch (data from the literature)*, (modified from Van Lennep & Komnick, 1970)

	Gland	Epithelial interdigitation	Concentration of secretion		Rate of secretion per g gland (μmole/min)		
			Na	K	Na	K	H_2O μl/min
Bird (*Larus argentatus*)	Nasal	Baso-lateral	718	24	430	14.4	600
Lizard (*Sauromalus obesus*)	Nasal	Lateral	64	611	1.6	15.2	25
Turtle (*Chelonia mydas*)	Lachrymal	Lateral	685	21	8.2	0.25	12
Elasmobranch (*Squalus acanthias*)	Rectal	Lateral	580	8.4	39.5	0.57	68

appear to have distinct functions, it is possible that the secretory tubules are in fact part of the duct system and that the alveoli with mucus-secreting cells have been lost during the evolutionary process.

The development of the ability to secrete a hypertonic solution at a high rate could have been achieved by changes in the osmotic permeability to water, an increase in the rate of pumping and by alterations in the geometry of the cell to permit 'standing gradients' to function and the rapid uptake of ions from the interstitial space. On this latter point Van Lennep & Komnick (1970) believe that marine birds may have evolved infolding of the basal membranes of the secretory cells in addition to the lateral folding and interdigitation seen in reptilian salt glands; they show clearly that the rate of secretion is lower in reptilian salt glands (Table 14.3).

Comparative physiology

We really have only dealt with the evolution of salt glands in the most general way. Just as interesting considerations would apply to the whole homeostatic system by which they are controlled but unfortunately we do not have the information. For example questions that must be answered are the nature and location of the receptors in the different groups, the secretory reflex, effects of hormones etc. The only clue we have at present is the claim that cation receptors rather than osmoreceptors appear to be responsible for triggering secretion in lizards (p. 256); this might be compatible with Sokol's idea that salt glands are ionic rather than osmotic regulators in lizards and, as we have suggested, in early reptiles as well.

Difficulties exist in obtaining comparative data, particularly in reptiles. It is very important to realize that rates of secretion are affected by environmental temperature and that this can, in other organs, result in marked changes in performance. For example, Shoemaker, Licht & Dawson (1966) found that in the lizard, *Trachydosaurus rugosus*, rates of urine production increased about 30-fold between 14 °C and 37 °C. Changes in glomerular filtration rate were mainly responsible but tubular sodium re-absorption also increased at higher temperatures. To study optimal performance it is essential that reptiles should be studied at the preferred body temperature (the temperature a particular species voluntarily attempts to attain in the wild or in a laboratory thermal gradient) which has been shown to be optimal in any one species for a number of physiological processes. However it is also important that the physiological mechanisms should be studied at lower temperatures since particularly in deserts and temperate regions body-temperature falls at night. For example it would be interesting to know whether nasal salt excretion in lizards continues at night when the body temperature is low in order to excrete the potassium ingested during the day. A discussion on the thermal requirements of reptiles in captivity has been published by one of us (Peaker, 1969).

REFERENCES

Abel, J. H. (1969). Electron microscopic demonstration of adenosine triphosphate phosphohydrolase activity in herring gull salt glands. *Journal of Histochemistry and Cytochemistry* **17**, 570–84.

Abel, J. H. & Ellis, R. A. (1966). Histochemical and electron microscopic observations on the salt secreting lachrymal glands of marine turtles. *American Journal of Anatomy* **118**, 337–58.

Abel, J. H., Rhees, R. W. & Haack, D. W. (1971). Studies on the presumptive corticotroph cells in the pituitaries of herring gulls. *Cytobiologie* **3**, 299–323.

Akester, A. R. (1967). Renal portal shunts in the kidney of the domestic fowl. *Journal of Anatomy* **101**, 569–94.

Allen, G. M. (1925). *Birds and their attributes* (Dover edition, 1962). Marshall Jones: New York.

Anderson, B. W. & Warner, D. W. (1969). Evidence from salt analysis for convergence of migratory routes and possible geographic variation in lesser scaup. *Bird-Banding* **40**, 198–207.

Anon (1667). *Observationes anatomicae selectiores Collegi privati Amstelodamensis.* Apud C. Commelinum. Amstelodami, 1667. This has been printed in the following: Cole, F. J. (1938). *Observationes anatomicae selectiores. Amstelodamensium 1667–1673.* University of Reading.

Ash, R. W. (1969). Plasma osmolality and salt gland secretion in the duck. *Quarterly Journal of Experimental Physiology* **54**, 68–79.

Ash, R. W., Pearce, J. W. & Silver, A. (1966). Factors influencing secretion from the salt glands of the domestic duck. *Journal of Physiology* **183**, 83–85P.

Ash, R. W., Pearce, J. W. & Silver, A. (1969). An investigation of the nerve supply to the salt gland of the duck. *Quarterly Journal of Experimental Physiology* **54**, 281–95.

Ball, J. N., Chester Jones, I., Forster, M. E., Hargreaves, G., Hawkins, E. F. & Milne, K. P. (1971). Measurement of plasma cortisol levels in the eel *Anguilla anguilla* in relation to osmotic adjustments. *Journal of Endocrinology* **50**, 75–96.

Ballantyne, B. (1967). Histochemical correlates of β-glucuronidase and sodium ion transport. *Journal of Physiology* **192**, 13–14P.

Ballantyne, B. & Fourman, J. (1967). Cholinesterases and the secretory activity of the duck supra-orbital gland. *Journal of Physiology* **188**, 32–3P.

Ballantyne, B. & Wood, W. G. (1967). A histochemical and biochemical investigation of β-glucuronidase activity in the quiescent and secreting supra-orbital gland of *Anas domesticus. Journal of Physiology* **191**, 89–90P.

Ballantyne, B. & Wood, W. G. (1968). ATP-ase and Na^+ transport: histochemical and biochemical observations on the avian nasal gland. *Journal of Physiology* **196**, 125–6P.

References

Ballantyne, B. & Wood, W. G. (1969). Mass and function of the avian nasal gland. *Cytobios* **4**, 337–45.

Ballantyne, B. & Wood, W. G. (1970*a*). Functional compensatory hypertrophy of the avian nasal gland. *Journal of Anatomy* **106**, 173–4.

Ballantyne, B. & Wood, W. G. (1970*b*). ATPase and Na^+–K^+ transport in the avian nasal gland. *Cytobios* **5**, 41–51.

Bang, B. G. & Bang, F. B. (1959). A comparative study of the vertebrate nasal chamber in relation to upper respiratory infections. *Bulletin of the Johns Hopkins Hospital* **104**, 107–49.

Barcroft, J. (1914). *The respiratory function of the blood.* Cambridge University Press: London.

Bayliss, W. M. (1923). *The Vaso-Motor System.* Longmans Green: London.

Beebe, W. (1924). *Galápagos. World's End.* Putnam: London & New York.

Bellairs, A. d'A. (1942). Observations on Jacobson's organ and its innervation in *Vipera berus. Journal of Anatomy* **76**, 167–77.

Bellairs, A. d'A. (1957). *Reptiles.* Hutchinson: London.

Bellairs, A. d'A. & Underwood, G. (1951). The origin of snakes. *Biological Reviews* **26**, 193–237.

Bellamy, D. & Phillips, J. G. (1966). Effects of the administration of sodium chloride solutions on the concentration of radioactivity in the nasal gland of ducks (*Anas platyrhynchos*) injected with [^{3}H]corticosterone. *Journal of Endocrinology* **36**, 97–8.

Bellamy, D., Phillips, J. G., Chester Jones, I. & Leonard, R. A. (1962). The uptake of cortisol by rat tissues. *Biochemical Journal* **85**, 537–45.

Bennett, G. (1834). *Wanderings in New South Wales, Batavia, Pedir Coast, Singapore and China* (in two volumes). Richard Bentley: London.

Bennett, G. (1834). On the nasal gland of the wandering albatross, *Diomedea exulans* Linn. *Proceedings of the Zoological Society of London* 1834, 151.

Bennett, H. S. (1963). Morphological aspects of extracellular polysaccharides. *Journal of Histochemistry and Cytochemistry* **11**, 14–23.

Benson, G. K. & Phillips, J. G. (1964). Observations on the histological structure of the supraorbital (nasal) glands from saline-fed and freshwater-fed domestic ducks. (*Anas platyrhynchos*). *Journal of Anatomy* **98**, 571–8.

Benson, G. K., Phillips, J. G. & Holmes, W. N. (1964). Observations on the histological structure of the nasal glands of the turtle. *Journal of Anatomy* **98**, 290.

Bentley, P. J., Bretz, W. L. & Schmidt-Nielsen, K. (1967). Osmoregulation in the diamond-back terrapin, *Malaclemys terrapin centrata. Journal of Experimental Biology* **46**, 161–7.

Bern, H. A. (1970). Discussion in *Hormones and the environment*, p. 604, eds Benson, G. K. & Phillips, J. G., (*Memoirs of the Society for Endocrinology*). Cambridge University Press: London.

Bern, H. A. & Nicoll, C. S. (1968). The comparative endocrinology of prolactin. *Recent Progress in Hormone Research* **24**, 681–720.

Bernard, C. (1858). De l'influence de deux ordres de nerfs qui déterminent les variations de couleur du sang veineux dans les organes glandulaires. *Compte rendu hebdomadaire des séances de l'Académie des sciences* **47**, 245–53.

Bernard, G. R. & Wynn, G. G. (1963*a*). Osmotic behavior of vertebrate 'salt gland' slices. *American Zoologist* **3**, 529–30.

References

Bernard, G. R. & Wynn, G. G. (1963*b*). Phosphatase activity of the secretory cells of some vertebrate 'salt glands'. *Association of Southeastern Biologists Bulletin* **10**, 24.

Bernard, G. R. & Wynn, G. G. (1964). Weight changes of vertebrate 'salt gland' slices exposed to saline or sucrose media. *Anatomical Record* **148**, 260.

Bhoola, K. D., Morley, J., Schachter, M. & Smaje, L. H. (1965). Vasodilatation in the submaxillary gland of the cat. *Journal of Physiology* **179**, 172–84.

Bindslev, N. & Skadhauge, E. (1971*a*). Salt and water permeability of the epithelium of the coprodaeum and large intestine in the normal and dehydrated fowl (*Gallus domesticus*). *In vivo* perfusion studies. *Journal of Physiology* **216**, 735–51.

Bindslev, N. & Skadhauge, E. (1971*b*). Sodium chloride absorption and solute-linked water flow across the epithelium of the coprodeum and large intestine in the normal and dehydrated fowl (*Gallus domesticus*). *In vivo* perfusion studies. *Journal of Physiology* **216**, 753–68.

Blaker, D. (1967). An outbreak of botulinus poisoning among waterbirds. *Ostrich* **38**, 144–7.

Blatt, C. M., Taylor, C. R. & Habal, M. B. (1972). Thermal panting in dogs: the lateral nasal gland, a source of water for evaporative cooling. *Science* **177**, 804–5.

Blösch, M. (1966). Die Aktivität der Salzdrüsen. Eine Undersuchung an freilebenden und gefangenen Silbermöwen. *Vogelwarte* **23**, 225–31.

Bonting, S. L. (1964). Sodium-potassium activated ATPase and active cation transport. In *Water and Electrolyte Metabolism*, vol. II, pp. 35–7, ed. de Gracff, J. & Leijnse, B. (*West-European Symposium of Clinical Chemistry* **3**).

Bonting, S. L. (1970). Sodium–potassium activated adenosinetriphosphatase and cation transport. In *Membranes and Ion Transport*, vol. I, pp. 257–363, ed. Bittar, E. E. Wiley-Interscience: London.

Bonting, S. L., Caravaggio, L. L., Canady, M. R. & Hawkins, N. M. (1964). Studies on sodium–potassium adenosinetriphosphatase. XI. The salt gland of the herring gull. *Archives of Biochemistry and Biophysics* **106**, 49–56.

Borut, A. & Schmidt-Nielsen, K. (1963). Respiration of avian salt-secreting gland in tissue slice experiments. *American Journal of Physiology* **204**, 573–81.

Botelho, S. Y., Brooks, F. P. & Shelley, W. B. (eds) (1969). *The exocrine glands.* University of Pennsylvania Press: Philadelphia.

Bradley, E. L. & Holmes, W. N. (1971). The effects of hypophysectomy on adrenocortical function in the duck (*Anas platyrhynchos*). *Journal of Endocrinology* **49**, 437–57.

Bradley, E. L. & Holmes, W. N. (1972*a*). The role of the nasal glands in the survival of ducks (*Anas platyrhynchos*) exposed to hypertonic saline drinking water. *Canadian Journal of Zoology* **50**, 611–17.

Bradley, E. L. & Holmes, W. N. (1972*b*). Corticotropic responses in the adenohypophysectomized duck (*Anas platyrhynchos*). *General and Comparative Endocrinology* **19**, 266–72.

Bradshaw, S. D. & Shoemaker, V. H. (1967). Aspects of water and electrolyte changes in a field population of *Amphibolurus* lizards. *Comparative Biochemistry and Physiology* **20**, 855–65.

Braysher, M. (1971). The structure and function of the nasal salt gland from the Australian sleepy lizard *Trachydosaurus* (formerly *Tiliqua*) *rugosus*: Family Scincidae. *Physiological Zoology* **44**, 129–36.

Burford, H. J. & Bond, R. F. (1968). Avian cardiovascular parameters: effect of intra-venous osmotic agents, relation to salt gland secretion. *Experientia* **24**, 1086–8.

Burgen, A. S. V. & Emmelin, N. G. (1961). *Physiology of the salivary glands.* Arnold: London.

Burger, J. W. & Hess, W. N. (1960). Function of the rectal gland in the spiny dogfish. *Science* **131**, 670–1.

Burock, G., Kühnel, W. & Petry, G. (1969). Über die inaktive Salzdrüse von Enten (*Anas platyrhynchus*). Histologische und histochemische Untersuchungen. *Zeitschrift für Zellforschung und mikroskopische Anatomie* **97**, 608–18.

Buxton, R., Hally, A. D. & Scothorne, R. J. (1966). A histological and histochemical study of the developing salt gland in the domestic duck. *Journal of Anatomy* **100**, 931.

Cade, T. J. & Greenwald, L. (1966). Nasal salt excretion in falconiform birds. *Condor* **68**, 338–50.

Caldwell, P. C. (1970). Models for sodium/potassium transport: a critique. In *Membranes and ion transport*, vol. I, pp. 433–61, ed. Bittar, E. E. Wiley-Interscience: London.

Carey, F. G. & Schmidt-Nielsen, K. (1962). Secretion of iodide by the nasal gland of birds. *Science* **137**, 866–7.

Carpenter, R. E. & Stafford, M. A. (1970). The secretory rates and the chemical stimulus for secretion of the nasal salt glands in the Rallidae. *Condor* **72**, 316–24.

Carr, A. (1952). *Handbook of turtles.* Cornell University Press: Ithaca, New York.

Chance, B., Lee, C., Oshino, K. & Van Rossum, G. (1964). Properties of mitochondria isolated from herring gull salt gland. *American Journal of Physiology* **206**, 461–8.

Chatwin, A. L., Linzell, J. L. & Setchell, B. P. (1969). Cardiovascular changes during lactation in the rat. *Journal of Endocrinology* **44**, 247–54.

Chester Jones, I. (1957). *The Adrenal Cortex.* Cambridge University Press: London.

Chester Jones, I., Bellamy, D., Chan, D. K. O., Follett, B. K., Henderson, I. W., Phillips, J. G. & Snart, R. S. (1972). Biological actions of steroid hormones in nonmammalian vertebrates. In *Steroids in Nonmammalian Vertebrates*, pp. 414–80, ed. Idler, D. R. Academic Press: New York.

Clark, R. B. & Kennedy, J. R. (1968). *Rehabilitation of oiled seabirds.* Department of Zoology, University of Newcastle-upon-Tyne.

Clark, R. B. & Kennedy, J. R. (undated). *How oiled seabirds are cleaned.* Department of Zoology, University of Newcastle-upon-Tyne.

Cloudsley-Thompson, J. L. & Mohamed, E. R. M. (1967). Water economy of ostrich. *Nature* **216**, 1040.

Cooch, F. G. (1961). Avian salt gland and botulism. *Canadian Wildlife Service Research Progress Reports*, p. 27.

Cooch, F. G. (1964). A preliminary study of the survival value of a functional salt gland in prairie Anatidae. *Auk* **81**, 380–93.

Cords, E. (1904). Beiträge zur Lehre vom Kopfnervensystem der Vogel. *Anatomische Hefte* **26**, 49–100 (plus plates 1–4 in same volume).

Cottle, M. K. W. & Pearce, J. W. (1970). Some observations on the nerve supply to the salt gland of the duck. *Quarterly Journal of Experimental Physiology* **55**, 207–12.

Cowan, F. B. M. (1967). Comparative studies on the cranial glands of turtles. *American Zoologist* **7**, 810.

References

Cowan, F. B. M. (1969). Gross and microscopic anatomy of the orbital glands of *Malaclemys* and other emydine turtles. *Canadian Journal of Zoology* **47**, 723–9.

Cowan, F. B. M. (1971). The ultrastructure of the lachrymal 'salt' gland and the Harderian gland in euryhaline *Malaclemys* and some closely related stenohaline emydines. *Canadian Journal of Zoology* **49**, 691–7.

Crocker, A. D. & Holmes, W. N. (1971*a*). Intestinal absorption in the duck (*Anas platyrhynchos*) maintained on freshwater and hypertonic saline. *Journal of Endocrinology* **49**, xviii–xix.

Crocker, A. D. & Holmes, W. N. (1971*b*). Intestinal absorption in ducklings (*Anas platyrhynchos*) maintained on fresh water and hypertonic saline. *Comparative Biochemistry and Physiology* **40A**, 203–11.

Cross, B. A. & Silver, I. A. (1962). Mammary oxygen tension and the milk-ejection mechanism. *Journal of Endocrinology* **23**, 375–84.

Crowe, J. H., Nagy, K. A. & Francis, C. (1970). Structure of lizard salt glands. *American Zoologist* **10**, 556.

Dale, H. H. & Gaddum, J. H. (1930). Reactions of denervated voluntary muscle, and their bearing on the mode of action of parasympathetic and related nerves. *Journal of Physiology* **70**, 109–44.

Daly, M. deB., Lambertson, C. J. & Schweitzer, A. (1954). Observations on the volume of blood flow and oxygen utilization of the carotid body in the cat. *Journal of Physiology* **125**, 67–89.

Darwin, C. (1889). *Journal of researches into the natural history and geology of the countries visited during the voyage of H.M.S. 'Beagle' round the world*, (new edition). Murray: London.

De Piceis Polver, P. (1968). Studio istomorfologico ed istochimico sulle ghiandole nasali di alcuni Lacertilii. *Bollettino di Zoologia* **35**, 375–6.

Diamond, J. M. (1965). The mechanism of isotonic water absorption and secretion. In *The state and movement of water in living organisms, Symposia of the Society for Experimental Biology* **19**, 329–47.

Diamond, J. M. & Bossert, W. H. (1967). Standing-gradient osmotic flow: a mechanism for coupling of water and solute transport in epithelia. *Journal of General Physiology* **50**, 2061–83.

Diamond, J. & Bossert, W. H. (1968). Functional consequences of ultrastructural geometry in 'backwards' fluid-transporting epithelia. *Journal of Cell Biology* **37**, 694–702.

Donaldson, E. M. & Holmes, W. N. (1965). Corticosteroidogenesis in the freshwater and saline-maintained duck (*Anas platyrhynchos*). *Journal of Endocrinology* **32**, 329–36.

Douglas, D. S. (1964). Extra-renal salt excretion in the Adélie penguin chick. In *Symposium de biologie antarctique*, pp. 503–8. Herman: Paris.

Douglas, D. S. (1966*a*). Low urine salt concentrations in salt-loaded gulls. *Physiologist* **9**, 171.

Douglas, D. S. (1966*b*). Secretion of the thiocyanate ion by the nasal gland of the Adelie penguin. *Nature* **209**, 1150–1.

Douglas, D. S. (1968). Hyponatraemic cloacal diuresis in salt-loaded gulls. *American Zoologist* **8**, 782.

Douglas, D. S. (1970). Electrolyte excretion in seawater-loaded herring gulls. *American Journal of Physiology* **219**, 534–9.

Douglas, D. S. & Neely, S. M. (1969). The effect of dehydration on salt gland performance. *American Zoologist* **9**, 1095.

Doyle, W. L. (1960). The principal cells of the salt-gland of marine birds. *Experimental Cell Research* **21**, 386–93.

Drost, R. (1931). Möwen trinken seewasser. *Ornithologische Monatsberichte* **39**, 119.

Dulzetto, F. (1965). La cosidetta 'ghiandola de sale' (ghiandola nasale) degli uccelli marini. *Atti della Società peloritana di scienze, fisiche, matematiche e naturali* **11**, 179–201.

Dulzetto, F. (1967). I meccanismi extrarenali nei casi di insufficienza fisiologica del rene. *Bollettino di Zoologia* **34**, 17–45.

Dunson, W. A. (1968). Salt gland secretion in the pelagic sea snake *Pelamis*. *American Journal of Physiology* **215**, 1512–17.

Dunson, W. A. (1969*a*). Reptilian salt glands. In *Exocrine glands*, pp. 83–101, eds Botelho, S. Y., Brooks, F. P. & Shelley, W. B. University of Pennsylvania Press: Philadelphia.

Dunson, W. A. (1969*b*). Electrolyte excretion by the salt gland of the Galápagos marine iguana. *American Journal of Physiology* **216**, 995–1002.

Dunson, W. A. (1970). Some aspects of electrolyte and water balance in three estuarine reptiles, the diamond-back terrapin, American and 'salt water' crocodiles. *Comparative Biochemistry and Physiology* **32**, 161–74.

Dunson, W. A., Packer, R. K. & Dunson, M. K. (1971). Sea snakes: an unusual salt gland under the tongue. *Science* **173**, 437–41.

Dunson, W. A. & Taub, A. M. (1966). Salt glands in sea snakes (*Laticauda*). *American Zoologist* **6**, 18.

Dunson, W. A. & Taub, A. M. (1967). Extrarenal salt excretion in sea snakes, *Laticauda*. *American Journal of Physiology* **213**, 975–82.

Duvdevani, I. (1972). Anatomy and histology of the nasal cavities and the nasal salt gland in four species of fringe-toed lizards, *Acanthodactylus* (Lacertidae). *Journal of Morphology* **137**, 353–64.

Ekström, J. & Holmberg, J. (1972). Effect of decentralization on the choline acetyltransferase of the canine parotid gland. *Journal of Physiology* **222**, 93–94P.

Ellis, R. A. (1965). DNA labelling and X-irradiation studies of the phosphatase-positive peripheral cells in the nasal (salt) glands of ducks. *American Zoologist* **5**, 648.

Ellis, R. A. & Abel, J. H. (1964*a*). Intercellular channels in the salt secreting glands of marine turtles. *Science* **144**, 1340–2.

Ellis, R. A. & Abel, J. H. (1964*b*). Electron microscopy and cytochemistry of the salt gland of sea turtles. *Anatomical Record* **148**, 278.

Ellis, R. A., DeLellis, R. A. & Kablotsky, Y. H. (1962). The effect of salt water intake on the development of the salt gland of the domestic duck. *American Zoologist* **2**, 406.

Ellis, R. A. & Ernst, S. A. (1967). The effect of salt-water intake on the specialization of the surface secretory cells in the nasal (salt) glands of domestic ducks. *Anatomical Record* **157**, 239–40.

Ellis, R. A., Goertemiller, C. C., DeLellis, R. A. & Kablotsky, Y. H. (1963). The effect of a salt water regimen on the development of the salt glands of domestic ducklings. *Developmental Biology* **8**, 286–308.

Ensor, D. M. & Phillips, J. G. (1970*a*). The effect of salt loading on the pituitary prolactin levels of the domestic duck (*Anas platyrhynchos*), and juvenile herring or lesser black-backed gulls (*Larus argentatus* or *Larus fuscus*). *Journal of Endocrinology* **48**, 167–72.

Ensor, D. M. & Phillips, J. G. (1970*b*). The effect of environmental stimuli on the circadian rhythm of prolactin production in the duck (*Anas platyrhynchos*). *Journal of Endocrinology* **48**, lxxi.

Ensor, D. M. & Phillips, J. G. (1972*a*). The effect of age and environment on extrarenal salt excretion in juvenile gulls (*Larus argentatus* and *L. fuscus*). *Journal of Zoology* **168**, 119–26.

Ensor, D. M. & Phillips, J. G. (1972*b*). The effect of dehydration on salt and water balance in gulls (*Larus argentatus and L. fuscus*). *Journal of Zoology* **168**, 127–37.

Ensor, D. M., Simons, I. M. & Phillips, J. G. (1973). The effect of hypophysectomy and prolactin replacement therapy on salt and water metabolism in *Anas platyrhynchos. Journal of Endocrinology* **57**, xi.

Ensor, D. M., Thomas, D. H. & Phillips, J. G. (1970). The possible role of the thyroid in extrarenal secretion following a hypertonic saline load in the duck (*Anas platyrhynchos*). *Journal of Endocrinology* **46**, x.

Ernst, S. A. (1971). Ultrastructural localization of K-dependent phosphatase activity in avian salt gland. *Anatomical Record* **169**, 310.

Ernst, S. A. (1972*a*). Transport adenosinetriphosphatase cytochemistry. 1. Biochemical characterization of a cytochemical medium for the ultrastructural localization of ouabain-sensitive, potassium-dependent phosphatase activity in the avian salt gland. *Journal of Histochemistry and Cytochemistry* **20**, 13–22.

Ernst, S. A. (1972*b*). Transport adenosinetriphosphatase cytochemistry. 2. Cytochemical localization of ouabain-sensitive, potassium-dependent phosphatase activity in the secretory epithelium of the avian salt gland. *Journal of Histochemistry and Cytochemistry* **20**, 23–38.

Ernst, S. A. & Ellis, R. A. (1969). The development of surface specialization in the secretory epithelium of the avian salt gland in response to osmotic stress. *Journal of Cell Biology* **40**, 305–21.

Ernst, S. A., Goertemiller, C. C. & Ellis, R. A. (1967). The effect of salt regimens on the development of ($Na^{+}K^{+}$)-dependent ATPase activity during the growth of salt glands in ducklings. *Biochimica et biophysica acta* **135**, 682–92.

Ernst, S. A. & Philpott, C. W. (1970). Preservation of Na-K-activated and Mg-activated adenosine triphosphatase activities of avian salt gland and teleost gill with formaldehyde as fixative. *Journal of Histochemistry and Cytochemistry* **18**, 251–63.

Fänge, R., Krog, J. & Reite, O. (1963). Blood flow in the avian salt gland studied by polarographic oxygen electrodes. *Acta physiologica scandinavica* **58**, 40–7.

Fänge, R., Schmidt-Nielsen, K. & Osaki, H. (1958). The salt gland of the herring gull. *Biological Bulletin. Marine Biological Laboratory, Woods Hole, Mass.* **115**, 162–71.

Fänge, R., Schmidt-Nielsen, K. & Robinson, M. (1958). Control of secretion from the avian salt gland. *American Journal of Physiology* **195**, 321–6.

Farber, S. J. (1960). Mucopolysaccharides and sodium metabolism. *Circulation* **21**, 941–53.

Fawcett, D. W. (1962). Physiologically significant specializations of the cell surface. *Circulation* **26**, 1105–25 (in discussion of this paper, 1126–32).

Fitzsimons, J. T. (1963). The effects of slow infusions of hypertonic solutions on drinking and drinking threshold in rats. *Journal of Physiology* **167**, 344–54.

Fletcher, G. L. & Holmes, W. N. (1968). Observations on the intake of water and electrolytes by the duck (*Anas platyrhynchos*) maintained on freshwater and on hypertonic saline. *Journal of Experimental Biology* **49**, 325–39.

Fletcher, G. L., Stainer, I. M. & Holmes, W. N. (1967). Sequential changes in adenosine triphosphatase activity and the electrolyte excretory capacity of the nasal-glands of the duck during the period of adaptation to salt-water. *Journal of Experimental Biology* **47**, 375–91.

Fourman, J. (1966). Cholinesterase and sodium transport in the supra-orbital gland of the duck. *Journal of Anatomy* **100**, 693.

Fourman, J. (1969). Cholinesterase activity in the supra-orbital salt secreting gland of the duck. *Journal of Anatomy* **104**, 233–9.

Frankel, A. I. (1970). Neurohumoral control of the avian adrenal: a review. *Poultry Science* **49**, 869–921.

Frey, E. K. & Kraut, H. (1928). Ein neues Kreislaufhormon unde seine Wirkung. *Archiv für experimentelle Pathologie und Pharmakologie* **33**, 1–56.

Friend, M., Haegele, M. A. & Wilson, R. (1973). DDE: interference with extra-renal salt excretion in the mallard. *Bulletin of Environmental Contamination and Toxicology* **9**, 49–53.

Frings, H., Anthony, A. & Schein, M. W. (1958). Salt excretion by the nasal gland of Laysan and black-footed albatrosses. *Science* **128**, 1572.

Frings, H. & Frings, M. (1959). Observations on salt balance and behavior of Laysan and black-footed albatrosses in captivity. *Condor* **61**, 305–14.

Frömter, E. & Diamond, J. (1972). Route of passive ion permeation in epithelia. *Nature* **235**, 9–13.

Furness, J. B. & Iwayama, T. (1972). The arrangement and identification of axons innervating the vas deferens of the guinea-pig. *Journal of Anatomy* **113**, 179–96.

Gabe, M. & Saint Girons, H. (1971). Polymorphisme des glandes nasales externes des sauriens. *Compte rendu hebdomadaire des séances de l'Académie des Sciences* **272**, 1275–8.

Gabe, M. & Saint Girons, H. (1973). Contribution à l'histologie des glandes externes de *Sphenodon punctatus* Gray (Reptilia, Rhynchocephalia). *Acta anatomica* **84**, 452–64.

Ganin, M. (1890). Einige Thatsachen zur Frage über das Jacobson'sche Organ der Vogel. *Zoologischer Anzeiger* **13**, 285–7.

Gaupp, E. (1888). Anatomische Untersuchungen über die Nervenversorgung der Mund- und Nasenhöhlendrüsen der Wirbeltiere. *Morphologisches Jahrbuch (Gegenbaur's Jarhbuch)* **14**, 436–89 plus plate XIX in same volume.

Gautvik, K. (1970*a*). Studies on kinin formation in functional vasodilatation of the submandibular salivary gland in cats. *Acta physiologica scandinavica* **79**, 174–87.

Gautvik, K. (1970*b*). The interaction of two different vasodilator mechanisms in the chorda-tympani activated submandibular salivary gland. *Acta physiologica scandinavica* **79**, 188–203.

Gautvik, K. (1970*c*). Parasympathetic neuro-effector transmission and functional vasodilatation in the submandibular salivary gland of cats. *Acta physiologica scandinavica* **79**, 204–15.

Gerzeli, G. (1967*a*). Sedi possibili e tentativi di stimolazione di una secrezione ionica extrarenale in *Lactera viridis. Bollettino di Zoologia* **34**, 120–1.

Gerzeli, G. (1967*b*). Osservazioni e considerazioni morfo-funzionali comparate sulle ghiandole lacrimali dei cheloni. *Archivio zoologico italiano* **L11**, 37–47.

References

Gerzeli, G. & De Piceis Polver, P. (1970). The lateral nasal gland of *Lacerta viridis* under different experimental conditions. *Monitore zoologico italiano* **4**, 191–200.

Gill, J. B. & Burford, H. J. (1968). Secretion from normal and supersensitive avian salt glands. *Journal of Experimental Zoology* **168**, 451–4.

Gill, J. B. & Burford, H. J. (1969). The effect of antidiuretic hormone on avian salt gland function. *Tissue & Cell* **1**, 497–501.

Glynn, I. M., Hoffman, J. F. & Lew, V. L. (1971). Some 'partial reactions' of the sodium pump. *Philosophical Transactions of the Royal Society, series B* **262**, 91–102.

Goertemiller, C. C. & Ellis, R. A. (1962). Secretory competence and the sodium and potassium content of the effluent of salt glands of domestic ducklings. *American Zoologist* **2**, 525.

Goertemiller, C. C. & Ellis, R. A. (1966). Specificity of sodium chloride in the stimulation of growth in the salt glands of ducklings. *Zeitschrift für mikroskopisch-anatomische Forschung* **74**, 296–302.

Goodge, W. R. (1961). A histological and histochemical study of the nasal gland of some non-marine birds. *American Zoologist* **1**, 357.

Gordon, M. S. (1968). Water and solute metabolism. In *Animal Function: Principles and Adaptations*, by Gordon, M. S., Bartholomew, G. A., Grinnell, A. D., Jörgensen, C. B. & White, F. N. Macmillan: New York.

Grenot, C. (1968). Sur l'excrétion de sels chez le lézard saharien: *Uromastix acanthinurus. Compte rendu hebdomadaire des séances de l'Académie des Sciences* **266**, 1871–4.

Gunnison, J. B. & Coleman, G. E. (1932). *Clostridium botulinum*, Type C, associated with western duck disease. *Journal of Infectious Diseases* **51**, 542–51.

Haase, P. & Fourman, J. (1970). The autonomic innervation of the avian salt gland. *Journal of Anatomy* **107**, 382–3.

Hajjar, R., Sattler, F., Anderson, B. G. & Gwinup, G. (1970). Definition of the stimulus to secretion of the nasal salt gland of the seagull. *Hormone and Metabolic Research* **2**, 35–7.

Håkansson, C. H. & Malcus, B. (1969). Secretive responses of the electrically stimulated nasal salt gland in *Larus argentatus* (herring gull). *Acta physiologica scandinavica* **76**, 385–92.

Håkansson, C. H. & Malcus, B. (1970). Secretive pressure exerted by the stimulated nasal salt gland in *Larus argentatus* (herring gull). *Acta physiologica scandinavica* **78**, 249–54.

Hally, A. D., Buxton, R. & Scothorne, R. J. (1966). The developing secretory capacity of the salt gland in the domestic duck. *Journal of Anatomy* **100**, 930.

Hand, A. R. (1972). Adrenergic and cholinergic nerve terminals in the rat parotid gland. Electron microscopic observations on permanganate-fixed glands. *Anatomical Record* **173**, 131–40.

Hanwell, A. & Linzell, J. L. (1973). The time course of cardiovascular changes in lactation in the rat. *Journal of Physiology* **233**, 93–109.

Hanwell, A., Linzell, J. L. & Peaker, M. (1970*a*). Avian salt-gland blood flow and the extraction of ions from the plasma. *Journal of Physiology* **207**, 83–4P.

Hanwell, A., Linzell, J. L. & Peaker, M. (1970*b*). Salt-gland function in the domestic goose. *Journal of Physiology* **210**, 97–9P.

Hanwell, A., Linzell, J. L. & Peaker, M. (1971*a*). Salt-gland secretion and blood flow in the goose. *Journal of Physiology* **213**, 373–87.

Hanwell, A., Linzell, J. L. & Peaker, M. (1971*b*). Cardiovascular responses to salt-loading in conscious domestic geese. *Journal of Physiology* **213**, 389–98.

Hanwell, A., Linzell, J. L. & Peaker, M. (1971*c*). The location and nature of the receptors for secretion by the salt gland of the goose. *Journal of Physiology* **216**, 28–9P.

Hanwell, A., Linzell, J. L. & Peaker, M. (1972). Nature and location of the receptors for salt-gland secretion in the goose. *Journal of Physiology* **226**, 453–72.

Hanwell, A. & Peaker, M. (1973). The effect of post-ganglionic denervation on functional hypertrophy in the salt gland of the goose during adaptation to salt-water. *Journal of Physiology* **234**, 78–80P.

Harriman, A. E. (1967). Laughing gulls offered saline in preference and survival tests. *Physiological Zoölogy* **40**, 273–9.

Harriman, A. E. & Kare, M. R. (1966*a*). Tolerance for hypertonic saline solutions in herring gulls, starlings and purple grackles. *Physiological Zoölogy* **39**, 117–22.

Harriman, A. E. & Kare, M. R. (1966*b*). Aversion to saline solutions in starlings, purple grackles and herring gulls. *Physiological Zoölogy* **39**, 123–6.

Harrison, F. A. & Paterson, J. Y. F. (1965). The specific activity of plasma cortisol in sheep after rapid intravenous injection of [1, 2-$^3H_2$]cortisol and its relation to the rate of cortisol secretion. *Journal of Endocrinology* **33**, 447–90.

Hart, W. M. & Essex, H. E. (1942). Water metabolism of the chicken with special reference to the role of the cloaca. *American Journal of Physiology* **136**, 657–68.

Heap, P. F. & Bhoola, K. D. (1969). Histochemical localization of kallikrein granules in the submaxillary gland of the guinea-pig. *Journal of Anatomy* **105**, 525–32.

Hebb, C. O. & Linzell, J. L. (1970). Innervation of the mammary gland. A histochemical study in the rabbit. *Histochemical Journal* **2**, 491–505.

Heidenhain, R. (1872). Über die Wirkung einiger Gifte auf die Nerven der glandula submaxillaris. *Pflügers Archiv für die gesamte Physiologie des Menschen und der Tiere* **5**, 309–18.

Heinroth, O. & Heinroth, M. (1928). *Die Vogel Mitteleuropas in allen Lebens- und Enticklungestufen photographisch aufgenommen und in ihrem Seelenleben bei der Aufzucht*, volume III. Bermühler: Berlin.

Helton, E. D. & Holmes, W. N. (1973). The distribution and metabolism of labelled corticosteroids in the duck (*Anas platyrhynchos*). *Journal of Endocrinology* **56**, 361–85.

Henderson, V. E. & Roepke, M. H. (1933). Über den lokalen hormonalen Mechanismus der Parasympathikusreizung. *Archiv für experimentelle Pathologie und Pharmakologie* **172**, 314–24.

Hilton, S. M. & Lewis, G. P. (1955*a*). The cause of the vasodilatation accompanying activity in the submandibular salivary gland. *Journal of Physiology* **128**, 235–48.

Hilton, S. M. & Lewis, G. P. (1955*b*). The mechanism of the functional hyperaemia in the submandibular salivary gland. *Journal of Physiology* **129**, 253–71.

Hilton, S. M. & Lewis, G. P. (1956). The relationship between glandular activity, bradykinin formation and functional vasodilatation in the submandibular salivary gland. *Journal of Physiology* **134**, 471–83.

Hirano, T. & Utida, S. (1968). Effects of ACTH and cortisol on water movements in isolated intestine of the eel, *Anguilla japonica. General and Comparative Endocrinology* **11**, 373–80.

Hokin, L. E. & Hokin, M. R. (1959). Evidence for phosphatidic acid as the sodium carrier. *Nature* **184**, 1068–9.

References

Hokin, L. E. & Hokin, M. R. (1960). Studies on the carrier function of phosphatidic acid in sodium transport. 1. The turnover of phosphatidic acid and phosphoinositide in the avian salt gland on stimulation of secretion. *Journal of General Physiology* **44**, 61–85.

Hokin, L. E. & Hokin, M. R. (1963*a*). The role of phosphatides in active transport with particular reference to sodium transport. In *Drugs and membranes*, pp. 23–40, ed. Hogben, C. A. M. (*Proceedings of the International Pharmacology Meeting, Stockholm*, vol. **4**).

Hokin, L. E. & Hokin, M. R. (1963*b*). Phosphatidic acid metabolism and active transport of sodium. *Federation Proceedings* **22**, 8–18.

Hokin, L. E. & Hokin, M. R. (1963*c*). On the lack of effect of acetylcholine on phosphoprotein metabolism in the salt gland of the sea gull. *Biochimica et biophysica acta* **71**, 462–3.

Hokin, L. E. & Hokin, M. R. (1963*d*). Biological transport. *Annual Reviews of Biochemistry* **32**, 553–78.

Hokin, M. R. (1963). Studies on a Na^+- and K^+-dependent ouabain-sensitive adenosine triphosphatase in the avian salt gland. *Biochimica et biophysica acta* **77**, 108–20.

Hokin, M. R. (1966). Respiration and ATP and ADP levels during Na^+ transport in salt gland slices. *Life Sciences* **5**, 1829–37.

Hokin, M. R. (1967). The Na^+, K^+ and Cl^- content of goose salt gland slices and the effects of acetylcholine and ouabain. *Journal of General Physiology* **50**, 2197–2209.

Hokin, M. R. (1969). Electrolyte transport in the avian salt gland. In *Exocrine glands*, pp. 73–83, eds Botelho, S. Y., Brooks, F. P. & Shelley, W. B. University of Pennsylvania Press: Philadelphia.

Hokin, M. R. & Hokin, L. E. (1964*a*). The synthesis of phosphatidic acid and protein-bound phosphorylserine in salt gland homogenates. *Journal of Biological Chemistry* **239**, 2116–22.

Hokin, M. R. & Hokin, L. E. (1964*b*). Interconversions of phosphatidylinositol and phosphatidic acid involved in the response to acetylcholine in the salt gland. In *Metabolic and physiological significance of lipids*, pp. 423–34, eds Dawson, R. M. C. & Rhodes, D. N. Wiley-Interscience: London.

Hokin, M. R. & Hokin, L. E. (1966). The formation and continuous turnover of a fraction of phosphatidic acid on stimulation of NaCl secretion by acetylcholine in the salt-gland. *Journal of General Physiology* **50**, 793–811.

Holmberg, J. (1971). The secretory nerves of the parotid gland of the dog. *Journal of Physiology* **219**, 463–76.

Holmes, W. N. (1965). Some aspects of osmoregulation in reptiles and birds. *Archives d'anatomie microscopique et de morphologie expérimentale* **54**, 491–513.

Holmes, W. N. (1972). Regulation of electrolyte balance in marine birds with special reference to the role of the pituitary-adrenal axis in the duck (*Anas platyrhynchos*). *Federation Proceedings* **31**, 1587–98.

Holmes, W. N. & Adams, B. M. (1963). Effects of adrenocortical and neurohypophysial hormones on the renal excretory pattern in the water-loaded duck. *Endocrinology* **73**, 5–10.

Holmes, W. N., Butler, D. G. & Phillips, J. G. (1961). Observations on the effects of maintaining glaucous-winged gulls (*Larus glaucescens*) on fresh-water and sea-water for long periods. *Journal of Endocrinology* **23**, 53–61.

Holmes, W. N., Chan, M.-Y., Bradley, J. S. & Stainer, I. M. (1970). The control of some endocrine mechanisms associated with salt regulation in aquatic birds. In *Hormones and the environment*, pp. 87–108, eds Benson, G. K. & Phillips, J. G. (*Memoirs, Society for Endocrinology* **18**). Cambridge University Press: London.

Holmes, W. N., Fletcher, G. L. & Stewart, D. J. (1968). The patterns of renal electrolyte excretion in the duck (*Anas platyrhynchos*) maintained on freshwater and on hypertonic saline. *Journal of Experimental Biology* **48**, 487–508.

Holmes, W. N., Lockwood, L. N. & Bradley, E. L. (1972). Adenohypophysial control of extrarenal excretion in the duck (*Anas platyrhynchos*). *General and Comparative Endocrinology* **18**, 59–68.

Holmes, W. N. & McBean, R. L. (1964). Some aspects of electrolyte excretion in the green turtle, *Chelonia mydas. Journal of Experimental Biology* **41**, 81–90.

Holmes, W. N. & Phillips, J. G. (1965). Adrenocortical hormones and electrolyte metabolism in birds. In *Proceedings of the IInd International Congress of Endocrinology, Excerpta Medica International Congress Series* **83**, 158–61.

Holmes, W. N., Phillips, J. G. & Butler, D. G. (1961). The effect of adrenocortical steroids on the renal and extra-renal responses of the domestic duck (*Anas platyrhynchus*) after hypertonic saline loading. *Endocrinology* **69**, 483–95.

Holmes, W. N., Phillips, J. G. & Chester Jones, I. (1963). Adrenocortical factors associated with adaptation of vertebrates to marine environments. *Recent Progress in Hormone Research* **19**, 619–72.

Holmes, W. N., Phillips, J. G. & Wright, A. (1969). The control of extrarenal excretion in the duck (*Anas platyrhynchos*) with special reference to the pituitary-adrenal axis. *General and Comparative Endocrinology, Supplement* **2**, 358–73.

Holmes, W. N. & Stewart, D. J. (1968). Changes in nucleic acids and protein composition of the nasal glands from the duck (*Anas platyrhynchos*) during the period of adaptation to hypertonic saline. *Journal of Experimental Biology* **48**, 509–19.

Holmes, W. N. & Wright, A. (1969). Some aspects of the control of osmoregulation and homeostasis in birds. In *Progress in Endocrinology*, pp. 237–48, eds Gual, C. & Ebling, F. J. G. (*Proceedings of the IIIrd International Congress of Endocrinology, Excerpta Medica International Congress Series* **184**).

Hsieh, T. H. (1951). The sympathetic and parasympathetic nervous systems of the fowl. Ph.D. thesis, University of Edinburgh.

Hughes, M. R. (1962). Studies on renal and extrarenal salt excretion in gulls and terns. Ph.D. thesis, Duke University, North Carolina.

Hughes, M. R. (1968). Renal and extrarenal excretion in the common tern *Sterna hirundo. Physiological Zoölogy* **41**, 210–19.

Hughes, M. R. (1969). Ionic and osmotic concentration of tears of the gull, *Larus glaucescens. Canadian Journal of Zoology* **47**, 1337–9.

Hughes, M. R. (1970*a*). Cloacal and salt-gland ion excretion in the seagull *Larus glaucescens* acclimated to increasing concentrations of sea water. *Comparative Biochemistry and Physiology* **32**, 315–25.

Hughes, M. R. (1970*b*). Flow rate and cation concentration in salt gland secretions of the glaucous-winged gull, *Larus glaucescens. Comparative Biochemistry and Physiology* **32**, 807–12.

Hughes, M. R. (1970*c*). Some observations on ion and water balance in the puffin, *Fratercula arctica. Canadian Journal of Zoology* **48**, 479–82.

Hughes, M. R. (1970*d*). Relative kidney size in nonpasserine birds with functional salt glands. *Condor* **72**, 164–8.

Hughes, M. R. (1972*a*). Hypertonic salt gland secretion in the glaucous-winged gull, *Larus glaucescens*, in response to stomach loading with dilute sodium chloride. *Comparative Biochemistry and Physiology* **41A**, 121–7.

Hughes, M. R. (1972*b*). The effect of salt gland removal on cloacal ion and water excretion in the growing kittiwake, *Rissa tridactyla. Canadian Journal of Zoology* **50**, 603–10.

Hughes, M. R. & Ruch, F. E. (1968). Sodium and potassium in the tears and salt gland secretion of saline acclimatized ducks. *Proceedings of the International Union of Physiological Sciences* **7**, 204 (*24th International Congress, Washington, D.C.*).

Hughes, M. R. & Ruch, F. E. (1969). Sodium and potassium in spontaneously produced salt-gland secretion and tears of ducks, *Anas platyrhynchos*, acclimated to fresh and saline waters. *Canadian Journal of Zoology* **47**, 1133–8.

Inoue, T. (1963). Nasal salt gland: independence of salt and water transport. *Science* **142**, 1299–1300.

Jacobson, L. L. (1813). Sur une glande conglomerée appartenante à la cavité nasale. *Bulletin de la Société philomathique de Paris* **3**, 267–9.

Janicki, R. H. & Kinter, W. B. (1971). DDT: disrupted osmoregulatory events in the intestine of the eel *Anguilla rostrata* adapted to seawater. *Science* **173**, 1146–8.

Jephcott, J. J. & Hally, A. D. (1970). The effects of tissue mass and functional demand on compensatory growth of the avian salt gland. *Journal of Anatomy* **106**, 405.

Jobert, C. (1869). Récherches anatomiques sur les glandes nasales des oiseaux. *Annales des sciences naturelles* (*Zoologie etc.*) **11**, 349–68.

Johnson, I. M. (1969). Electrolyte and water balance of the red-tailed hawk, *Buteo jamaicensis. American Zoologist* **9**, 587.

Kalmbach, E. R. & Gunderson, M. F. (1934). Western duck sickness: a form of botulism. *United States Department of Agriculture, Technical Bulletin* **411**.

Karlsson, K.-A., Samuelsson, B. E. & Steen, G. O. (1969). Sphingolipid composition of the avian salt gland. *Biochimica et biophysica acta* **176**, 429–31.

Karlsson, K.-A., Samuelsson, B. E. & Steen, G. O. (1971). Lipid pattern and Na^+-K^+-dependent adenosine triphosphatase activity in the salt gland of the duck before and after adaptation to hypertonic saline. *Journal of Membrane Biology* **5**, 169–84.

Karpinski, E., Barton, S. & Schachter, M. (1971). Vasodilator nerve fibres to the submaxillary gland of the cat. *Nature* **232**, 122–4.

Kasbekar, D. K. & Durbin, R. P. (1965). An adenosine triphosphatase from frog gastric mucosa. *Biochimica et biophysica acta* **105**, 472–82.

Keynes, R. D. (1969). From frog skin to sheep rumen: a survey of transport of salts and water across multicellular structures. *Quarterly Reviews of Biophysics* **2**, 177–281.

Komnick, H. (1963*a*). Elektronenmikroskopische Untersuchungen zur funktionellen Morphologie des Ionentransportes in der Salzdrüse. I. Teil: Bau und Feinstruktur der Salzdrüse. *Protoplasma* **56**, 274–314.

Komnick, H. (1963*b*). Elektronenmikroskopische Untersuchungen zur funktionellen Morphologie des Ionentransportes in der Salzdrüse. II. Teil: Funktionelle Morphologie der Blutgefaße. *Protoplasma* **56**, 385–419.

Komnick, H. (1963*c*). Elektronenmikroskopische Untersuchungen zur funktionellen Morphologie des Ionentransportes in der Salzdrüse. III. Teil: Funktionelle Morphologie der Tubulusepithelzellen. *Protoplasma* **56**, 605–36.

Komnick, H. (1964). Elektronenmikroskopische Untersuchungen zur funktionellen Morphologie des Ionentransportes in der Salzdrüse von *Larus argentatus.* IV. Teil: Funktionelle Morphologie der Epithelzellen des Sammelkanals. *Protoplasma* **58**, 96–127.

Komnick, H. (1965). Funktionelle Morphologie von Salzdrüsenzellen. In *Sekretion und Exkretion*, pp. 289–314. Springer-Verlag: Berlin.

Komnick, H. & Kniprath, E. (1970). Morphometrische Untersuchungen an der Salzdrüse von Silbermöwen. *Cytobiologie* **1**, 228–47.

Komnick, H. & Komnick, U. (1963). Elektronenmikroskopische Untersuchungen zur funktionellen Morphologie des Ionentransportes in der Salzdrüse von *Larus argentatus.* V. Teil: Experimenteller Nachweis der Transportwege. *Zeitschrift für Zellforschung und mikroskopische Anatomie* **60**, 163–203.

Krista, L. M., Carlson, C. W. & Olson, O. E. (1961). Some effects of saline waters on chicks, laying hens, poults and ducklings. *Poultry Science* **40**, 938–44.

Kühnel, W. (1972). On the innervation of the salt gland. *Zeitschrift für Zellforschung und mikroskopische Anatomie* **134**, 435–8.

Kühnel, W., Burock, G. & Petry, G. (1969). Enzymhistotopochemische Studien an inaktiven Salzdrüsen von Hausenten (*Anas platyrhynchus*). I. Histiogramme einiger Oxydoreduktasen. *Histochemie* **19**, 235–47.

Kühnel, W., Petry, G. & Burock, G. (1969). Enzymhistotopochemische Studien ab inaktiven Salzdrüsen von Hausenten (*Anas platyrhynchus*). II. Cytochemische lokalisation einiger Hydrolasen. *Zeitschrift für Zellforschung und mikroskopische Anatomie* **99**, 560–9.

Kuijpers, W. & Bonting, S. L. (1969). Studies on (Na^+-K^+)-activated ATPase. XXIV. Localisation and properties of ATPase in the inner ear of the guinea pig. *Biochimica et biophysica acta* **173**, 477–85.

Lange, R. & Staaland, H. (1965). Anatomy and physiology of the salt gland in the grey heron, *Ardea cinerea. Nytt magasin for zoologi* **13**, 5–9.

Lanthier, A., Pépin, A. & Sandor, T. (1965). Secretion of the nasal gland in ducks. *Union médicale du Canada* **94**, 90.

Lanthier, A. & Sandor, T. (1967). Control of the salt-secreting gland of the duck. 1. Osmotic regulation. *Canadian Journal of Physiology and Pharmacology* **45**, 925–36.

Ledsome, J. R. & Linden, R. J. (1968). The role of left atrial receptors in the diuretic response to left atrial distension. *Journal of Physiology* **198**, 487–503.

Le Maire, M. (1971). Un phenomene d'ecochimie: les glandes á sel des oiseaux marins. *Aves* **8**, 18–20.

Lemire, M., Deloince, R. & Grenot, C. (1970). Etude des cavités et glandes nasales du lézard Fouette-Queue, *Uromastyx acanthinurus* Bell, *Compte rendu hebdomadaires des séances de l'Académie des Sciences* **270**, 817–20.

Levine, A. M., Higgins, J. A. & Barrnett, R. J. (1972). Biogenesis of plasma membranes in salt glands of salt-stressed domestic ducklings: localization of acyltransferase activity. *Journal of Cell Science* **11**, 855–73.

Linzell, J. L. (1955). Some observations on the contractile tissue of the mammary glands. *Journal of Physiology* **130**, 257–67.

Linzell, J. L. (1959). The innervation of the mammary glands in the sheep and goat with some observations on the lumbo-sacral autonomic nerves. *Quarterly Journal of Experimental Physiology* **44**, 160–78.

References

Linzell, J. L. (1968). The magnitude and mechanisms of the uptake of milk precursors by the mammary gland. *Proceedings of the Nutrition Society* **27**, 44–52.

Linzell, J. L. (1971). Techniques for measuring nutrient uptake by the mammary glands. In *Lactation*, ed. Falconer, I. R. Butterworths: London.

Linzell, J. L. & Peaker, M. (1971). Mechanism of milk secretion. *Physiological Reviews* **51**, 564–97.

Lockett, M. F. (1965). A comparison of the direct renal actions of pituitary growth and lactogenic hormones. *Journal of Physiology* **181**, 192–9.

Lockett, M. F. & Nail, B. (1965). A comparative study of the renal actions of growth and lactogenic hormones in rats. *Journal of Physiology* **180**, 147–56.

Lunn, T. & Hally, A. D. (1967). The effect of saline loading on the development of the prefunctional salt gland in the domestic duck. *Journal of Anatomy* **101**, 834.

Macchi, I. A., Phillips, J. G. & Brown, P. (1967). Relationship between the concentration of corticosteroids in avian plasma and nasal-gland function. *Journal of Endocrinology* **38**, 319–29.

Macchi, I. A., Phillips, J. G., Brown, P. & Yasuna, M. (1965). Relationship between nasal gland function and plasma corticoid levels in birds. *Federation Proceedings* **24**, 575.

Maetz, J. (1971). Fish gills: mechanisms of salt transfer in fresh water and sea water. *Philosophical Transactions of the Royal Society, series B* **262**, 209–49.

Marples, B. J. (1932). Structure and development of nasal glands of birds. *Proceedings of the Zoological Society of London* (1932), 829–44.

Martin, D. W. & Diamond, J. M. (1966). Energetics of coupled active transport of sodium and chloride. *Journal of General Physiology* **50**, 295–315.

Matthews, L. H. (1959). Salt excretion in marine birds. *Nature* **183**, 202.

McFarland, L. Z. (1959). Captive marine birds possessing a functional nasal gland (salt gland). *Nature* **184**, 2030–1.

McFarland, L. Z. (1960*a*). Histological observations on the avian nasal (salt) glands. *Anatomical Record* **137**, 376–7.

McFarland, L. Z. (1960*b*). Salt excretion from the nasal glands of captive penguins. *Anatomical Record* **138**, 366.

McFarland, L. Z. (1960*c*). Salt excretion from the nasal glands from various species of the Pelicaniformes. *Anatomical Record* **138**, 366.

McFarland, L. Z. (1963*a*). The effect of salt intake on the weight of the lateral nasal (salt) gland and other organs of Brandt's cormorant (*Phalacrocorax penicillatus*). *Anatomical Record* **145**, 259.

McFarland, L. Z. (1963*b*). Observations on the haematology and blood volume of captive western gulls. *Proceedings of the XVIth International Congress of Zoology*, vol. **2**, 86.

McFarland, L. Z. (1964*a*). Static blood volume of the nasal salt gland and other organs of the sea gull. *American Zoologist* **4**, 421–2.

McFarland, L. Z. (1964*b*). Minimal salt load required to induce secretion from the nasal salt-glands of sea gulls. *Nature* **204**, 1202–3.

McFarland, L. Z. (1965). Influence of external stimuli on the secretory rate of the avian nasal salt gland. *Nature* **205**, 391–2.

McFarland, L. Z., Martin, K. D. & Freedland, R. A. (1965). The activity of selected soluble enzymes in the avian nasal salt gland. *Journal of Cellular and Comparative Physiology* **65**, 237–41.

McFarland, L. Z. & Sanui, H. (1963). Sodium and potassium binding by microsomes from the nasal salt gland, Harder's gland, kidney and liver of sea gulls. *Proceedings of the Society for Experimental Biology and Medicine* **113**, 105–7.

McFarland, L. Z. & Warner, R. (1966). Blood volume of the nasal salt glands and other glands and organs of sea gulls. *Nature* **210**, 1389–90.

McLelland, J. & Pickering, E. C. (1969). The effects of exposure to increasing salt loads on the plasma osmolarity and the nasal gland of *Gallus domesticus. Research in Veterinary Science* **10**, 518–22.

McLelland, J., Moorhouse, P. D. S. & Pickering, E. C. (1968). An anatomical and histochemical study of the nasal gland of *Gallus gallus domesticus. Acta anatomica* **71**, 122–33.

McWhirter, N. & McWhirter, R. (1972). *The Guinness Book of Records.* Guinness Superlatives: London.

Meier, A. H., Farner, D. S. & King, J. R. (1965). A possible endocrine basis for migratory behaviour in the white-crowned sparrow, *Zonotrichia leucophrys gambelii. Animal Behaviour* **13**, 453–65.

Meischke, M. H. A. (1967). Observations of salt excretion by Pacific gulls (*Larus pacificus, Larus novaehollandiae*). *Ardea* **55**, 269.

Mihálik, P. V. (1932). Über die Glandula lateralis nasi der Vögel. *Ergebnisse der Anatomie und Entwicklungsgeschichte* **29**, 399–448.

Minnich, J. E. (1968). Maintenance of water and electrolytic balance by desert iguana, *Dipsosaurus dorsalis. American Zoologist* **8**, 782.

Minnich, J. E. (1970). Water and electrolyte balance of the desert iguana, *Dipsosaurus dorsalis*, in its natural habitat. *Comparative Biochemistry and Physiology* **35**, 921–33.

Minnich, J. E. (1972). Excretion of urate salts by reptiles. *Comparative Biochemistry and Physiology* **41A**, 535–49.

Murphy, R. C. (1936). *Oceanic birds of South America.* American Museum of Natural History: New York.

Murrish, D. E. & Schmidt-Nielsen, K. (1970). Exhaled air temperature and water conservation in lizards. *Respiration Physiology* **10**, 151–8.

Natochin, Y. V. & Krestinskaya, T. V. (1961). Succinic dehydrogenase and the active transfer of sodium in the osmoregulatory organs of vertebrates. *Sechenov Physiological Journal of the U.S.S.R.* **47** (translated version), 1437–46 (74–80 in Russian edition).

Nechay, B. R., Larimer, J. L. & Maren, T. H. (1960). Effects of drugs and physiologic alterations on nasal salt excretion in sea gulls. *Journal of Pharmacology and Experimental Therapeutics* **130**, 401–9.

Neutra, M. & Leblond, C. P. (1969). The Golgi apparatus. *Scientific American* **220**, 100–7.

Nicoll, C. S. & Bern, H. A. (1972). On the actions of prolactin among the vertebrates: is there a common denominator? In *Lactogenic hormones*, pp. 299–317, eds Wolstenholme, G. E. W. & Knight, J. Ciba Foundation Symposium. Churchill Livingston: Edinburgh.

Nitzsch, C. L. (1820). Über die Nasendrüse der Vögel. *Deutsches Archiv für die Physiologie* (*Meckel's Archiv*) **6**, 234–69.

Norris, K. S. & Dawson, W. R. (1964). Observations on the water economy and electrolyte excretion of chuckwallas (Lacertilia, *Sauromalus*). *Copeia* (1964), 638–46.

References

Ohmart, R. D. (1972). Physiological and ecological observations concerning the salt-secreting nasal glands of the roadrunner. *Comparative Biochemistry and Physiology* **43A**, 311–16.

Owen, M. & Kear, J. (1972). Food and feeding habits. In *The swans*, 58–77, ed. Atkinson-Willes, G. Michael Joseph: London.

Paterson, J. Y. F. & Harrison, F. A. (1967). The specific activity of plasma cortisol in sheep during continuous infusion of [1,2-$^3H_2$]cortisol, and its relation to the rate of cortisol secretion. *Journal of Endocrinology* **37**, 269–77.

Patterson, J. Y. F. & Harrison, F. A. (1968). The specific activity of plasma cortisol in sheep after intravenous infusion of [1,2-$^3H_2$]cortisol, and its relation to the distribution of cortisol. *Journal of Endocrinology* **40**, 37–47.

Paynter, R. A. (1971). Nasal glands in *Cinclodes nigrofumosus*, a maritime passerine. *Bulletin of the British Ornithologists' Club* **91**, 11–12.

Peaker, M. (1969). Some aspects of the thermal requirements of reptiles in captivity. In *International Zoo Yearbook* **9**, 3–8, ed. Lucas, J., Zoological Society of London. Academic Press: London.

Peaker, M. (1971*a*). Intracellular concentrations of sodium, potassium and chloride in the salt gland of the domestic goose and their relation to the secretory mechanism. *Journal of Physiology* **213**, 399–410.

Peaker, M. (1971*b*). Effects of arginine vasotocin on the salt gland of the goose, *Anser anser. Journal of Endocrinology* **49**, xxvi–xxvii.

Peaker, M. (1971*c*). Avian salt glands. *Philosophical Transactions of the Royal Society, series B* **262**, 289–300.

Peaker, M. (1971*d*). Salt balance of oiled sea-birds. *Ibis* **113**, 536.

Peaker, M., Hanwell, A. & Linzell, J. L. (1971). Location of receptors for salt gland secretion in the goose. *Proceedings of the International Union of Physiological Sciences (25th International Congress, Munich)* **9**, 444.

Peaker, M. & Phillips, J. G. (1969). The role of prolactin and other hormones in the adaptation of the duck (*Anas platyrhynchos*) to saline conditions. *Journal of Endocrinology* **43**, lx.

Peaker, M., Phillips, J. G. & Wright, A. (1970). The effect of prolactin on the secretory activity of the nasal salt-gland of the domestic duck (*Anas platyrhynchos*). *Journal of Endocrinology* **47**, 123–7.

Peaker, M., Peaker, S. J., Hanwell, A. & Linzell, J. L. (1973). Sensitivity of the receptors for salt-gland secretion in the domestic duck and goose. *Comparative Biochemistry and Physiology* **44A**, 41–6.

Peaker, M., Peaker, S. J., Phillips, J. G. & Wright, A. (1971). The effects of corticotrophin, glucose and potassium chloride on secretion by the nasal salt gland of the duck, *Anas platyrhynchos. Journal of Endocrinology* **50**, 293–9.

Peaker, M. & Stockley, S. J. (1973). Lithium secretion by the salt gland of the goose. *Nature* **243**, 297–8.

Peaker, M. & Stockley, S. J. (1974). The effects of lithium and methacholine on the intracellular ionic composition of goose salt gland slices: relation to sodium and chloride transport. *Experientia* **30**, 158–9.

Peaker, M., Wright, A., Peaker, S. J. & Phillips, J. G. (1968). Absorption of tritiated water by the cloaca of the domestic duck (*Anas platyrhynchos*). *Physiological Zoölogy* **41**, 461–5.

Peters, A. (1890). Beitrag zur Kenntniss der Harder'schen Drüse. *Archiv für mikroskopische Anatomie* **36**, 192–203.

Phillips, J. G. (1968). Studies on the nasal gland of birds. *Archives d'anatomie, d'histologie et d'embryologie* **51**, 533–7.

Phillips, J. G. & Bellamy, D. (1962). Aspects of the hormonal control of nasal gland secretion in birds. *Journal of Endocrinology* **24**, vi–vii.

Phillips, J. G. & Bellamy, D. (1963). Adrenocortical hormones. In *Comparative Endocrinology*, vol. **1**, 208–57, ed. von Euler, U. S. & Heller, H. Academic Press: New York & London.

Phillips, J. G. & Bellamy, D. (1967). The control of nasal gland function with special reference to the control by adrenocorticosteroids. *Proceedings of the IInd International Symposium on Hormonal Steroids. Excerpta Medica International Congress Series* **132**, 1065–9.

Phillips, J. G. & Ensor, D. M. (1972). The significance of environmental factors in the hormone mediated changes of nasal (salt) gland activity in birds. *General and Comparative Endocrinology, supplement* **3**, 393–404.

Phillips, J. G., Holmes, W. N. & Butler, D. G. (1961). The effect of total and subtotal adrenalectomy on the renal and extra-renal response of the domestic duck (*Anas platyrhynchos*) to saline loading. *Endocrinology* **69**, 958–69.

Philpott, C. W. & Templeton, J. R. (1964). A comparative study of the histology and fine structure of the nasal salt secreting gland of the lizard *Dipsosaurus. Anatomical Record* **148**, 394–5.

Pittard, J. B. & Hally, A. D. (1973). The effect of denervation on genotypic and compensatory growth of the immature avian salt gland. *Journal of Anatomy* **114**, 303.

Poulsen, J. H. (1973). An attempt to elicit salivary secretion by changing the intracellular sodium and potassium concentrations without applying neurotransmitters. *Acta physiologica scandinavica* **87**, 51–52A.

Poulson, T. L. (1965). Countercurrent multipliers in avian kidneys. *Science* **148**, 389–91.

Prange, H. D. & Schmidt-Nielsen, K. (1970). The metabolic cost of swimming in ducks. *Journal of Experimental Biology* **53**, 763–77.

Reynaert, H., Peeters, G., Verbeke, R. & Houvenaghel, A. (1968). Further studies of the physiology of plasma kinins and kallikreins in the udder of ruminants. *Archives internationales de Pharmacodynamie et Therapie* **176**, 473–5.

Rhees, R. W., Abel, J. H. & Frame, J. R. (1972). Effect of osmotic stress and hormone therapy on the hypothalamus of the duck (*Anas platyrhynchos*). *Neuroendocrinology* **10**, 1–22.

Richardson, K. C. (1966). Electron microscopic identification of autonomic nerve endings. *Nature* **210**, 756.

Russell, F. S. (1958). Salt excretion in marine birds. *Nature* **182**, 1755.

Sandor, T. & Fazekas, A. G. (1973). Corticosteroid 'receptor(s)' in the nasal gland cytosol of the domestic duck. *Acta endocrinologica* **177** (*supplement*), 251.

Santiago-Calvo, E., Mule, S., Redman, C. M., Hokin, M. R. & Hokin, L. E. (1964). The chromatographic separation of polyphosphoinositides and studies on their turnover in various tissues. *Biochimica et biophysica acta* **84**, 550–62.

Schildmacher, H. (1932). Über den Einfluß des Salzwassers auf die Enticklung der Nasendrüsen. *Journal für Ornithologie* **80**, 293–9.

References

Schmidt, K. P. (1957). Reptiles (except turtles) (annotated bibliography of species occurring in salt-water). *Memoirs. Geological Society of America* **67**, 1213–16.

Schmidt-Nielsen, K. (1959). Salt glands. *Scientific American* **200**, 109–16.

Schmidt-Nielsen, K. (1960). Salt-secreting gland of marine birds. *Circulation* **21**, 955–67.

Schmidt-Nielsen, K. (1963). Osmotic regulation in higher vertebrates *Harvey Lectures* **58**, 55–93.

Schmidt-Nielsen, K. (1964). *Desert animals.* Clarendon: Oxford.

Schmidt-Nielsen, K. (1965). Physiology of salt glands. In *Sekretion und exkretion,* 269–88. Springer-Verlag: Berlin.

Schmidt-Nielsen, K., Borut, A., Lee, P. & Crawford, E. (1963). Nasal salt excretion and the possible function of the cloaca in water conservation. *Science* **142**, 1300–1.

Schmidt-Nielsen, K. & Fänge, R. (1958*a*). The function of the salt gland in the brown pelican. *Auk* **75**, 282–9.

Schmidt-Nielsen, K. & Fänge, R. (1958*b*). Salt glands in marine reptiles. *Nature* **182**, 783–5.

Schmidt-Nielsen, K. & Fänge, R. (1958*c*). Extra-renal salt excretion. *Federation Proceedings* **17**, 142.

Schmidt-Nielsen, K. & Kim, Y. T. (1964). The effect of salt intake on the size and function of the salt gland in ducks. *Auk* **81**, 160–72.

Schmidt-Nielsen, K., Jörgensen, C. B. & Osaki, H. (1957). Secretion of hypertonic solutions in marine birds. *Federation Proceedings* **16**, 113–14.

Schmidt-Nielsen, K., Jörgensen, C. B. & Osaki, H. (1958). Extrarenal salt excretion in birds. *American Journal of Physiology* **193**, 101–7.

Schmidt-Nielsen, K. & Sladen, W. J. L. (1958). Nasal salt secretion in the Humboldt penguin. *Nature* **181**, 1217–18.

Schwarz, D. (1962). Untersuchungen zur biologischen Bedeutung der Salzdrüsen bei freilebenden Sturmöwen (*Larus canus* L.) (I. Sommer 1961). *Journal für Ornithologie* **103**, 180–6.

Schwarz, D. (1966). Das Verhalten von Lariden gegenüber Salzlösungen von verschiedener Konzentration. *Beiträge zur Vogelkunde* **11**, 359–67.

Schwarz, D. & Nehls, H. W. (1967). Untersuchungen zur biologischen Bedeutung der Salzdrüsen bei freilebenden Sturmöwen (*Larus canus* L.) (II. Winter 1961–Winter 1965). *Journal für Ornithologie* **108**, 335–40.

Schwarz, D. & Spannof, L. (1961). Zur Frage der NaCl-Ausscheidung durch die sogennante Salzdrüse bei Vögeln. *Nátúrwissenschaften* **48**, 414.

Scothorne, R. J. (1958*a*). On the anatomy and development of the nasal cavity in the gannet (*Sula bassana* L.). *Journal of Anatomy* **92**, 648.

Scothorne, R. J. (1958*b*). Histochemical study of the nasal (supraorbital) gland of the duck. *Nature* **182**, 732.

Scothorne, R. J. (1959*a*). The nasal glands of birds: a histochemical and histological study of the inactive gland of the domestic duck. *Journal of Anatomy* **93**, 246–56.

Scothorne, R. J. (1959*b*). Further studies of the salt-secreting ('nasal') glands of birds. *Journal of Anatomy* **93**, 588.

Scothorne, R. J. (1959*c*). On the response of the duck and the pigeon to intravenous hypertonic saline solutions. *Quarterly Journal of Experimental Physiology* **44**, 200–7.

Scothorne, R. J. (1959*d*). Histochemical study of succinic dehydrogenase in the nasal (salt-secreting) gland of the Aylesbury duck. *Quarterly Journal of Experimental Physiology* **44**, 329–32.

Scothorne, R. J. (1960). Alkaline and acid phosphatases in the salt glands of birds. *Journal of Anatomy* **94**, 581.

Scothorne, R. J. & Hally, A. D. (1960). The fine structure of the nasal gland in the Aylesbury duck. *Journal of Anatomy* **94**, 581–2.

Secker, J. (1938). Suprarenals and the transmission of the sympathetic nerves in the cat. *Journal of Physiology* **94**, 259–79.

Secker, J. (1948). The influence of ovarian hormones on the transmission of the activity of the sympathetic nerves of the cat. *Journal of Physiology* **107**, 265–71.

Serventy, D. L. (1971). Biology of desert birds. In *Avian biology*, vol. **1**, pp. 287–339, eds Farner, D. S. & King, J. R. Academic Press: New York.

Shaw, P. A. (1929). Duck disease studies 1. Blood analyses in diseased birds. *Proceedings of the Society for Experimental Biology and Medicine* **27**, 6–7.

Shoemaker, V. H. (1972). Osmoregulation and excretion in birds. In *Avian biology*, vol. II, pp. 527–74, eds Farner, D. S. & King, J. R. Academic Press: New York.

Shoemaker, V. H., Licht, P. & Dawson, W. R. (1966). Effects of temperature on kidney function in the lizard *Tiliqua rugosa*. *Physiological Zoology* **39**, 244–52.

Shoemaker, V. H., Nagy, K. A. & Bradshaw, S. D. (1972). Studies on the control of electrolyte excretion by the nasal gland of the lizard *Dipsosaurus dorsalis*. *Comparative Biochemistry and Physiology* **42A**, 749–57.

Simon, B., Kinne, R. & Knauf, H. (1972). The presence of a HCO_3^- ATPase in glandula submandibularis of rabbit. *Pflügers Archiv* **337**, 177–84.

Simon, B., Kinne, R. & Sachs, G. (1972). The presence of a HCO_3^- ATPase in pancreatic tissue. *Biochimica et biophysica acta* **282**, 293–300.

Skadhauge, E. (1967). *In vivo* perfusion studies of the cloacal water and electrolyte resorption in the fowl (*Gallus domesticus*). *Comparative Biochemistry and Physiology* **23**, 483–501.

Skadhauge, E. (1968). The cloacal storage of urine in the rooster. *Comparative Biochemistry and Physiology* **24**, 7–18.

Skadhauge, E. & Schmidt-Nielsen, B. (1967). Renal function in domestic fowl. *American Journal of Physiology* **212**, 793–8.

Smith, D. P. (1970). On the regulation of monovalent cation concentrations in the fluid secreted by salt glands of domestic ducks. *Journal of Anatomy* **107**, 383.

Smith, D. P. (1972*a*). Ethacrynic acid and nasal salt excretion in the duck (*Anas platyrhynchos*). *Cytobios* **5**, 217–18.

Smith, D. P. (1972*b*). An investigation of interrelated factors affecting the flow and concentration of salt-induced nasal secretion in the duck, *Anas platyrhynchos*. *Comparative Biochemistry and Physiology* **43A**, 1003–17.

Smith, D. P., Fourman, J. M. & Haase, P. (1971*a*). The effect of an intravenous injection of isotonic saline or sucrose on the secretory activity and butyrylcholinesterase content of the salt glands of *Anas domesticus*. *Cytobios* **3**, 49–55.

Smith, D. P., Fourman, J. M. & Haase, P. (1971*b*). The secretory activity and butyrylcholinesterase content of the salt glands of *Anas domesticus* in response to an intravenous injection of hypertonic saline. *Cytobios* **3**, 57–64.

Smith, H. W. (1953). *From fish to philosopher*. Little Brown: Boston.

Sokol, O. M. (1967). Herbivory in lizards. *Evolution* **21**, 192–4.

References

Spannhof, L. & Jürss, K. (1967). Untersuchungen zur Genese einiger Enzyme in den Salzdrüsen junger Sturmöwen. *Acta biologica et medica germanica* **19**, 137–44.

Staaland, H. (1967*a*). Temperature sensitivity of the avian salt gland. *Comparative Biochemistry and Physiology* **23**, 991–3.

Staaland, H. (1967*b*). Anatomical and physiological adaptations of the nasal glands of Charadriiformes birds. *Comparative Biochemistry and Physiology* **23**, 933–44.

Staaland, H. (1968). Excretion of salt in waders, Charadrii, after acute salt loads. *Nytt magasin for Zoologi* **16**, 25–8.

Stainer, I. M., Ensor, D. M., Phillips, J. G. & Holmes, W. N. (1970). Changes in glycolytic enzyme activity in the duck (*Anas platyrhynchos*) nasal gland during the period of adaptation to salt water. *Comparative Biochemistry and Physiology* **37**, 257–63.

Stewart, D. J. (1972). Secretion by salt gland during water deprivation in the duck. *American Journal of Physiology* **223**, 384–6.

Stewart, D. J. & Holmes, W. N. (1970). Relation between ribosomes and functional growth in the avian nasal gland. *American Journal of Physiology* **219**, 1819–24.

Stewart, D. J., Holmes, W. N. & Fletcher, G. L. (1969). The renal excretion of nitrogenous compounds by the duck (*Anas platyrhynchos*) maintained on freshwater and on hypertonic saline. *Journal of Experimental Biology* **50**, 527–9.

Storer, R. W. (1971). Classification of birds. In *Avian Biology*, Vol **1**, 1–18. eds Farner, D. S. & King, J. R. Academic Press: New York.

Stresemann, E. (1928). 1-Fünfte Klasse der Craniota zweite und zugleich letzte Klasse der Sauropsida, Aves = Vögel. In *Kukenthal-Krumbach Handbuch der Zoologie*, vol. 7 (2), W. de Gruyter: Berlin.

Sturkie, P. D. (1965). *Avian Physiology*, 2nd edition. Bailliere, Tindall & Cassell: London.

Sturkie, P. D. (1970). Circulation in aves. *Federation Proceedings* **29**, 1674–9.

Taub, A. M. & Dunson, W. A. (1966). A new gland in sea snakes (Squamata, Reptilia). *American Zoologist* **6**, 265.

Taub, A. M. & Dunson, W. A. (1967). The salt gland in a sea snake (*Laticauda*). *Nature* **215**, 995–6.

Technau, G. (1936). Die Nasendrüse der Vögel. *Journal für Ornithologie* **84**, 511–617.

Templeton, J. R. (1963). Nasal salt secretion in terrestrial iguanids. *American Zoologist* **3**, 530.

Templeton, J. R. (1964*a*). Nasal salt gland in terrestrial lizards. *Texas Reports on Biology and Medicine* **22**, 206.

Templeton, J. R. (1964*b*). Nasal salt excretion in terrestrial lizards. *Comparative Biochemistry and Physiology* **11**, 223–9.

Templeton, J. R. (1966). Responses of the lizard nasal salt gland to chronic hypersalaemia. *Comparative Biochemistry and Physiology* **18**, 563–72.

Templeton, J. R. (1967). Nasal salt gland excretion and adjustment to sodium loading in the lizard, *Ctenosaura pectinata*. *Copeia* (1967), 136–40.

Templeton, J. R. (1972). Salt and water balance in desert lizards. In *Comparative physiology of desert animals*, pp. 61–77, ed. Maloiy, G. M. O. (*Symposia of the Zoological Society of London* **31**). Academic Press: London.

Templeton, J. R., Murrish, D., Randall, E. & Mugaas, J. (1968). The effect of aldosterone and adrenalectomy on nasal salt excretion of the desert iguana, *Dipsosaurus dorsalis*. *American Zoologist* **8**, 818–19.

Templeton, J. R., Murrish, D. E., Randall, E. M. & Mugaas, J. N. (1972*a*). Salt and water balance in the desert iguana, *Dipsosaurus dorsalis*. I. The effect of dehydration, rehydration and full hydration. *Zeitschrift für vergleichende Physiologie (Journal of Comparative Physiology)* **76**, 245–54.

Templeton, J. R., Murrish, D. E., Randall, E. M. & Mugaas, J. N. (1972*b*). Salt and water balance in the desert iguana, *Dipsosaurus dorsalis*. II. The effect of aldosterone and adrenalectomy. *Zeitschrift für vergleichende Physiologie (Journal of Comparative Physiology)* **76**, 255–69.

Thesleff, S. & Schmidt-Nielsen, K. (1962). An electrophysiological study of the salt gland of the herring gull. *American Journal of Physiology* **202**, 597–600.

Thomas, D. H. & Phillips, J. G. (1973). The kinetics of exogenous corticosterone and aldosterone in relation to different levels of food and NaCl intake by domestic ducks. *Journal of Endocrinology* **57**, xiv.

Tormey, J. McD. (1966). Significance of the histochemical demonstration of ATPase in epithelia noted for active transport. *Nature* **210**, 820–2.

Ungar, G. & Parrot, J.-L. (1936). Sur la présence de la callicréine dans la saliva, et la possibilité de son intervention dans la transmission chimique de l'influx nerveux. *Compte rendu des séances de la Société de biologie* **122**, 1052–5.

Ussing, H. H. (1949). The distinction by means of traces between active transport and diffusion. The transfer of iodide across the isolated frog skin. *Acta physiologica scandinavica* **19**, 43–56.

Ussing, H. H. (1960). The alkali metal ions in isolated systems and tissues. In *Handbuch der experimentellen Pharmakologie* **13**, 1–195, eds Eichler, O. & Farah, A. Springer-Verlag: Berlin.

Vader, W. (1971). Dippers feeding on marine invertebrates. *British Birds* **64**, 456–8.

Van Lennep, E. W. & Komnick, H. (1970). Fine structure of the nasal salt gland in the desert lizard *Uromastyx acanthinurus*. *Cytobiologie* **2**, 47–67.

Van Rossum, G. D. V. (1964*a*). The effects of succinate and other substrates on the pyridine nucleotides of slices of rat liver and avian salt gland. *Biochimica et biophysica acta* **86**, 198–201.

Van Rossum, G. D. V. (1964*b*). Measurements of respiratory pigments and sodium efflux in slices of avian salt gland. *Biochemical and Biophysical Research Communications* **15**, 540–5.

Van Rossum, G. D. V. (1964*c*). Observations on the fluorescence emitted by slices of rat liver and avian salt gland. *Biochimica et biophysica acta* **88**, 507–16.

Van Rossum, G. D. V. (1965*a*). Observations on respiratory pigments in slices of avian salt gland and rat liver 1. Effects of inhibitors and uncouplers. *Biochimica et biophysica acta* **110**, 221–36.

Van Rossum, G. D. V. (1965*b*). Observations on respiratory pigments in slices of avian salt gland and rat liver II. Evidence for reversal of electron transfer. *Biochimica et biophysica acta* **110**, 237–51.

Van Rossum, G. D. V. (1966). Movements of Na^+ and K^+ in slices of herring-gull salt gland. *Biochimica et biophysica acta* **126**, 338–49.

Van Rossum, G. D. V. (1968). Relation of the oxidoreduction level of electron carriers to ion transport in slices of avian salt gland. *Biochimica et biophysica acta* **153**, 124–31.

Verney, E. B. (1947). The anti-diuretic hormone and the factors which determine its release. *Proceedings of the Royal Society, series B* **135**, 25–106.

References

Vieria, F. L., Caplan, S. R. & Essig, A. (1972). Energetics of sodium transport in frog skin. 1. Oxygen consumption in the short-circuited state. *Journal of General Physiology* **59**, 60–76.

Watson, G. E. & Dovoky, G. J. (1971). Identification of *Diomedea leptorhyncha* Coues 1866, an albatross with remarkably small salt glands. *Condor* **73**, 487–9.

Webb, M. (1957). The ontogeny of the cranial bones, cranial sympathetic nerves, together with a study of the visceral muscles of *Struthio. Acta zoologica fennica* **38**, 81–203.

Weiss, P. (1952). Self-regulation of organ growth by its own products. *Science* **115**, 487–8.

Werle, E. & Roden, P. (1936). Über das Vorkommen von Kallikrein in den Speicheldrusen und im Mundspeichel. *Biochemische Zeitschrift* **286**, 213–19.

Werle, E., Hochstrasser, K. & Trautschold, I. (1966). Studies of bovine plasma kininogen, ornitho-kallikrein, and ornitho-kinin. In *Hypotensive Peptides*, eds Erdös, E. G., Back, N. & Sicuteri, F. Springer-Verlag: New York.

Wetmore, A. (1915). Mortality among waterfowl around Great Salt Lake, Utah. *United States Department of Agriculture. Technical Bulletin* **217**.

Wetmore, A. (1918). The duck sickness in Utah. *United States Department of Agriculture. Technical Bulletin* **672**.

Wetsheloff, M. (1900). *Beiträge zur Kenntnis der Nasendrüsen bei den Vögeln.* Inaugural Dissertation. E. Ebering: Berlin.

Wood, W. G. & Ballantyne, B. (1968). Sodium ion transport and β-glucuronidase activity in the nasal gland of *Anas domesticus. Journal of Anatomy* **103**, 277–87.

Wright, A., Phillips, J. G. & Huang, D. P. (1966). The effect of adenohypophysectomy on the extrarenal and renal excretion of the saline-loaded duck (*Anas platyrhynchos*). *Journal of Endocrinology* **36**, 249–56.

Wright, A., Phillips, J. G., Peaker, M. & Peaker, S. J. (1967). Some aspects of the endocrine control of water and salt electrolytes in the duck (*Anas platyrhynchos*). *Proceedings of the IIIrd Asia and Oceania Congress of Endocrinology*, vol. II, pp. 322–7, ed. Litonjua, A. Manila.

Zaks, N. G. & Sokolova, M. M. (1961). Ontogenetic and specific features of the nasal gland in some marine birds. *Sechenov Physiological Journal of the U.S.S.R.* **47** (translated version), 120–7.

Zerahn, K. (1956). Oxygen consumption and active sodium transport in the isolated and short-circuited frog skin. *Acta physiologica scandinavica* **36**, 300–18.

Zierler, K. L. (1961). Theory of the use of arteriovenous concentration differences for measuring metabolism in steady and non-steady states. *Journal of Clinical Investigation* **40**, 2111–25.

ADDENDUM

AVIAN SALT GLANDS

Chapter 1

Sanson (1972) has studied, in the Fulmar (*Fulmarus glacialis*), the ultrastructure of the cells that occur between the bases of the secretory cells; their function remains unknown. The histology of the gland in the penguin, *Spheniscus demersus* has been studied by Oelofsen (1973).

Chapter 4

Ballantyne (1974) has published in full similar findings to those of Ballantyne & Fourman (1967) on acetylcholinesterase in the salt gland of the duck.

Peaker (1975) has reviewed the nervous control of secretion, blood flow and adaptive changes.

Chapter 6

Martin & Philpott (1973) have shown that horseradish peroxidase may penetrate some of the tight junctions between secretory cells in the duck. They also added lanthanum to the fixative and found that it too penetrated the tight junctions. However the possibility that fixation affected the junctions and allowed lanthanum to enter must be considered before it can be accepted that most of the tight junctions in the secretory epithelium are not physiologically tight. They have also found that anionic protein-carbohydrates are associated with the plasmalemma of the secretory cells in the duck; evidence was obtained that the anionic nature of the surface is due to sialic acid and that the amount of this substance in the glands increases during adaptation (Martin & Philpott, 1974).

Fourman (1968) has indicated in an abstract that butyrylcholinesterase is located on membranes and in vesicles between the mitochondria and the basal membrane of the secretory cells. Ballantyne (1974) has extended the findings of Ballantyne & Fourman (1967) on butyrylcholinesterase in the gland. Karlsson, Samuelsson & Steen (1974) have extended their work on lipid composition in relation to ATPase, and Hughes (1974) has studied the water content of the gland during adaptation to salt water in the Black Swan (*Cygnus atratus*); it did not change.

Addendum

Chapter 8

Lanthier & Sandor (1973) have found that 18-hydroxycorticosterone enhances the secretory rate, but they did not investigate whether the effect was direct or indirect, for example, on the kidney or on plasma composition. It is not known whether this steroid is secreted by the adrenal *in vivo*. The abstract by Sandor & Fazekas (1974) on corticosteroid binding is similar to that by the same authors in 1973.

Ensor (1974) has found that prolactin release from the pituitary, as well as secretion, is blocked when birds in which the vagi are treated with local anaesthetic are salt-loaded, but that secretion can still be induced by electrical stimulation of the salt gland. Thus the afferent pathway of the secretory reflex arc also appears to be responsible for the release of prolactin in response to an osmotic stimulus. The role of prolactin in adaptation has been reviewed (Ensor, 1975).

Chapter 9

Martin & Philpott (1973) have studied the ultrastructure of the secretory cells during adaptation in the duck; their findings are similar to those of other authors.

Chapter 11

Hughes & Blackman (1973) have observed salt secretion in the Brolga (*Grus rubicundus*), an Australian crane that drinks fresh or salt water. Lavery (1972) has studied the size of the salt glands in relation to habitat in the Australian Grey Teal (*Anas gibberifrons gracilis*). Oelofsen (1973) has found no difference in the composition of nasal fluid from penguins (*Spheniscus demersus*) kept at 18 °C or 28 °C; secretory rate was not reported.

The salt glands have been implicated in the effect of chlorinated hydrocarbon insecticides in 'egg-shell thinning'. Cooke (1973) in reviewing this effect stated, 'Since an excess of chloride in the diet causes chickens to lay eggs with thin shells...Risebrough, Davis & Anderson (1970) suggested that pesticides may inhibit proper functioning of the salt gland in marine species, so reducing chloride excretion and causing a decrease in shell thickness. They also pointed out that this would explain why some species, such as the brown pelican, which largely rely on extrarenal means for chloride excretion, should be particularly affected. This is an interesting theory and affords yet another site of action for a carbonic anhydrase inhibitor [which these substances are]. It is, however, still only a tentative theory. One objection is that chloride salts in the diet do not consistently reduce shell thickness.'

REPTILIAN SALT GLANDS

Chapter 12

Cowan (1973) has produced further anatomical evidence that the lachrymal, and not the Harderian gland, is the salt gland in chelonians. He has also studied extra-renal excretion and lachrymal-gland weight during adaptation of the terrapin, *Malaclemys*, to sea water (Cowan, 1974). Evans (1973) in a balance study on the Loggerhead Turtle (*Caretta caretta*) has concluded that 60 per cent of the gross efflux of sodium occurs via the salt glands.

The work by Burns & Pickwell on the cephalic glands of sea snakes has been published (Burns & Pickwell, 1972). Dunson & Dunson (1973) have found that the posterior sublingual gland functions as a salt gland in *Acrochordus granulatus* (called *Chersydrus granulatus* in main text); the structure of the gland is similar to that of the true sea snakes. Since the Acrochordidae is not considered to be closely related to the true sea snakes (Hydrophiidae) these findings could suggest that this type of salt gland may have evolved independently in the two groups. Dunson & Dunson (1974) have made further interesting studies on sea snakes including gland weight, secretory rate, ultrastructure and ATPase activity. The fresh-water 'sea snake', *Hydrophis semperi*, from Lake Taal does have a functional salt gland but the authors discuss the possibility that this lake was connected to the sea until comparatively recent times. Some sea snakes have been found which secrete very little fluid from their salt glands.

Dunson (1974) has also found that the nasal gland of the Australian Rusty Monitor Lizard (*Varanus semiremax*), which lives in coastal mangrove swamps, secretes a fluid rich in sodium and chloride but relatively low in potassium.

Chapter 13

Lemire & Grenot (1973) have investigated the structure of the nasal gland in the lizard *Agama mutabilis*.

Minnich & Shoemaker (1972) have found that the salt gland of the lizard *Uma scoparia*, in the wild, secretes mainly potassium and chloride; Nagy (1972) has reached a similar conclusion following studies on a wild population of *Sauromalus obesus*. Deavers (1972) has shown that *Uma notata* also has a functional salt gland.

REFERENCES

Ballantyne, B. (1974). Choline ester hydrolase activity in the avian nasal gland. *Cytobios* **9**, 39–53.

Burns, B. & Pickwell, G. V. (1972). Cephalic glands in sea snakes (*Pelamis, Hydrophis* and *Laticauda*). *Copeia* (1972), 547–59.

Cooke, A. S. (1973). Shell thinning in avian eggs by environmental pollutants. *Environmental Pollution* **4**, 85–152.

Cowan, F. B. M. (1973). The homology of cranial glands in turtles: with special reference to the nomenclature of 'salt glands'. *Journal of Morphology* **141**, 157–70.

Cowan, F. B. M. (1974). Observations on extrarenal excretion by orbital glands and osmoregulation in *Malaclemys terrapin*. *Comparative Biochemistry and Physiology* **48A**, 489–500.

Deavers, D. R. (1972). Water and electrolyte metabolism in the arenicolous lizard *Uma notata notata*. *Copeia* (1972), 109–22.

Dunson, W. A. (1974). Salt gland secretion in a mangrove monitor lizard. *Comparative Biochemistry and Physiology* **47A**, 1245–55.

Dunson, W. A. & Dunson, M. K. (1973). Convergent evolution in sublingual salt glands in the marine file snake and the true sea snakes. *Journal of Comparative Physiology* **86**, 193–208.

Addendum

Dunson, W. A. & Dunson, M. K. (1974). Sea snake salt glands: relation between gland weight, fluid concentration, flow rate, cell ultrastructure and Na^+–K^+ ATPase activity. *American Journal of Physiology* **227**, 430–8.

Ensor, D. M. (1974). The effect of diuretics and vagus blockade on the release of avian prolactin. *Journal of Endocrinology* **63**, 45–46P.

Ensor, D. M. (1975). Prolactin and adaptation. In *Advances in avian physiology*, ed. Peaker, M. Zoological Society of London Symposium. Academic Press: London (in press).

Evans, D. H. (1973). The sodium balance of the euryhaline marine loggerhead turtle, *Caretta caretta. Journal of Comparative Physiology* **83**, 179–85.

Fourman, J. (1968). Electron microscopic localization of non-neuronal cholinesterase in the mammalian kidney and the duck salt gland. *Journal of Anatomy* **103**, 585.

Hughes, M. R. (1974). Water content of the salt glands and other avian tissues. *Comparative Biochemistry and Physiology* **47A**, 1089–93.

Hughes, M. R. & Blackman, J. G. (1973). Cation content of salt gland secretion and tears in the Brolga, *Grus rubicundus* (Perry) (Aves: Gruidae). *Australian Journal of Zoology* **21**, 515–18.

Karlsson, K.-A., Samuelsson, B. E. & Steen, G. O. (1974). The lipid composition and Na^+–K^+-dependent adenosine-triphosphatase activity of the salt (nasal) gland of Eider duck and Herring gull. *European Journal of Biochemistry* **46**, 243–58.

Lanthier, A. & Sandor, T. (1973). The effect of 18-hydroxycorticosterone on the salt-secreting gland of the duck (*Anas platyrhynchos*). *Canadian Journal of Physiology and Pharmacology* **51**, 776–8.

Lavery, H. J. (1972). The grey teal at saline drought-refuges in north Queensland. *Wildfowl* **23**, 56–63.

Lemire, M. & Grenot, C. (1973). La structure nasale du lézard Saharien *Agama mutabilis* Merrem (agamidae). *Compte rendu hebdomadaire des séances de l'académie des Sciences*, series D **277**, 2719–22.

Martin, B. J. & Philpott, C. W. (1973). The adaptive response of the salt glands of adult Mallard ducks to a salt water regime: an ultrastructural and tracer study. *Journal of Experimental Zoology* **186**, 111–22.

Martin, B. J. & Philpott, C. W. (1974). The biochemical nature of the cell periphery of the salt gland secretory cells of fresh and salt water adapted mallard ducks. *Cell and Tissue Research* **150**, 193–211.

Minnich, J. E. & Shoemaker, V. H. (1972). Water and electrolyte turnover in a field population of the lizard, *Uma scoparia. Copeia* (1972), 650–9.

Nagy, K. A. (1972). Water and electrolyte budgets of a free-living desert lizard, *Sauromalus obesus. Journal of Comparative Physiology* **79**, 39–62.

Oelofsen, B. W. (1973). The influence of ambient temperature on the function of the nasal salt gland and the composition of cloacal fluid in the penguin, *Spheniscus demersus. Zoologica Africana* **8**, 63–74.

Peaker, M. (1975). Recent advances in the physiology of the salt gland. In *Advances in avian physiology*, ed. Peaker, M. Zoological Society of London Symposium. Academic Press: London (in press).

Risebrough, R. W., Davis, J. & Anderson, D. W. (1970). Effects of various chlorinated hydrocarbons. *Oregon State University, Environmental Health Science*, series 1, 40–53.

Addendum

Sandor, T. & Fazekas, A. G. (1974). Corticosteroid-binding macromolecules in the nasal gland of the domestic duck. *General and Comparative Endocrinology* **22**, 348–9.

Sanson, R. (1972). Fine structure and location of basal cells in the salt gland of the fulmar, *Fulmarus glacialis* L. (Aves, Procellariiformes). *Norwegian Journal of Zoology* **20**, 105–9.

AUTHOR INDEX

References to the Addendum are indicated by bold type

Author index

Author index

SUBJECT INDEX

Subject index